THICKER THAN WATER

STRODE STUDIES IN EARLY MODERN LITERATURE AND CULTURE

Michelle M. Dowd, series editor

Thicker Than Water

BLOOD, AFFINITY, AND HEGEMONY IN EARLY MODERN DRAMA

LAUREN WEINDLING

THE UNIVERSITY OF ALABAMA PRESS
Tuscaloosa

The University of Alabama Press
Tuscaloosa, Alabama 35487-0380
uapress.ua.edu

Typeface: Arno Pro

Cover image: Illuminated manuscript, *Gratian's Decretum*;
© The Fitzwilliam Museum, Cambridge
Cover design: Lori Lynch

Cataloging-in-Publication data is available from the Library of Congress.
ISBN: 978-0-8173-2148-2 (cloth)
ISBN: 978-0-8173-6101-3 (paper)
E-ISBN: 978-0-8173-9436-3

FOR MICHAEL, I CHOOSE YOU

CONTENTS

ACKNOWLEDGMENTS

When my committee chair, Heather James, first told me that my project was about blood back in 2014, I didn't believe her. But, as I soon learned, one should always believe Heather James; and blood became a far more capacious, productive, and unwieldy thing than I ever imagined.

Both halves of my first chapter have appeared as earlier versions; I am grateful to the publishers for permission to use this material: *Cahiers du dix-septième siècle 18* (2017) and *Early Modern Literary Studies* 21.2 (2020). Most of the ideas and arguments in this book were developed and presented for workshops, seminars, and classes. Thus, I would like to thank the audiences and organizers of numerous seminars for the Shakespeare Association of America as well as the Early Modern Conversions Project. I would also like to thank my students from Auburn University, Toronto Metropolitan University (recently renamed), and the graduate students in my Shakespeare seminar at York University for engaging with, or at least entertaining, many of the readings found in these pages.

A whole host of institutions deserve my gratitude for ensuring that I had a roof over my head, vital library access, food to eat, and wine to drink while completing this work: the Provost's Office at the University of Southern California, the USC-Huntington Early Modern Studies Institute, the Gold Family Fellowship, and the Centre for Renaissance and Reformation Studies at the University of Toronto. My fourteen-pound corgi-Chihuahua, Raymond, also thanks you for feeding him over the years.

I would like to thank everyone at the University of Alabama Press: the series editor, Michele Dowd, copy editors Susan Harris and Jon Berry, and editor-in-chief, Dan Waterman, who kindly shepherded me through my first publication experience. I am, moreover, truly grateful for the thoughtful commentary of the readers, Ariane Balizet and Eric Langley, whose suggestions helped this book attain its best form. Any persistent problems are entirely my own.

I have had the good fortune of finding many wonderful mentors and colleagues throughout my time working on this book, greatly benefiting from their suggestions and feedback: Penelope Geng, Anna Rosensweig, Bruce Smith, Antonia Szabari, and Devin Toohey. Special mention is due to the following: Heather James, to whom I'm deeply indebted for her feedback and patience such that this project didn't devolve into something reminiscent of *Titus*'s bloody hole; Rebecca Lemon, whose marginalia always found its way to the bleeding wounds of this manuscript and gave me the tools to patch them; Margaret (Tita) Rosenthal for her generous line-editing and the countless hours we spent discussing Italian drama; and finally, the USC Early Modern Writing Group (Megan Herrold, Amanda Ruud, and Betsy Sullivan), whose eyes have graced this entire project, countless times, in some of its bloodiest and most mangled iterations. Without their support, both intellectual and emotional, this book simply would not exist.

To my family, thank you for letting me stay up into the wee hours of the morning as a child, so long as I was reading. This practice inspired a love affair that would last a lifetime. And last but not least, to my partner, Michael, thank you for your unshakable faith not only in this project, but also in me. I could not do this without you.

EMBLEM 28 WITHER

Would God, I could as feelingly infuse
A good effect of what this *Emblem* shewes,
As I can tell in words, what *Moralls* bee,
The life of that, which here you pictur'd see.
Most *Lovers,* minde their *Penny,* or their *Pleasure;*
Or, painted *Honors;* and, they all things measure,
Not as they are, but as they helpfull seeme,
In compassing those toyes, they most esteeme.
Though many with to gaine a faithfull *Friend,*
They seldome seeke one, for the noblest end:
Not know they (should they finde what they had sought)

How *Friendship* should be manag'd, as it ought.
Such, as good *Husbands* cover, or good *Wives*
(The deare companions of most happy lives)
Wrong Courses take to gaine them; yet, contemne
Their honest love, who rightly counsell them:
And, left, they unawares the Marke may hit,
They blinde their *judgements,* and befoole their *wit.*
He, that will finde a *Friend,* must seeke out one
To exercise unfeigned *love* upon;
And, *mutuall duties,* must both yield, and take,
Not for himselfe; but, for his Friendship sake.
Such, as do rightly *marry,* neither be
With *Dowries* caught, nor wooe a *Pedigree;*
Nor, merely come together, when they wed,
To reape the youthfull pleasures of the Bed:
But, seeke that fitnesse, and, that *Sympathy,*
Which maketh up the perfectest *Amity.*
A *paire,* so match'd: *like Hands that wash each other,*
As *mutuall helpes,* will sweetly live together.

THICKER THAN WATER

Introduction

AN IDEOLOGY OF BLOOD

"Blood's thicker than Water."

—A. RAMSAY, *A Collection of Scots Proverbs*... (vii.13.1737)

Although the proverb of this epigram was not recorded in its contemporary form until the eighteenth century, its sentiment far predates Ramsay's collection. John Lydgate's Middle English poem, *Troy Book* (1412), states, "For naturely blod will ay of kynde / Draw vn-to blod, wher he may it fynde" (book 3, 2071–72).[1] While recounting a fight between Ajax and Hector, here the narrative pauses to explain how the two were naturally drawn to each other in battle to find a suitable opponent. Blood, in Lydgate's account, not only explains affinity among consanguineous kin but dictates an innate attraction between those similar in "kind," of *like* noble blood. Readers of early modern drama will be rather familiar with this belief, from the trope of the *cri du sang*, the recognition of long-lost kin featured in romances, to friendships featuring like pairs such as Leontes and Polixenes in *The Winter's Tale*, and, finally, to romantic attraction founded in similar constitutions and, thus, a natural affinity, like the aforementioned protagonists' children Florizel and Perdita.

And where natural affinity based in a similarity of blood is not obvious, one is presumed. One particularly notable example comes from Shakespeare's *The Merchant of Venice*. Salarino, a friend of the Christian Lorenzo, tells Shylock that his daughter, Jessica, eloped with a Christian because "there is more difference between thy flesh and hers than between jet and ivory, more between your bloods than there is between red wine and Rhenish" (3.1.35–38).[2] Salarino here comprehends the romantic pairing by appealing to Shylock's purported bodily difference from Jessica, which, Salarino believes, substantiates their divergent characters. In his mind, despite their consanguinity, only a material difference could explain why Jessica has not inherited Shylock's temperament. Salarino's proclamation is partly metaphorical since these more obvious physical differences stand in for a physiological difference that one cannot see. The text gives no indication other than this quote that there is a visible difference between Shylock's and Jessica's complexions,[3] and we can safely assume that their bloods would not resemble white and red wines if spilt on the stage. And yet, what Salarino's exclamation implies is that in order for the world to function as he believes it does—that is, in order for Jessica to be a woman worthy of Lorenzo's love and for their coupling to make sense—Jessica must be *different* from Shylock in order to be sufficiently *like* Lorenzo, of his kind. Yet Salarino's proclamation effectively produces this distinction. This exchange betrays the ultimately rhetorical function of assumptions that are presumed to be physiological. Throughout this book, I argue that this play and others of this period stage limit cases for this belief in plots of love, courtship, and marriage, thereby demonstrating that blood functions not as a biological basis for affinities but *discursively*. Early modern drama highlights that *talk* of blood carefully defined groups, cast as kinds or species, and established hierarchies by asserting limits to affine and, especially, romantic relationships. This discursive system, moreover, was used to claim power, justify privilege, and effectively reproduce hegemony. And yet, though this system is discursive, it nevertheless has real, and often violent, consequences.

I call this system an *ideology* of blood because it is a collection of then-unquestioned beliefs about the world.[4] Patricia Canning aptly defines *ideology* as "a particular social, political, or cultural story or collection of stories[,] . . . a series of ideas that we make plausible in order to make sense of our conditions of existence, however obscure those conditions may seem to us, or indeed, others. That said, ideology is not only a mental phenomenon; it has very real manifestations because it is primarily a social *practice*. . . . Thus one of the fundamental ways by which ideologies are constructed, perpetuated and even opposed is through discourse" (1–2).

Steven Mullaney argues that though many scholars object to using the term *ideology* in early modern studies on the basis of anachronism, early modern people had the concept.[5] And for the ideology of blood that I trace here, conceptions of blood's physiological characteristics inflected how early modern people understood personal identity, social position, and affective relationships. Namely, it dictated to which groups a person belonged by telling them with whom they should feel affinity. Alternatively put, like racial ideologies that would follow, this ideology provided a way to construct differences between, and the worth of, individual persons or groups.[6] This collection of beliefs then justified the existence of a social hierarchy by proclaiming its foundation in an effectively unchangeable bodily material, that is, in blood, and further circumscribed social groups by explaining how blood and its concomitant characteristics could circulate among these persons.[7]

This ideology of blood has a few key tenets or logics regarding how blood prescribes affinities between persons, thereby limiting access to or allegiance with particular social groups. First, those who share the same blood (i.e., kin) have a natural affinity for one another determined by said consanguinity, or, "like likes like."[8] By extension, those persons who have similar blood—are similar in "kynd" like Lydgate's Ajax and Hector marking similarities in race, ethnicity, class, and virtues or character traits—likewise have a natural affinity for one another. And this tenet provides a biological foundation for appropriate romantic relationships and marriage matches. In short, every Jack will find his Jill; chivalrous, noble men will love virtuous, noble women, peasants others of their kind, birds other birds, and fish fully satisfied with their like in the ocean.[9] Second, in becoming "one flesh" in marriage, bloods mix or mingle together, bridging prior differences to create a new admixture and bloodline. This mixture of blood engendered by sexual consort dictates and completes an erotic, romantic, and spiritual love between the two persons, while simultaneously founding a new blended family unit in the creation of progeny, or new kin. The process then starts all over again, and the next generation acts in accordance with the same principles.

Now a commonplace in early modern studies, anxieties about identity and community membership ran rampant in a period fraught with political and religious struggles, new contact with global populations, and the simultaneous freedom and threat of newfound social mobility in a burgeoning market economy.[10] In light of this rapidly changing social landscape, the ideology of blood provided a means to solidify and stabilize the boundaries of particular social and political groups. Or, if blood embodied the virtues of nobility, then it justified the nobility's preservation of power and the hegemonic status quo as this system, quite literally, reproduced itself via endogamy. In this historical moment, invoking bloodlines maintained differences, distinctions, and boundaries between people and justified the privilege of some over others; but it did so by appealing to likeness and affinity. As Laurie Shannon notes in her discussion of Renaissance friendship, "The insistent emphasis on sexual and social sameness is a systematic response to that most acute form of early modern difference: the hierarchical difference of degree [. . .] a difference in 'kind'" (2).[11] In short, these differences in kind were sustained, in part, by outlining not only whom one is like but also whom one *will* like.

Thicker Than Water investigates how the drama of early modern Europe interrogated blood's ideology. Drama, by virtue of the multiple voices it presents and its definitionally performative character, provides a forum for examining these pan-European assumptions about blood. For, by performing this ideology in a theatrical space, plays could demonstrate blood's rhetorical status. Characters like Salarino exhibit that this vocabulary does not track incontrovertible truths but rather functions as part of a cultural discourse, one that can be manipulated in the pursuit of power, influence, and (sometimes) love in all social relationships. Theater undoubtedly played a role in ideological production, especially given the influence of the royal censor; it is likewise the case that some dramatic genres, like comedy, pushed to foreclose social alternatives by staging a return to the status quo. However, as Jean Howard demonstrates in her seminal work, *The Stage and Social Struggle*, clearly "drama enacted ideological contestation as much as it mirrored or reproduced anything that one could call the dominant ideology of a single class, class faction, or sex," and it thus creates space for the voices of marginalized groups (7).[12] While theater sometimes affirmed the values of patriarchy, racial hegemony, and aristocracy, it could simultaneously represent a variety of competing interests that lingered, providing a natural place for skepticism and noncompliance with prevailing cultural assumptions.[13]

I thus take for granted, as Stephen Greenblatt does, a similarity between literary (or in my case theatrical) characters and early modern persons.[14] Namely, drama stages theatrical characters negotiating the ideologies and discourses of contemporary society, struggling within or bristling against them, just as the social persons of their audience do outside of the playhouse. As theorist Paolo Valesio describes in "Practice of Literary Semiotics," social persons produce signs in their social system, though they are equally conditioned by them, and, as a result, these discourses can be both a source of power and of constraint. At a distance from the action onstage, and thus with a measure of detachment, the audience observes the variety of ways in which characters use, invoke, and manipulate the discourses around blood to achieve (or fail to achieve) their desired ends; the plays thereby encourage a measure of self-awareness about this ideology beyond the stage. Theatrical performance thus uncovers and renders visible the linguistic strategies of the powerful.[15] The plays examined here expose the political function of blood's ideology by staging limit cases for its logic in plots of love, courtship, and marriage. In them, some characters become creative within the confines of blood's ideology abiding by its rules and using its vocabulary. For instance, Romeo hopes to establish a new blended bloodline to dissolve the enmity supposedly inherited from both Capulet and Montague blood; Giovanni invokes blood's discourse to justify a love that goes beyond a seemly affinity for his sister; Helena hopes that by secretly bedding her beloved Bertram, thereby mingling their bloods, she can establish a love between them; and Jessica marries Lorenzo in the hopes that she will materially divorce herself from her father's Jewish blood. Yet, as these examples show, these plays likewise reveal that investment in blood's discourse—agreeing to play the game by the established rules—can have disappointing, tragic, and even disastrous, consequences just as often as it facilitates the achievement of an individual's desired ends.

Other marginal characters, however, see the limits of these discourses and search for an alternative. They realize that the game is fundamentally flawed, rigged even, and ask whether it is worth playing. These persons do not gain advantage from the existing social system and consequently are most open to *thinking otherwise.*[16] Disadvantaged and subjects of discrimination, these characters fall into two somewhat predictable camps, namely women and the lower classes (usually servants or clowns), sometimes both at once.[17] And their voices can be heard in this dramatic space, if not outside the playhouse. These marginal voices dare to ask whether blood must truly determine a person's nature, their allegiances, their loves. Could we smell roses by other names and is blood the result if we do not (Juliet of *Romeo and Juliet* and Elvire of *Le Cid*)? Might we signify bodies however we please regardless of biology, or might all persons be basically alike (Trasimaco of *La Sorella,* Pickadille of *No Wit, No Help Like a Woman's,* and Putana of *'Tis Pity She's a Whore*)? Are affinities or kin truly based in consanguinity (Lavatch of *All's Well That Ends Well* and well everyone in *La mandragola*)? And lastly, could love be found in difference rather than in similitude (for the lovers of both *Othello* and *The Merchant of Venice*)? All of these voices ultimately point to the same answer, which is, to paraphrase the King of *All's Well*, blood *does* confound distinction.

In sum, I hold that the European drama considered here has a few lines of social critique. First, the early modern period's natural philosophy and other cultural discourses about blood—both appealing to the aphoristic quality of "like likes like" and presuming that blood has agency—provide a convenient narrative that substantiates the elite's claim to power, privilege, and authority. Second, since it is an incredibly powerful and capacious ideology with adverse and violent consequences, this drama subjects the ideology to scrutiny by showcasing its limits.[18] And thirdly, this drama suggests that we might benefit from hearing and sympathizing with the communities that would challenge these prevailing assumptions. Yet, despite these critical gestures, these plays likewise intimate that finding alternative vocabularies to society's present ideology (*thinking otherwise*) is unbelievably difficult, often ineffectual, and might require great sacrifices. In short, these dramas call for the dismantling of this ideology beneath our deepest forms of social organization and oppression, but they recognize that the path to social change will be difficult and hard-fought.

RHETORICAL SELF-FASHIONING, BODILY DETERMINISM, AND LIKENESS

This study of blood participates in a long history of early modern work on identity or character, either as a product of performance like Greenblatt's "self-fashioning" or in relation to (determined by) bodies.[19] As Salarino demonstrates, discourses of blood largely resemble performative aspects of identity. Unlike costume, rhetorical mastery, or Castiglione's *sprezzatura*, however, talk of blood disguises itself as bodily determinism even as it belongs to the social construction of identity. Studies of bodily determinism in relation to early modern "character" have taken a few different tracks. One aims to show the origins of "inwardness" or a psychological interior such as Katherine Eisaman Maus's work on theater,[20] Jonathan Sawday's on early modern dissection,[21] Michael Schoenfeldt's on early modern poetry's citation of the humoral body,[22] and Eric Langley's discussion of sympathy and contagion.[23] Another strain of criticism focuses on specific discourses that reduce or materialize character/identity in bodies: most notably Gail Kern Paster's work on the humors, Maurizio Calbi's study of specific types of early modern bodies (e.g., the aristocratic female or male homosocial body), or David Hillman and Carla Mazzio's edited collection examining particular body parts.[24] In thinking about a specific bodily material, this project then builds upon the latter of these two strains, focusing on blood as an extremely capacious signifier that materialized all aspects of a person: personality, character, identity, innate affective predispositions, and even a "soul."

Yet my own discussion of this bodily material is not confined to the realm of natural philosophy—the Galenic geohumoral framework illuminated by Paster and others—which can be limiting, for the assumptions and beliefs about bodies press beyond the confines of natural philosophy and bleed into (pardon the pun) numerous cultural discourses. This project embraces a more disparate framework, attending to discussions of blood that betray the period's anxiety over the foundations of identity and group membership. These discourses range from treatises on nobility, to advice literature on marriage, to discussions of breastfeeding, myth, and even magic.[25] As Howard Marchitello asserts in *The Machine in the Text*, "Science cannot be construed as an essentially autonomous feature within early modern culture, [and] what we take today as early modern science is more properly understood as a set of desires widely dispersed among disparate cultural practices and institutions in early modern Europe" (15). I concur that no single early modern discourse asserts all of the assumptions underpinning blood's ideology simultaneously. Instead, the cultural imaginary of the early modern period likely conflated assumptions about blood from a Galenic geohumoral framework with a whole host of other assumptions related to identity and character, kinship and allegiance, the *commixtio sanguinis* model of marriage, and both affine and romantic love.

Moreover, it is unlikely that the Galenic geohumoral framework was unilaterally supported or fully believed. Steven Mullaney in his study of Reformation emotions frames it this way, "I remain skeptical, unable to understand how an etiological theory of the passions could become the basis for a phenomenology of emotions. . . . Such contemporaneous theories have a great deal to tell us about the history of ideas but their application to the way things felt—a phenomenology of Elizabethan emotions—seems, to say the least, problematic" (21). With regards to the discourses of blood and affinity then, I maintain that the distance between what early modern persons supposedly felt according to these theories of natural philosophy (i.e., instinctual sympathy for kin, etc.) and what they did, in fact, feel likely prompted many early modern persons to intuit that blood might *not* determine affinity or love after all. The early modern stage emphasized and highlighted this intuition, calling audiences to interrogate the ideology and social structure that it supported. I tend to agree with Julie Solomon that "rather than supporting the idea that Shakespeare simply endorses the materialist psychology of his age, Shakespeare's ironic representations of humoral explanation . . . suggest that early modern psychology was not simply reducible to a material base" ("You've Got to Have Soul" 198).[26] Yet I, nevertheless, agree with Paster, Rowe, Floyd-Wilson, and others that the Galenic-determinist viewpoint was both ubiquitous and a formidable force in this cultural imaginary.

This project likewise participates in studies on the role of likeness or proximity in early modern conceptions of identity. Marjorie Rubright, for instance, discusses the role of similarity, affiliation, and proximity in the context of Anglo-Dutch relations and the formation of English national identity in *Doppelgänger Dilemmas*. Valerie Traub notes the role of erotic sameness that paradoxically underlies the "naturalness of heterosexuality" (*Renaissance of Lesbianism* 261). John Dollimore, writing on sexual dissidence, importantly reminds us that the "adjacent becomes threatening in a way that the excluded never does" (52). These studies illuminate the ways in which differences are fabricated and emphasized when the "other" is too much alike. Scholars have traced this psychological process with a variety of groups, be it Amazons (Kathryn Schwarz in *Tough Love*), Catholics (Frances Dolan in *Whores of Babylon*), the Irish (Andrew Murphy in *But the Irish Sea*), or the Moors (Emily Bartels's *Speaking of the Moor*). Eric Langley underlines the latent perceived threat in concepts of sympathy, likeness, and correspondence since these concepts likewise characterize the dispersal of disease.[27] And finally, Uvrashi Chakravarty highlights the vexed relationship between sameness and difference in the evolution of the concept of family. She reminds us that the original denotation of the word *family* meant *household,* which included servants. But with the increase of servants of different races or ethnicities, these "blood strangers," though still emotionally intimate members of the family unit, were deemed strangers nonetheless; this new vocabulary then clearly delimits the concept of kinship as consanguinity ("More Than Kin"). Yet another instance of this psychological phenomenon, blood's elitist ideology could be explained as a response to the anxiety that mankind might have a common genealogy, a belief that would erase the naturalized foundation for class hierarchy. *Thicker Than Water* thus shares in this story about how discourses of likeness and liking (in both the affine and erotic sense) facilitate and perpetuate difference.

I now turn to a historical overview in order to illustrate blood's ubiquity as well as its utility as a vocabulary for character, identity, and affinity. I pull evidence from both primary and secondary historical sources from all over Western Europe (England, France, Italy, Germany, etc.). In so doing, I aim to demonstrate a shared cultural sensibility as well as the enormous reach of this discourse. Blood's ideology appears to be invoked by all of these cultures in similar circumstances for similar reasons—which would explain a shared dramatic, critical response to this ideology in my countries of concern.

BLOODY DISCOURSES, A HISTORICAL BACKGROUND

Blood as More than Humor

Blood was first and foremost a bodily fluid studied by natural philosophers. Material blood in a Galenic framework was one of the four humors (blood, phlegm, and yellow and black bile); the particular mixture of the four in the body determined the complexion or temperament of an individual. In theory, a physician could "discover" a person's dominant humor by examining how that person's body responded to the elements and disease. For example, bodies predisposed to overheating had a large amount of yellow bile in their systems, which could manifest itself in an aggressive personality. The healthy person, in both mind and body, had the ideal balance or mixture of all four humors. Yet blood was always viewed as superior to the other three. For Galen, blood was the most "natural" of the humors, and even though all were necessary for a well-balanced temperament, the other humors tended toward imbalance or pathological excess (Hankinson 219).[28] Men of sanguine temperaments, for instance, were prone to live longest; and while the other humors were only visible during illness, blood could also be a marker of health (Camporesi 17; Balizet *Blood and Home*14). Moreover, as stated in Levinus Lemnius's *De miraculis occultis naturae* (1559), "in force and value, blood lies far above the other three humours, as long as it remains pure, clean and clear," for if too thick, people become fierce, cruel, and even inhumane, typical of "rougher professions" (quoted in Santing 422–23).

Yet early modern thinkers likewise distinguished blood as humor from blood *as such* since, rather confusingly, as explained by Lemnius (c. 1505–68) in *The Touchstone of Complexions*, blood found in veins was the medium for all four humors (Santing 427).[29] The perfect application of the body's "innate heat" produced blood the humor, while either deviant scorching or cooling produced the others (Balizet *Blood and Home* 13). Blood was thus the in-between state of all other bodily fluids; or by way of illustration, the cow eats grass, which becomes cow blood, that becomes the milk that the human drinks to become part of their own blood and temperament, seemingly justifying the old parable "you are what you eat" (E. Keller 103–4).[30] Reproductive bodily fluids (semen and menstrual blood) were also variants of blood; and because early modern people believed that conception was the combination of semen and menstrual blood—both of which contained all four humors necessary for creation—blood was the source of life (Camporesi 84–86; Hankinson 218). This second "blood" is roughly equivalent to a person's individual or unique temperament, made more plausible because the number of humoral admixtures were theoretically infinite. The complexion or temperament indicated by or embodied in one's blood was thus unique and highly individual, effectively corresponding with one's soul or "essence."

This equivalence of complexion/temperament with individual identity and life force contributed to the notion of blood as having spiritual content and transmitting inherent qualities (Guerreau-Jalabert 61). This belief has a classical precedent, for the Greeks believed that the *thumos*, the organ of feeling or inner voice, resided in blood (Cherpack 10); Homer's Odysseus, for example, gives those in Hades blood to drink so that he may speak with them. Yet blood's connection to the soul or life breath likewise has origins in the Old Testament's discussion of sacrifice, further solidified by the Christian belief that the divine blood of Christ brought humanity's salvation (Delille 136–37; Camporesi 14).[31] In Germany, blood was etymologically connected to a blooming flower, representing the living human as well as life more generally (Linke 44). As Piero Camporesi outlines, in the early modern imagination, "Blood was the seat of the soul—that invisible, elusive principle (abstract, but not without its inexpressible physical concreteness)—that was deemed to ebb and flow in hiding, swelling and diffusing in the oily liquid of life" (32).[32] This "hemocentric" theory in natural philosophy may have originated with the pre-Socratic philosopher Empedocles, but it gained popularity with its partial acceptance in Aristotle's *De partibus animalium* (Conticelli 55, 60). Of course, Galen later expresses this conception of the soul in his work, *The Capacities of the Soul Depend on the Mixtures of the Body*. Although the nature of this "dependence" is not entirely clear,[33] Galen proffers that the "soul actually *is* precisely this: the mixture of the body" (386). This belief follows from his starting assumption that young children must differ from each other in the substances of their souls precisely because he believed their diverse capacities (intellectual or otherwise) could not be the result of education (375–76). According to his theory, the soul joined with blood via the spirits "a kind of vapour, streaming or smoking from the thinnest and most active part of the blood" (Sugg 3). Numerous early modern scholars reiterated this principle, including Thomas Adams; Thomas Bilson, the bishop of Winchester; Helkiah Crooke in his *Microcosmographia* (1615); and Spanish scholar and martyr of the Reformation Michael Servetus (Sugg 18; Balizet *Blood and Home* 12).

Consequently, the health and qualities of one's soul were inextricably linked to the health of one's body. Galen states in his treatise *Character Traits*, "The knowledge of the soul . . . from which comes its fineness, knows the elements of which the body is composed and from which the affections of the soul are generated, composed and increased. The knowledge of this is followed by discovery of their treatments" (162). Galen advocated for bodily management that would improve the soul for "the mixtures themselves are consequent on the original formation and on well-humoured daily regimes, and these things mutually increase each other. So, to be sure, people who become sharp-spirited because of the hot mixture then fire up their innate heat by their sharpness of spirit; and those who are well-balanced in their mixtures, having well-balanced motions of the soul, are assisted towards good humour" (*Capacities* 408–9). Lemnius's *Touchstone of Complexions*, a descendant of this Galenic tradition, explicitly frames itself as a manual for every person to know their individual "habit and constitution." Lemnius judges, "It right needful also to have a diligent eye and respect to the body, leaste [otherwise] it should be a burthen to the Soule, and hinder it from matters of more wayte and worthines" (chapter 1). The period envisioned humoral management as a whole, and phlebotomy especially, as a way to govern and amend one's soul, preparing oneself for goodness and salvation by ritually cleansing the humors at the close of seasonal cycles (Camporesi 27). Bloodletting theoretically purified the blood as a means by which "faded humors, the 'passions'—those agents of evil and of physiopsychological imbalance [e.g., sins of carnality, lust, sloth, and churlishness]—could be constrained to spurt away" (Camporesi 28). Nicholas Gyer, both a theologian and clergyman, discusses the benefits of bloodletting in *The English Phlebotomy* (1592): "First it maketh glad those that are pensive. Secondly, it appeaseth such as are angry. Anger is especially caused through mixture of much yellow choler with blood. And sadness, by commixture of much melancholy with blood" (quoted in Belling "Act of Flesh and Blood" 38). In short, in this Galenic worldview, there was no discernible difference between the spiritual and the medicinal, the physiological and the theological.

Blood then did not simply operate as a metaphor in this period, but the difference between signifier and signified all but collapsed in reference to blood and other bodily fluids. Scholars of early modern bodies Paster, Rowe, and Floyd-Wilson argue that one cannot separate the body and soul in this period, in an explicitly pre-Cartesian moment ("Shakespeare and Embodiment").[34] As Charles Taylor argues about the Galenic framework, "Black bile doesn't just cause melancholy; melancholy somehow resides in it. The substance embodies the significance" (quoted in Paster *Humoring the Body* 5). Although works like William Harvey's *De Motu Cordis* (1628) on the circulation of the blood posed an empirical challenge to older scientific discourses,[35] assumed connections between theological and medical epistemology persisted in anatomical treatises well into the seventeenth century (Balizet *Blood and Home* 12). In sum, blood persisted as more than metaphor despite changing scientific and philosophical discourses.

Blood's Nobility: Lineage and Character Traits Materialized

In the previously outlined formulation, blood literally carried characteristics or traits passed down from one generation to the next; it was the repository of properties and virtues that could be reproduced indefinitely. Nobility was not merely a social station (although it included material advantages such as lands and rights), it was also a natural predisposition carried in blood indicating gentility, bravery, refinement, politeness, and the like.[36] As Francis Markham asserted in his handbook for noblemen, *Booke of Honour* (1625), blood had "severall degrees" (46). In a Galenic framework, the blood of different degrees even looked different; common blood was supposedly thinner and less refined, a product of sweating during labor, accounting for "their relative coarseness of feeling, poverty of manners, and spiritual simplicity" (Berkeley 7). A Germanic belief took this notion of degrees to an extreme, holding that members of certain blood lineages had superior, even supernatural abilities derived from their blood "such as the power to ensure good crops, victory in battle, and the ability to heal certain diseases" (Linke 54). As Jean Feerick importantly reminds us, family and social station inherited via blood were not social constructs but far more akin to how we understand race today, "trac[ing] deep divisions among people to the body's inner recesses, mystifying and essentializing those markers" ("'Rude Uncivill Blood'" 65). After all, John Florio's Italian-English dictionary *Worlde of Wordes* (1598) defines *razza* or race as "a kind, a brood, a blood, a stock, a pedigree" (545).

Additionally, discussion of bloodlines or lineage led to privileging the preservation of the material bloodline over and above its individual members. As historian of the family Lawrence Stone explains, "Allegiance to lineage and kinship creates a system which gives priority to the permanent interests of the 'house,' its maintenance, continuity, function and economic well-being for century after century. The living members, the kin, are regarded as no more than the trustees of the lineage's name, property and blood" (29). Esteeming blood over person not only serves as the foundation for the blood feud's logic but also explains some of the seemingly fetishistic treatment of blood throughout texts of this period—including many of the otherwise puzzling motivations of characters in early modern drama (Corneille's Chimène, Shakespeare's Bertram and Helena, and Ford's Giovanni, for starters).

Consanguinity, Inheritance, and Mixed-Blood Progeny

The notion that blood embodies or materializes certain character traits has a profound effect on theories of procreation, taking for granted "the key idea that like engenders like" (Sabean 153). According to Germanic folklore, one could even verify consanguinity by dropping some blood on remains; if the bones absorbed the blood, then the two parties shared a bloodline (Cherpack 10). The conventional conception of European kinship is "based on the idea that an individual is composed of a substance—more or less explicitly thought of as blood—that comes in equal parts from two parents" (Sabean and Teuscher 5).[37] Evidence for belief in equal blood mixing comes from incest proscriptions in canon law wherein one counted degrees of separation on both sides of the family tree.[38] Galen's theory of generation supports this assumption since both male and female sperm, two forms of blood, contribute to the fetus, suggesting a roughly equal composition (Sabean and Teuscher 5–9). Moreover, observations of physiological similitude seemed to favor this formulation, for clearly offspring never replicated a single parent perfectly; thus "the similarity with respect to any one part was said to correspond to the quantity of semen from each parent for that part" (Russell 25).

There were, nonetheless, conflicting views about the amount of blood inherited from each parent, especially since many early modern scholars were eager to emphasize patrilineal descent. For example, some argued that semen was the primary transmitter of "pure" blood (Sabean 153). Or, in an argument for women breastfeeding their own children, a mother's milk, although a form of her blood, was "constituted by her husband through the act of generation" (Sabean and Teuscher 10).[39] Hence, while most scholars of the period claimed that fathers, not mothers, passed down nobility, notions of blood-mixing clearly remained. Stefano Guazzo (1574), for instance, held that mixed marriages would debase blood, producing less courageous children (Delille 129). Moreover, most discussion of breastfeeding in advice literature emphasized the mother's blood and the qualities or characteristics it contained.[40] Since breast milk came from the menstrual blood, which had nurtured the fetus, this blood-milk continued to shape children and their qualities. This meant that if the mother herself could not breastfeed, then the nursemaid should closely resemble the mother. Advocated by advice literature like Lorenzo Gioberti's *Errori Popolari,* this choice was "preferable in any event to animal milk—goat's milk or cow's milk—which might degrade the little man-to-be and push him in the direction of the brutes" (Klapsich-Zuber 161). Often authors cite racial discourses to make this point more forcefully. For example, Matteo Palmieri says in his dialogue *Della vita civile,* "Nature in all its perfection nourished the baby in the womb with his mother's blood, and now this blood has moved upward to her breasts. No other food can be as good, and if the substitute is too far removed from the mother's characteristics then the child will become like a stranger, just as the white lamb suckled by a black sheep turns dark and the lamb given suck by a goat not only develops coarse, tough hair but sees its body desiccated, its voice and habits changed" (quoted in Bell 128–29). In this context, then, marriage to someone of similar blood was deemed immensely important for the maintenance of familial lines and the preservation of their virtues in progeny.

Unitas Carnis and Commixtio Sanguinis: Mixing Blood in Marriage

A child's blood was not the only matter at issue in the rite of marriage, however. The saying that two would become "one flesh" or *unitas carnis* (Gen.

2:24) appears to have been taken literally. And if not effected in the moment of marriage's sacrament,[41] then this process was presumed to take place during the sexual intercourse of marriage's consummation.[42] Just as Adam and Eve were literally of the same flesh—Eve pulled from Adam's rib in the first marriage in Eden[43]—so too becoming "one flesh" was not limited to the generation of matter (i.e., a child) in sexual reproduction but was the by-product of matrimonial union. Paul's letter to the Ephesians (5:25–28) provides biblical support for a one-flesh model for "husbands should love their wives as they do their own bodies. He who loves his wife loves himself" (quoted in Bell 236). Calvin in his commentary on 1 Corinthians also advocated this model as a way to emphasize the necessity of marriage, saying that "a 'man without a wife was half a man', and that only together did 'l'homme et la femme font l'homme tout entier' [man and woman makes man completely whole]" (Lipscomb 351, my translation).[44] And ideally, this blended unit of one flesh forged *unanimitas*, or the bringing together of individual, and sometimes contrary, desires and needs into a single, harmonious will (Bell 226).

This discussion of shared flesh changes to one of blood possibly due to Old Testament references connecting blood to the soul (Delille 136–37). In the fifteenth century, people began to speak of lineages and bloods mixing in marriage. Certainly blood seems central to the symbolism of the wedding ceremony since "the placement of the wedding ring on the fourth finger of the left hand . . . took its meaning from the traditional belief that a vein ran from the ring finger directly to a woman's heart" (Balizet *Blood and Home* 31). Yet, because the vocabulary of sexual intercourse changed from *copula carnalis* to commixtio sanguinis (the mixing of blood), this new designation provoked greater attention to deciphering how this might work. Philosophers like Benedict Wincler in his *Principiorum iuris* (1615) gave a Galenist rendition of marriage, which explicitly equated semen with blood and posited that a woman might also need to ejaculate a blood of sorts to complete the process (Sabean 146, 166n12).[45] Moral theologian Jean Pontas in the *Dictionnaire de cas de conscience* (1715) goes so far as to claim that affinity is established by the *commixtio seminum*, and thus, according to criminal law, the man must ejaculate or release seed into the vagina in order for the marriage to be fully consummated; moreover, since bloods have been mixed, this affinity does not end with the death of either spouse but would persist between the families (Sabean 166–67n12).[46] The eponymous character of Corneille's *Agésilas Tragedy* confronts his brother about his niece's marriage based upon this very assumption, insisting that his blood (their family blood) would pass from the father to his son-in-law via his daughter (3.3.1:391).

This belief in the mixture of bloodlines prompted a degree of ambiguity between consanguineous and affinal relations—that is, kin by blood and kin by marriage—in which there was theoretically little distinction. If blood mixed in marriage, then even relations by marriage were consanguineous. Old Norse has a linguistic trace of this literal belief wherein "kinsmen by marriage are referred to not only as 'sib-folk' but also as 'minglers of blood'" and most familial terms in German do not distinguish between relatives by birth and those by marriage (Linke 55). Henricus Greve, a commentator on the *Arbor consanguinitatis* by Ioannes Andreae, held that affinity is a *purely* material matter; affinity "is made by blood, not by words" (quoted in Teuscher 96). The ambiguity/imprecision of kinship vocabulary mirrors this loose system of consanguineous connections. As historian David Cressy articulates,

> The language of kinship in England was (and is) limited and loose. The basic relational terms—uncle, father, daughter, cousin, etc.—were used without precision or consistency.... Female offspring and the wives or widows of sons or stepsons could each be addressed as "daughter." "Daughter-in-law" could mean a son's wife (more commonly simply called 'daughter') or a stepdaughter, a spouse's daughter by a former marriage. "Uncle" could indicate any male relative of an ascending generation, and "nephew" one of a descending generation, without strictly referring to a parent's brother or a sibling's son. (66)

This belief that blood-mixing purportedly occurred between spouses, for one, explains why marriage interdictions aimed at preventing incest include both consanguineal *and* affinal kin (Guerreau-Jalabert 72). In marriage and its consummation, one's body becomes a vessel for this new mixture of bloods, and one's person or identity becomes susceptible to change. For instance, this new blood-mixing was the means by which non-Christian spouses were converted or baptized in marriage. As Paul states in 1 Corinthians, "For the unbelieving husband is sanctified by the wife, and the unbelieving wife is sanctified by the husband: else were your children unclean"

(7:14). In fact, in Corneille's drama *Polyeucte*, "both father and daughter are converted to Christianity following the martyrdom of the son-in-law/spouse, [for] the blood of the spouse falls upon his wife with a salvific effect" (quoted in Sabean 157). However, the body might likewise be subject to contamination or corruption, which is the impetus behind the genre of marital advice literature written for an elite audience. As Torquato Tasso mentions in *Householder's Philosophie*, "Matrimonie maketh equall many differences" (quoted in Cook 41).

"Like Likes Like"

Luckily then, a person's blood presumably dictated affinity for others of their "kind." Love's purported basis in similitude helped to mitigate the potential threats of blood-mixing. English cleric and Calvinist Thomas Taylor pronounced, affective bonds can be discerned between "every Creature [who] hath agreement, and sympathy with his kinde [as] things thrive best among their like" (105). Moreover, philosopher Erasmus, in his colloquy "Amicitia or Friendship," speaks to natural, instinctual sympathies or antipathies based on similarity or difference in character. The speaker claims that "[I] had mortal Aversion to a Lyar. . . . I abhorred him more than a Viper, and knew no Reason for it, but only a certain hidden Instinct in Nature" (314). One would assume then, that natural sympathies would be no different for prospective romantic partners as they theoretically were for friends; and blood that carried nobility would instinctually prompt love for virtuous nobles in order to produce more virtuous and noble children.[47]

Most early modern scholars dispensing advice about marriage emphasized equity in lineage or status not merely for practical or tactical reasons but because it assumed a parity of values, temperament, and compatibility resulting in love. For instance as early as the twelfth century, Andrew the Chaplain—a clergyman on the fringe of the court of King Philip II of France (1179–1223)—includes a moralization of hierarchy and class difference in his treatise *Tractatus de amore* (c. 1180). He insists upon rankings in exclusive, ascending levels with boundaries that should not be crossed, just as species should not mix because "social groups had common mental and physical characteristics" (241). In an example from early modern Italy, Leonardo Fioravanti's *Dello specchio di scientia universale* (1564) insists that "bride and groom should be of equal blood. This rule may not bring you total happiness, but at least it will result in your avoiding the worst disasters and scandals that arise from an unhappy marriage" (quoted in Bell 215). Likewise, Robert Crofts in *The Lover, or Nuptiall Love* (1638) preaches, "Choose such as are of equall yeeres, birth, fortunes, and degree, of good parentage, and kindred, of such a countenance, complexion and constitution, as best agrees to our love and disposition" (quoted in Cook 40). Here, references to compatibility are not limited to those of personality or status conceived as a social marker, rather words like *complexion* and *constitution* have a biological valence and love supposedly arises from this biological compatibility.

This advice also frequently references the dangers of material incompatibility. Philip Gawdy, esquire of the body to both Elizabeth and James I, wrote in a letter, "It is thought very fytt and conuenyent that euery man now sholde marry within his owne element" (quoted in Cook 45). The term *element* here can simply mean "the surroundings in which one feels at home" or "source of origin" (*OED* 12), but it likewise indicates the matter of substances involved in one's composition (*OED* 2). Ludovico Dolce's *Della institutione delle donne* (1545) makes an explicit comparison between pairing similar personality types with pairing similar species. He advises, "Good companionship is most likely among equals, so do not try to marry your daughter to someone either superior or inferior to yourself in rank. A household where there is no love quickly falls to ruin. The second rule is that the prospective couple should be of similar personality and habits. You can't expect a wolf to live together happily ever after with a lamb, and the same goes for people" (quoted in Bell 218). Dire warnings accompany the act of pairing like with unlike in Daniel Rogers's *Matrimoniall Honour* (1642), which says, "When a poore party meets with a rich, a well-bred one with a rude and illiberal, a curteous with a froward, a bountiful with a miserly, a noble with a base; one from the court with another from the cart or the shop; a proper and personable, with a deformed, crooked or dwarfe, what a disproportion doth it cause, and a kind of loathsomnesse?" (quoted in Cook 40). Rogers cautions that pairing unlikes results in hatred between the couple; however, his vocabulary of "disproportion" and "loathsomnesse" similarly intimate his own feelings of distaste or disgust for these unlike pairings. And his list, which spans from an incompatibility in personality to social and bodily ones, has the effect of conflating all of these considerations. John Ferne's character Bartholus in *The Blazon of Gentrie* (1586) voices this position even more adamantly. He admonishes that even if the woman of lower class is worthy on other grounds such as beauty or wealth,

> yet for all this, heere is a disparagment, an is ins the vnequall coupling in yoke of the cleane Oxe, and the vncleane Asse, an iniurie not onely done to the person of the yong Gentleman, but eeke a dishonor to the whole house from which he is descended, and that vppon his reason in that nobilitie, shoulde bee ioyned with ignobilitie: so that the issue of their kinsman, discended as it were, of *Abraham and Agar,* should seeme but halfe noble, nay but halfe a man monstrous in kinde, and degenerated from the rest of his family: and hereby the house suffereth a lamentable diminution and rebatement of her Noblenes, even in that ligne, which ought to be the chiefest honor to the same. (9–10)

According to Bartholus, half-noble mongrels harm the noble gentleman as well as his entire house or bloodline. It is best for the bird not to love a fish, however beautiful she may be, in order to avoid creating half men "monstrous in kinde." Even if (and this is a big if) these descriptions were fundamentally metaphorical, these are very plainly not innocent metaphors. Indeed, as Ania Loomba remarks, "sexual intercourse between members of different groups was the kind of crossover that generated the greatest anxiety. Since sperm was widely understood as man's purest blood, sexual activity was an exchange of blood" with profound consequences ("Delicious Traffick" 213).

Affinity or sympathy between those of similar blood only increased for those who shared the *same* blood, a philosophy that substantiated love and loyalty among family members as well as romantic love between spouses having been mixed in the process of commixtio sanguinis.[48] In fact, the "mingling of blood is sometimes supposed to be a direct cause of mutual sympathy. . . . Even in Europe there are traces of the belief that a few drops of blood transferred from one person to another inspire the recipient with friendly feelings" (Edvard Wesermarck, quoted in Cherpack 10). "Natural love" between parents and their progeny was a commonplace. Gasparo of Castiglione's *Il Cortegiano,* for instance, reiterates the importance of a woman's chastity to ensure knowledge of true offspring, for if not, "the bond would be dissolved that binds the whole world through the blood, and through each man's natural love for what he engenders" (176).[49] Richard Younge's *The Cure for Misprision* (1646) likewise insists, "There can be no amity, where there is no sympathy, yea, though they be man and wife, Parent and child, [as] if they be not alike, they will not like" (4). The well-known cri du sang (cry of blood), an established commonplace of the romance genre, presupposed that this natural affinity would be so strong between consanguineous kin that one could miraculously identify long-lost relatives due to a body recognizing its blood. Pietro Pomponazzi in his *De Incantationibus* (1556) describes the physiological basis for this phenomenon whereby the blood from one person emits vapors that then enter the bloodstream of the other person through their pores (Cherpack 11). In another strange application of this assumption, Francesco Barbaro, in his treatise "On Wifely Duties," suggests sharing a wet-nurse as a means to foster sibling-like affinity between playmates on account of the shared consumption of the same breast-milk, that is, blood (Bell 129).[50] Moreover, consanguinity seemingly had the capacity to carry affinity (or familial duty) across oceans. Edward Johnson, a migrant to Massachusetts and author of *Wonder Working Providence* (c. 1630), idealizes "kindred blood that binds the bowells of affection in a true lover's knot" when articulating how colonists naturally miss their kinsmen (quoted in Cressy 47).[51]

From a lexical standpoint, the etymology of *affinity* serves as a case study for the ambivalent notions yoked together in the period's conceptions of the relationship between blood, kinship, and affection. The Latin etymon *affinitāt-, affinitās* has a set of divergent definitions, though they understandably connect. In classical Latin, it designates a relationship by marriage and in postclassical Latin, it also designates neighborhood or adjacency. The Anglo-Norman etymon however—*afineté* to later become French *affinité*—began to absorb a far more conflicting set of definitions. At first, it meant "from a neighboring region" (1160), which seems to give rise to meanings of connection, mutual dependence, or alliance. About a century later, it designated a relationship by marriage (c.1260), often used in contrast to consanguineous relations, such as those seen in debates around incest prohibitions in canon law. It then extended to our contemporary sense of likeness concerning a relationship between people based in their similar characters and tastes (1283) but then also more generally a similarity or resemblance (1286) that later included structural resemblances between species of animals, plants, or minerals (1538). Connections based on similarity then understandably were applied to friendly relationships (c.1355). This definition expanded to signify our sense of liking or attraction to a person or thing, natural inclination, or sympathy (as early as 1600). This trend continued to include spiritual (nonfamilial) connections between godchildren and their godparents (early fifteenth century) (*OED* "affinity, n.").

And yet, despite the long history of the word's application to nonfamilial relationships, in explicit contrast to those of consanguinity, *affinity* in the early fourteenth century was applied to family or kin, explicitly designating a blood relation. Affinity's etymology is, to put it mildly, curious.[52] However, this confusion makes a great deal more sense when one considers the discussion of blood and kinship relations in natural philosophy. If blood mixes upon consummation, then there *is* no difference between affine relations and consanguineous ones. And the definition of *affinity* thus tracks the discussions of likeness in character among those similar in "kind" as well as the assumed natural attraction between those who are alike in character.

In addition to familial love or affinity resulting from consanguinity, blood had a well-established connection to romantic or erotic love. In a variation of the Venus and Adonis myth, Adonis's flower was originally white, only stained red by blood from Venus's foot, which she pierced with a thorn in her rush to his aid (Linke 50–51). Similarly in a German myth—recorded in 1790 but evidently circulating prior—the red rose, a symbol of romantic love, was red on account of Amor's blood. The story goes that looking for a flower that would suit the celebration of his mother, Venus, Amor only found a silver rose. When he went to pluck it, the thorns cut his finger, and the rose was permanently stained with his blood (Linke 51). Outside the realm of myth, natural philosophy held that the process of falling in love involved blood, as for instance described by Giambattista Della Porta's *Magia Naturalis* and Marsilio Ficino's *De Amore*. Although one fell in love through sight, the spirits "caught" through the eyes transmute into blood that then dictated affection or sympathy (Sugg 45).[53] One could then explain attraction in "purely material (and perhaps selfish) terms. You are attracted to another (at one level) because they have quite literally taken something of yourself" (Sugg 42).[54] Also, those that had sanguine temperaments were most susceptible to the passion of love and lovesickness, as explained by Arnaldo da Villanova, the first medical doctor to write a treatise on love sickness titled *De amore qui heroicus nominatur* (c.1300) (Ciavolella 69). Pierre Le Loyer likewise notes in his work *Des spectres* (1586), "The most furious love which physicians know is Heroic love which is wont to overtake heroes and men of valour in whom the sensual element abounds because they have a greater share of earthly spirits and a body filled only with hot and wicked blood" (quoted in Céard 43).

The fact that blood was a key ingredient in superstitious and magical practices around love potions provides further evidence of a widespread belief in a material basis for love and affection. Laurent Joubert's *Erreurs Populaires* (1578) mentions the popular superstition that "women in some places command the midwives attending the birth of a daughter to save the 'vedille' or 'nombril' (navel cord), so that in later years they can make a powder of it to give to the man whom they would like to fall instantly in love with their daughters" (Wack 20). The logic went that the umbilical cord, carrying blood from the infant girl in utero, would dictate the affection of her future beloved if ingested (Ruggiero *Binding Passions* 115–16). Evidence of another practice comes from a confession made to the Holy Office in 1584: "Giulia, the daughter of Ludovico di Verona, . . . offered to spill the blood of her intended lover two-fifths for herself and three-fifths for the eclectic mix of damned and divine spirits that she called on to help her" (Ruggiero *Binding Passions* 247n50). In theory, the mixture of their two bloods together—a parody of commixtio sanguinis resulting from marriage's sexual consummation—would create the elixir necessary to call on the spirits who would cast the love spell.

Sex as marriage's consummation or the completion of the commixtio sanguinis was thus presumed to inspire romantic love. For if similar blood inspired liking among those of the same class or kind, and consanguinity inspired affinity and familial love, then as logic would dictate, so much greater must the love be between those who have the exact same blood after their bloods have mixed. Clergyman Daniel Rogers's conduct book *Matrimoniall Honour* (1642) explains, "There ariseth in the spirit of an husband (who is not degenerate,) a merveilous natural and tender instinct of sympathie towards his wife [who is] one with him in all things, one in flesh . . . in the superlative union of marriage, wherein . . . union breedes love, sympathy and compassion" (237). And a practitioner of magnetic physic, Christopher Irvine goes so far as to assert that "'neither love nor compassion [can] be generated without the commixture of the spirits' whereby 'the spirit of the body, being married to the qualities of that body, is communicated to another body, . . . generating a certain compassion'" (9). Jansenist theologian Antoine Arnauld even warns in his *Lettres* (1727) that conjugal love can turn into brutal or excessive passion if consanguineous kin added this conjugal mingling of blood (Sabean 163). This physiological process of producing love was attended by a linguistic ambivalence; as Lawrence Stone notes, "the distinction between love and sex was not clearly defined, the latter word to describe the purely physical act not being generally employed. The phrase 'to make love' is a relic of this early linguistic ambiguity" (644). And even though Catholic theologians did not generally emphasize sex as the means for inspiring spousal affection, they still held the basic premise that sex inspired love and emotional bonding (Brundage 9). Sex as the foundation of matrimonial love—the latter phrase providing a euphemism for the former—was further endorsed by texts such as Flaminio de' Nobili's *Trattato dell'amore umano* (1556) and Giovan Battista Giraldi Cinzio's introduction to his collection of *Hecatommithi* (1565) (Giannetti 210). In sum, whether it be familial love determined by consanguinity or erotic love founded in newly mingled blood, there was a clear precedent for the basic assumption that like not only engendered like but inherently liked its like.

Bloody Discourses of Stability

Upon completing this picture of blood's ideology, it is easy to see how this discourse functioned to both establish and reproduce hegemony. Indeed, the preoccupation with the quality of one's blood lineage appeared to intensify in a time of social change, aiming to solidify social, economic, and political power. Although this fixation began in the Middle Ages,[55] there is evidence that the discourse of agnatic lineage increased in the early modern era to stabilize the boundaries of particular groups. In Simon Teuscher's study of treatises on kin diagrams called *arbores consanguinitatis,*[56] he describes the phenomenon as follows: "Speaking of blood rather than of flesh appears to liquidize the bodily substances of kinship and to facilitate processes of mixing and dilution. Nevertheless, the notion of being of the same blood paved the way for exclusive conceptions of belonging. Thus being of one blood, or of the right blood, became a precondition of belonging to the nobilities and the patriciates at the passage to the early modern period" (101). Fellow historian Gerard Delille seconds, "Naturalist theory, which rested on the idea that qualities are transmitted through 'inheritance' and so describe, classify, and perpetuate particular and definite human groups, began to assert its position vigorously in a debate that blended the definition of nobility with social orders, social mobility, the place and function of kinship, the nature and role of blood" (129).

This trend seemed to spread throughout Europe. German historian Georg Kugler, for instance, notes the obsession with blood in the Hapsburg dynasty, producing numerous incestuous marriages. Given how often lands shifted hands in the period, there was a need to invest power in something else like blood, which cannot be taken or annexed in the same way (Kugler 122).[57] Early Venetian humanists defended the discourse of nobility carried by blood alongside the closure of the Venetian Republic's noble order in 1403. This closure prevented the purchasing of titles and thus the possibility of rich commoners' participation in government (Pullan 157). One example of this defense, Francesco Barbaro's *On Wifely Duties* (1416), asserts aristocrats' duty to preserve their lineages through marriage in order to maintain the inheritance of ruling offices in the Venetian Republic; having inherited the right to rule, they likewise inherited the responsibility to provide future rulers (D'Elia 28). Historian Anthony D'Elia observes a greater emphasis on genealogy in wedding orations composed by the humanists of Renaissance Florence as well. These orations effectively acted as propaganda—justifying and solidifying the power of ruling elites by claiming the inheritability of attributes (65–72).[58] The work of John F. Padgett likewise demonstrates that endogamy was a guiding principle in Italy in an effort to stabilize social position among the elite, even though it did not always work to prevent social mobility in republican Florence.[59]

France shared this impulse to invoke blood purity in this period. Arlette Jouanna's *L'idee de race en France* lists seventy-seven authors who debated a nobility of race (understood as hereditary bloodline) versus a nobility of virtue between 1550 and 1615. And André Devyver has shown that the self-conscious equivalence of nobility with blood purity increased after 1560 when the Count of Rochefort called for the end of ennoblements in an attempt to limit the increasing number of *parvenus* among them. Rochefort claimed the ancient prerogatives of blood as the medium of psychic and moral virtues, including bravery and strength, which were inherited from the ancient conquering Franks, and thus justified their monopoly on military power (7).[60] Similarly, Cardin Le Bret as *avocat general* insisted in the Cour des Aides (1598) that "nobility depends less on the law than on nature, that is to say, on the law of nations and on a series of degrees of consanguinity" (quoted in Descimon 97). This myth of Frankish blood gained more force toward the end of Louis XIV's reign, in part due to the nobility's continued impoverishment but also because of the decline that they suffered in prestige as a function of the court at Versailles, augmented by the work of Boulainvilliers.[61] This vocabulary incited a great deal of debate and gave rise to a legal definition of nobility.[62]

In England, too, works like John Ferne's *The Blazon of Gentrie* (1586) proffers advice for the English nobility in the opening epistle, disparaging those that stand upon "the vaine ostentation of noblenes and bloud, thinke themselues without vertue most truely noble" and insisting upon a nobility of character (*Epistle to the Gentlemen of the Inner Temple*).[63] Yet the text also contends that the highest brand of nobility *depends* on the body in addition to merit and virtues: "For as gentleness of blood, with ungentleness of mind, can not make the person perfectlie noble: so on the other part, the worthy merits of any man, borne of vnnoble parents, can not make him a perfect Gentleman, although thereby he may condignly deserue by lawe, to possesse both ensigne and title of Gentlenesse" (23). The epistle likewise reminds its readers that the Innes of Court at their inception only permitted gentlemen of blood, "for to the mind, inclining towards vertue, it auayleth [avails] much to be borne of gentle bloud" (24); and the most perfect of gentle blood required at least five lines of descent. Finally, the text provides an extremely long list of privileges that these bodies afford to gentlemen in contrast to the commonwealth (87).[64] In sum, one might characterize Ferne's message as follows: nobles, your bodies afford you the potential for virtue and a great deal of privilege, please start acting accordingly. Clearly, Ferne's text is symptomatic of a moment when the nobility were anxious about their status and thus asserted it more forcefully.

There were, nevertheless, difficulties built into these discourses for those who desired to solidify and delimit power in particular bodies. Blood discourses could stabilize qualities or character traits, but blood itself was also subject to change and flux according to these formulations. For one, there was the possibility of degeneration, namely that a bloodline could be progressively watered down. Sir Thomas Elyot's *The Governor* (1531) discusses a variety of material practices aimed to help noble families avoid this degeneration of noble blood and encourage their children's noble temperament (Feerick *Strangers in Blood* 16). Moreover, as part of the humoral discourse in which the body remains an open system, a person could be susceptible to change with their climate.[65] With the beginnings of the exploration era and colonial transplantation, this susceptibility became a greater concern, one discussed in Juan Huarte's *The Examination of Men's Wits* (Englished 1594)

(Feerick *Strangers in Blood* 17–19).[66] Yet, despite these difficulties with blood discourses, talk of blood prevailed, popping up everywhere from appeals to sympathy, loyalty, and shared traits among kin; to justifications for loving or hating another; to concerns for the integrity of a person's body or personality. The elite cited blood for the purposes of social, economic, and political hegemony, but this vocabulary was, as we have seen, far-reaching and capacious. Blood was thought to dictate many qualities of a person, contributing to a person's social status as well as their sense of self and with whom they belonged. The drama of this period then brings blood's multifaceted discourse into play (pun intended) and, likewise, open to scrutiny.

BLOOD ON OUR HANDS: CRITICAL COMPANIONS AND METHODS

In addition to larger critical discussions of identity, bodily determinism, and likeness, this project participates in a resurgence of interest in early modern blood over the last couple of decades. Roland Greene examines blood as one of his *Five Words*, a new philological project tracing its "semantic envelope" and outlining its conceptual transformation from "religious, chivalric, and cosmological values into a new one with scientific, social and racial ones" (10).[67] While Greene studies blood's overarching change in signification, other critics have isolated more specific aspects of the blood discourse, a conversation to which this project contributes. Sujata Iyengar studies blood in the context of race studies; she highlights that "race" as bloodline encapsulated family origin and lineage (especially in relation to class) and likewise referenced inherited physical differences (especially skin pigmentation), both of which delineated differences in "kind" or species that should not reproduce (13–14). Similarly, Ania Loomba's *Shakespeare, Race, and Colonialism* analyzes stories of miscegenation between those different in blood or kind in Shakespeare in order to map the origins of race in the early modern period. Jean Feerick's *Strangers in Blood* likewise examines blood to underscore a transitional moment for racial discourses from terms of blood to those of skin color; she maintains that the latter served as a means to draw attention away from class difference within the nation to racial differences without it. Examining discourses of lovesickness and "hot blood," Carol Thomas Neely builds on this reading, arguing that concepts of humoral complexion and climactic heat provided "the ground for emergent categories of essentialized racial and national bodily difference—in particular, difference which valorizes English bodies and discredits Mediterranean ones" ("Hot Blood" 56).[68] Janet Adelman's *Blood Relations* foregrounds blood as a religious/ethnic discourse in relation to Shakespeare's *The Merchant of Venice*. In it, she contends that the play's preoccupation with blood speaks to the early modern racialization of Jews, which was the product of anxiety about the blood lineage shared by the two religions.[69] And Ariane M. Balizet studies blood in the domestic space and its relationship to gendered identities. In particular, she illustrates that Renaissance dramatists used domestic, material blood (female menstruation, bloodletting, bleeding hymens, bloody childbirth) to speak to ideologies of kinship and nation as emblematized by the ideal, middle-class, Protestant marriage. This work voices "anxieties regarding masculine identity [and authority] vis à vis domesticity as well as the misogynistic fear of women's bodies and spaces that fueled such anxieties" (*Blood and Home* 4–5). Finally, Johnson and Decamp's edited collection *Blood Matters* calls for a greater attention to blood's ubiquity with an explicitly interdisciplinary and interperiod (1400–1700) approach.[70]

All of these works attend to blood's ideology and its participation in the production of hegemonic structures of one sort or another, on which I build here. In so doing, and as should be evidenced by the historical section of this introduction, I take particular interest in discourses around social class and station, work also taken up recently by Jean Feerick and Katherine Craik.[71] Consequently, although *Thicker Than Water* addresses the feminist and racial effects of blood logics as they influence marginal status (especially in the final chapter), which likewise concern Iyengar, Feerick, Adelman, Balizet, and others, it largely emphasizes class difference and hierarchy as a product of discourses regarding affinity. Regardless of the type of difference that blood's ideology produces, ultimately, I am interested in *how* these differences are created and sustain both hierarchy and hegemony, for these are the mechanisms that are likewise at work in contemporary racial ideologies. That is to say, I agree with Patricia Akhimie, for I find value in "exposing the workings of racist thinking that link a social process of differentiation (always connected to existing structures of power) to the naturalization of such difference—the stigmatization of bodily markers imagined to be biological, inheritable, socially meaningful, and indelible" (11).

Thicker Than Water examines plots of courtship, love, and marriage as drama's testing ground for blood's ideology since these are points of tension for negotiating endogamous social bonds. This investigation will reveal how blood's ideology permitted early modern persons to invest blood with both meaning and, supposedly, agency to determine affinal relations. And though the period's natural philosophy presumed that these logics merely described facts, this drama makes clear how early modern persons perpetuated and manipulated these discourses for particular ends. Consequently, though I attend to the purported agency of bodily matter (as do some ecocritical or new materialist scholars), this agency is conceived *by humans and their social constructs*.[72] I have chosen, therefore, to shift the conversation to ask *what* these assumptions allowed or permitted. What kinds of problems did they create, and how did dramatists expose those problems or attempt to negotiate a way out of them? In short, though this discourse is pervasive and widely influential, it is nonetheless *a discourse* and one that early modern persons could either invoke or search to think beyond.

Finally, in examining drama not only of England but also of France and Italy, I hope to call attention to blood's ideology as a pan-European phenomenon. As evidenced by the historical overview here, not only did these countries have similar working assumptions about blood from natural philosophy and a myriad of other cultural discourses, the dramatists from these different national traditions responded to this ideological system in a similar fashion. As this study shows, they engaged in similar modes of critique, responding to both parallel cultural imaginaries and shared source materials to then present the logical limit cases for this ideology.

This project thus aligns with that of many other scholars revitalizing source study in early modern drama.[73] In the introduction to the edited collection *Rethinking Shakespeare Source Study*, Dennis Britton and Melissa Walter announce this renewed attention to source study, underlining that this methodology is not incompatible with New Historicist concerns, the investigation of power relations and subjectivities, or transnational interests. Moreover, the analysis of source materials need not view a (re)interpretation of the English, usually Shakespearean, text as the exclusive payoff of this work or devolve into bardolatry as was sometimes the case with traditional source study.[74] Like much of this scholarship, I take Louise George Clubb's "theatergrams" as a starting point (*Italian Drama in Shakespeare's Time*).[75] I organize my comparative groupings according to a particular formal unit—which provide, what I call, a "limit case"—such as a plot device, premise, or theme, even if one text did not strictly "influence" the other.[76]

Yet my own formal units develop more robust source studies connections than do Clubb's; I, for instance, make an effort to trace some textual basis for commonality in each comparative chapter (chapters 1 through 3), even though finding "influence" is not the primary objective. Furthermore, these formal units of comparison have been activated for (New) Historicist ends. Although unlike some New Historicist methodologies—which primarily look at the influence of nonliterary texts or allusions and foreground the near-contemporaneous or microhistory[77]—my interest is to sketch a more general, wide-ranging, and pervasive pan-European phenomenon: blood's ideology as a naturalized discourse that supported the elite's hegemony.

One model for blending a formalist methodology with historical considerations is Douglas Bruster's "bricolage,"[78] but in lieu of the more neutral "quotation," I offer the metaphor of *kinship* or *consanguinity*.[79] This metaphor is by no means a new one, for scholars of genre theory have borrowed the concept of family resemblance from Wittgenstein for a long time.[80] David Fishelov productively developed the metaphor by insisting on prototypical or core members of a genre family with a "common ancestry" to capture the more robust connections between its members. Traditional source studies, too, have invoked family metaphors, especially the father-son relationship. Namely, the borrowing text has "inherited" some topos, motif, theme, style, or diction; some characteristic of the father is present in its literary progeny.[81]

Yet, taking a cue from the early modern conception of blood and the work of Fishelov, one might see the father-son relationship as only one type of kin relation among many. Certainly, direct inheritance plays a role in the way that texts speak to each other, but this may not be the only way. Some texts could be called siblings or cousins. Blood offers a more fluid model of relationality, since consanguinity presupposes similarity not limited to the relationship between patriarchs and their progeny. Consequently, though one text has not directly influenced another in my pairings, just as the physical blood shared between kin was said to determine some shared characteristic, so consanguinity provides a particularly compelling metaphor for textual relationships that have a similar question or preoccupation. Or, otherwise stated, my application of this kinship metaphor offers a means to compare texts from the same "genealogy" not diachronically but synchronically.[82] Using this metaphor to characterize a group of texts with common stakes offers a more pliable comparative methodology, allowing for both comparisons across national boundaries and a long early modern time frame—from the very early sixteenth century and the Italian theater of Machiavelli to the French theatrical renaissance of Corneille in the mid-seventeenth century. By using formalist tools to enlarge the scope and increase the flexibility of source study, I hope to productively discuss "kin" plays. In so doing, I aim to show that despite cultural or national boundaries, placing these plays in conversation sometimes reveals a greater similarity between them than with their respective sources due to a shared cultural critique of blood's ideology. Inspired by similar sources that present a particular limit case for one of blood's logics, these plays when examined together suggest a shared testing of an ideology's limitations in the early modern period. They expose the workings of an oppressive system that reproduces hegemony by proclaiming inherent affinities for similar persons, and thus, this oppressive system will continue to oppress due to the "natural" reproduction of the same.

CHAPTER SUMMARIES: PUSHING BLOOD'S IDEOLOGY TO ITS LIMITS

This book is comprised of four chapters that demonstrate how drama interrogated the beliefs of blood's ideology by probing particular limit cases. As noted previously, not only do love plots provide occasion to examine the premise that blood determines affinity for prospective matches but they also examine whether, according to this logic, the rite of marriage transforms the individuals involved. Each chapter compares plays that ask a particular question. Can a person fall in love with their predetermined enemy as in the case of a blood feud? Would incest be the logical conclusion of the assumption that a person should love one most similar to themselves? Do certain bloodlines, in fact, warrant divine protection? Could a person force love and connection by way of mixing bloods in sexual consort (via a bed trick)? Or, could one use a stand-in stud since blood plays no role in prescribing familial relations? Does the process of commixtio sanguinis actually dissolve supposed differences in kind? And lastly, can love be found in difference rather than similarity?

Chapter 1, "Feuding Bloods and Bloody Romance," examines two plays wherein the protagonists' romantic love combats with familial sympathies dictated by a blood feud: Shakespeare's *Romeo and Juliet* and Corneille's *Le Cid*. Romantic love is here the occasion for the characters to reexamine blood's ideology, either to embrace its terms or imagine some alternative picture. Although according to most measures, the lovers are extremely well suited—to borrow Shakespeare's phrase, both are "alike in dignity" (prologue, line 1)—the logic of blood insists that blood carries predetermined affinities as well as enmities. Moreover, it assumes that one would adopt a spouse's blood, and thus their character, in marriage. In both cases, tragedy results from the characters' inability to escape the parameters and vocabulary of this discourse; honoring consanguinity in this way necessitates violence. Romeo and Juliet hope to establish affective relations between the warring families via their marriage. Yet the protagonists' appropriation of the blood feud's framework means likewise assuming its violent consequences; in this case, the families' enmity is only finally purged by spilling blood from Juliet's body. Or maybe worse yet for Corneille's Chimène, the prospective blood-mixing with her beloved Rodrigue upon their marriage would render her guilty of patricide. And though the King may insist on a romantic conclusion for the pair, his own motives are equally bloody. Namely, he is dependent upon the aristocracy spilling Moorish blood to solidify his own claims to power. Disadvantaged female characters by virtue of their gender, age, or station suggest that by rejecting these principles altogether the romantic love could survive—for Juliet, nothing is in a name, or, as for the maidservant Elvire, mounting bloodshed should dissuade these actors from further pursuing their bloody logics. Their resistance, however, is ultimately squashed by the tragic propulsion of blood's ideology.

Chapter 2, "When Blood Cries: Incest and Bloody Affinity," considers three sibling incest plays: two comedies, Giambattista della Porta's *La Sorella* and its English analog, Thomas Middleton's *No Wit, No Help Like a Woman's*, and one tragedy, John Ford's *'Tis Pity She's a Whore*. Despite their vastly different endings and whether the incest is realized, all of them employ the premise of sibling incest to ask what happens when the notion that "like likes its most like" is taken to its natural conclusion. Does the love between consanguineous family members always threaten to bleed into an all-encompassing, erotic love since both are founded upon shared blood? In posing these questions, these plays expose the ugly truths behind maintaining a closed sociopolitical system via endogamy. The comedies that avert incest appear to endorse the status quo of blood's importance: solidifying consanguineous ties by establishing affinity among kin as well as ensuring bodily difference between romantic partners in the name of avoiding incest. But beneath this endorsement of the ideology is the suggestion that the bodily distinctions mean very little, thereby exposing the ultimately rhetorical nature of this discourse that naturalizes and solidifies existing power structures. Moreover, *No Wit* brings to the fore the true motivations beneath blood's ideology, namely wealth, which, by virtue of its ability to circulate, can be divorced from noble bloodlines. *'Tis Pity's* tragedy, on the other hand, explicitly stages the violent consequences that result from this logic that consanguinity prescribes affinity. Otherwise stated, by rendering literal the already violent logic of this ideology, the play asks its audience to direct their moral shock and disgust not toward Giovanni but toward incest as the perfect instantiation of an oppressive, claustrophobic system.

"Bed Tricks and Bloodlines, or All's Well That Ends with a Baby" compares two plays featuring allusions to biblical bed tricks and specifically Boccaccio's Giletta (*Decameron* 3.9).[83] These bed tricks are orchestrated, and apparently justified, in the name of producing a vessel for a specific bloodline: Shakespeare's *All's Well That Ends Well* and Machiavelli's *La mandragola*. The biblical bed trick, wherein some venerable bloodline divinely sanctions the trick's deception, provides the test case for blood having intrinsic worth since it implies that God endorses certain naturalized hierarchies.[84] These plays first undermine this tradition because blood does not deem any character of particular worth; and, second, they ask whether blood inspires affinity or whether one might relinquish the importance of consanguinity for the sake of pragmatism. Shakespeare's class-conscious protagonist Bertram refuses to entertain loving a person "unlike" himself in kind. And for Helena, producing a child for him is secondary to her hope that sharing blood dictates affinity, fully believing that consummating their marriage (even if in secret) would win his love by overcoming their supposed biological incompatibility. Yet the play's unromantic ending ultimately disproves their logic. And though the Countess and the King search for alternative ways of conceptualizing personal worth, alliance, and love beyond blood, they are limited by their privileged position within this ideological system. Only Lavatch the clown offers a formidable critique by insisting that blood should be a great equalizer among persons, not a marker of difference in kind. By contrast, Machiavelli's skeptical play takes the material in an even more critical direction. *La mandragola* demonstrates that even though a child is crucial for the continuation of a family's lineage—and thus the foolish husband, Messer Nicia, willingly allows his beautiful wife, Lucrezia, to bed another man for its conception—the particularities of that child's body, and subsequently the notion of preserving an all-important bloodline, is of no real import when social signification ultimately determines identity. In sum, the biblical bed trick trope emphasizes the extent to which God will supposedly venture for blood, for its divine sanction is rhetorically convenient for those persons whose blood embodies their power and privilege. These playwrights, either implicitly or explicitly, place pressure on this limit case, thus undermining its foundational assumption.

The final chapter treats two overtly bloody Shakespeare plays, *Othello* and *The Merchant of Venice*, both of which feature relationships that cross the boundaries of kind on ethnic or racialized lines. This chapter thus departs from those prior in both focus and comparative method for two reasons. First, I aim to illustrate the translation or application of this ideology of blood to the realm of race and thus contribute to the larger picture of racist ideologies that persist today. As Akhimie aptly puts it, "Race is not so much distinguishable from as contiguous with other systems of differentiation in its reliance on a strategy of naturalizing social difference through biological narratives such as blood, lineage, phenotype, or geohumoralism" (13). Second, I believe it most constructive to bring this lens of blood's discourse, one way of producing and policing social boundaries, to the place where most of the debate on early modern race has occurred: Shakespeare's Venetian plays. These plays ask if love (both romantic and familial) only accommodates likeness as the ideology of blood presumes. *Othello* stages love between Othello and Desdemona, but the couple apparently conceptualizes and rationalizes their love by seeking similarity and concord between them. The play's tragedy thus tracks the enduring power of blood's ideology as Othello falls prey to the belief that Desdemona could have never truly loved him due to a biological incompatibility. *Merchant*, on the other hand, highlights blood's rhetorical function, invoked by opposing sides around the issue of Jessica's ambiguous, Jewish identity and the possibility of her conversion to Christianity. One camp speaks of her remarkably similar body and blood to provide proof that she was already Christian-like, thereby explaining Lorenzo's attraction to her. Others insist on Jessica's persistent consanguinity with her father. And Jessica herself appeals to the logic of commixtio sanguinis (like Helena in *All's Well*) in order to sever her blood's connection with her Jewish father by becoming a fully Christian wife. Yet none of these characters successfully appeal to this ideology because Jessica cannot be absorbed into either network of affinity. Nevertheless, by staging the romantic relationship between Jessica and Lorenzo wherein the pair appears to embrace their differences, *Merchant* entertains a love that falls outside of blood's discourse, one that can potentially disrupt the status quo.

The epilogue closes by offering another vision of love outside of blood's ideology. The Italian pastoral, Maddalena Campiglia's *Flori* (1588), seemingly offers the most radical alternative to this hegemonic system since its characters have little concern for either reproduction or endogamy. They rather embrace the arbitrariness of love objects. Here, love does not arise from bodily likeness; instead, lovers are merely placeholders for the projection of feelings and desires. The fact that marginalized voices here take center stage, placed there by a similarly marginal female author, provides one possible mechanism for dismantling the existing bloody system.

1

Feuding Bloods and Bloody Romance

"'Tis but thy name that is my enemy."

—*Romeo and Juliet* (2.1.80)

With this line, Juliet launches the iconoclastic speech at her window, one which flies in the face of the tragedy's premise; this premise is based in the enduring belief in an "ancient grudge" between the two families, an enmity determined by their bloodline, a necessary by-product of honoring consanguinity. Juliet's suggestion flouts convention because, even though, as David Crouch notes, the blood feud was exceptional in England as early as the twelfth century, the logic behind it persisted in the early modern period. In *Romeo and Juliet,* characters invest in blood's ideology that inspires (and justifies) this feud. For them, shared blood signals shared character traits, shared responsibility for any crimes committed, and, lastly, shared affinity for kin and enmity toward those who have dishonored said bloodline. As historian Lawrence Stone reminds us, kin were believed to be mere "trustees" or vessels of a given bloodline, taking precedence over its individual members (29). The allegiances or affinities of this larger consanguineous network would (or at least should) dictate the affections and feelings of its constituents. Being alike, they will like according to their shared constitution.

Yet, according to Juliet, the young lovers have no reason to be separated by this familial rancor: it is not Romeo's blood but only his name, "Montague," that is her enemy. Juliet's revolutionary suggestion is that Romeo might renounce his role as "trustee" of his kinship group. This kinship is not fundamentally based in consanguinity and thus need not predetermine his emotional relation to her or the Capulets. In the infamous balcony scene,

Juliet establishes herself as a tragic heroine seeking to resist the terms in which her family and society see the blood feud: she recasts it as a game of names. Blood, she speculates, is not a defining substance but rather a rhetorical tool, one used to perpetuate hostilities that continue to endanger the lives of both Capulets and Montagues. Juliet emerges as a figure who, by virtue of her youth, sex, and thus marginal status, sees beyond the assumptions and limitations of her society. However, Juliet's iconoclastic voice and reason are eventually overwhelmed by the discourse of blood that permeates Verona.

Can romantic love amid a blood feud—be it "ancient" as is the case for Shakespeare or freshly sown in Corneille's *Le Cid*—challenge the presumption that consanguinity determines affinity? Both Shakespeare's play and its kin play from across the channel, Pierre Corneille's *Le Cid,* activate the potential of this plot premise to interrogate blood's ideology; and they do so in a historical moment when the elite appeal to this ideology more forcefully in an effort to maintain their hegemony. First, both plays speak to a world in which this tale is not simply a tragedy of "star-crossed lovers" but a conflict between the young protagonists' love and contemporary assumptions about blood's role in affinities, responsibilities, privilege, and power. Many, if not most, characters fully invest in and ascribe to blood's ideology as reflective of the natural world—though both dramas call upon the audience to view these appeals as rhetorical rather than factual. Tragedy in both plays stems from honoring consanguinity and filial love in ways prescribed by this ideology: requiring more blood spilt of one's enemy and thus perpetuating violence. In Shakespeare's Verona, this discourse creates bloodthirsty Tybalts who genuinely believe that their bodies cannot do otherwise but hate and destroy another family's bloodline to honor their own. In *Le Cid,* Chimène's refusal to marry her love, Rodrigue, may forever interrupt the promised romantic ending; yet her complaint carries enormous weight if the theories of shared blood are taken seriously, as she does. If one's blood physically determines one's character, and blood is mixed in marriage, then husband and wife ultimately come to share the traits, and thus the crimes, of one another. In Chimène's case, this crime would be patricide. Chimène feels the pull of her father's blood on the ground, as though *in* her body, and she persists in demanding retribution for her father's murder to save not only herself but also her entire lineage from "eternal reproach."

Second, both plays present characters who attempt to invoke this blood discourse toward a different end—supporting the central romantic love or, at the very least, peace. In *Romeo and Juliet,* Romeo and Friar Lawrence hope that once married, the mixing of the two bloodlines will end the blood feud by establishing new consanguineous and thus affine ties between the families. And although their scheme is "successful," it does not escape the ideology's violent consequences. The social reintegration of the two families comes at the cost of blood spilt in a Galenic framework. Juliet's bleeding body in the Capulet tomb, her body the vessel of this new blended bloodline, acts as the mechanism for a humoral purge of the familial enmity.[1] In *Le Cid* too, both the Infante and King Fernand look to interrupt the blood feud's destructive logic, but they likewise fail to escape this discourse fully. While this play has often been read as the political struggle between older aristocratic appeals to independent power based in bloodlines and the rising absolutist monarchy, *Le Cid* highlights that despite these political differences, both sides of this divide use a vocabulary of blood to establish their identities. King Fernand's royal bloodline and his dependence upon the aristocracy to spill Moorish blood means that he cannot completely squelch the aristocracy's concern for their own blood as a source of authority. Chimène's "point of honor" holds value regardless of the problems that it poses for Fernand's kingdom (at least not without charges of hypocrisy). Both the aristocratic blood feud and the monarch's national security require blood and, consequently, violence.

Third, both plays make space for an alternative outside of these blood-soaked discourses: a small crack in which the audience can recognize the rhetorical status of blood and, therefore, the possibility for an escape from these violent consequences. In this case, a few women, marginalized and easily disregarded due to their gender as well as their age or social position, voice this alternative even though they ultimately fail to change the course of these tragedies. As mentioned, Juliet, being both young and a woman, endeavors to reject the principle of inherited allegiances, flirting with the notion that the difference between her and her love might be in name only and not a physiological difference. But unable to fully elaborate this alternative picture, she, persuaded by Romeo with Friar Lawrence, becomes overwhelmed by ideological pressures. In Corneille's play as well, Chimène's servant, Elvire, alone suggests that to escape from the violence of blood discourse and have a romantic conclusion necessitates a wholesale rejection of this ideology. This escape requires that one reject the bloody vocabulary of both the nobility and the monarchy, insisting on a space outside of the existing dialectic. One cannot merely command that the blood feud cease, as the king declares in the interest of his own power, but one must explicitly *recognize* the violent by-products of defining others, or oneself, by blood. Of course, Elvire's moment is fleeting and her message disregarded.

COMPARING KIN TEXTS, EARLIER ITERATIONS OF THE ROMEO AND JULIET PLOT

A study that places Shakespeare's *Romeo and Juliet* alongside Corneille's *Le Cid*, though founded in a seemingly natural comparison due to their common plot premise, would be inconceivable according to traditional source study. This might be why this comparison, as far as I can tell, has not yet been pursued.[2] Shakespeare's play could not have influenced Corneille's in any direct fashion because Shakespeare was unknown to seventeenth-century France. The plays of the French Renaissance were translated into English and performed at court after the Restoration, but there appeared to be very little material traveling from England to France, likely due to religious and political tensions between the countries as well as aesthetic disagreements. After all, English drama did not faithfully obey the classical rules of Aristotle such as the three unities, nor was it overly concerned with decorum or *bienséance* mandated by the Académie Française.[3] Apparently Shakespeare's work became known to the French reading public in 1776 with Le Tourneur's translations, although Voltaire also knew of him, having learned English during his exile purportedly in order to read Shakespeare (1726–28).[4] Additionally, the playwright Jean-François Ducis rewrote many of Shakespeare's tragic plots.[5] But regardless of how he finally got there, it is clear that Shakespeare did not make it across the channel until the eighteenth century.

Yet, if we look to the plays' shared subject matter—a young couple torn between romantic love and allegiances determined by consanguinity—we can trace this premise to some probable sources common to both playwrights. And in a cultural moment when Europe's nobility, in a desperate attempt to maintain power, forcefully asserted blood's import, its power to determine allegiances and affinities, this plot would appear apt for both authors. To begin, Kenneth Muir's suggested sources for Shakespeare's *Romeo and Juliet* are not limited to Arthur Brooke's prose version (1562). He also identifies the central plot in several Italian sources: Luigi da Porto's text (1530),[6] Matteo Bandello's novella (1554), and Luigi Groto's dramatized rendition titled *Hadriana* (1578).[7] These Italian sources, it turns out, have French "cousins." One such cousin is a French "translation" of Bandello, one of six tales in Pierre Boaistuau's *Historiques tragiques* (1559).[8] Given Boaistuau's infamy as the editor who botched Margueritte de Navarre's *Heptaméron*, Corneille's familiarity with this text is not terribly far-fetched. Additionally, Adrien Sevin had rewritten the da Porto version with a pseudo-Greek setting in 1542,[9] and Sevin's adaptation, according to many other scholars, likely influenced Corneille's *Le Cid*.[10] After all, Corneille's Spanish source text, Guillén de Castro's *Mocedades del Cid* (1618), focuses on the complicated action and violence *not* on the psychological struggles of the two lovers in the clash between love and duties to one's lineage.[11] In short, the many formal similarities between the English and French renditions of this theme point to a shared literary lineage, one which, in this instance, suggests similar preoccupations. It seems to me that both plays employ this plot at a time when assumptions about the force of consanguinity are not only pervasive but also primed for interrogation.

Both plays, it turns out, make similar departures from the sources outlined above. The most significant of which is an increased emphasis on blood; or, put simply, the blood feuds were made literally bloodier. Earlier versions tend to downplay the family feud except as background, a necessary obstacle for the young lovers. The blood feud operates as premise for the tragic plot but not as its central preoccupation. This holds true for Bandello's novella, "La sfortunata morte di due infelicissimi amanti che l'uno di veleno e l'altro di dolore morirono, con vari accidenti" (The unfortunate death of two unhappy lovers who die, one of poison the other of sorrow, with various accidents) (*Le Novelle* ix).[12] Here, it is not the blood feud as such that has made the families so angry but rather the fact that so many have died.[13] The "sanguinolente nemicizia" (the bloody feud) is one not necessarily inherited via blood but one created by bloody violence. The blood feud's status as premise rather than emphasis holds for Boaistuau as well. The introductory summary to the tale depicts the tragic conclusion as the inevitable outcome of passionate love, in which lovers die "par une trop excessive joye" (by a too excessive joy) (61). It is one tale among many that proves that one can die "par les furieuses flammes du trop ardent amour" (on account of the furious flames of a too ardent love) and by a reversal of fortune (61).[14] When

Boaistuau does describe the blood feud, he gives it a distinctly social origin, the result of inevitable competition between families of similar rank.[15]

Since in both tales the blood feud provides a necessary obstacle for the romantic, tragic narrative, the hope that marriage will make peace between the two families is framed as an afterthought.[16] Bandello's Fra Lorenzo desires to conduct the marriage because "con questo mezzo si persuadeva poter pacificare insieme i Capelletti e i Montecchi" (with this means one will persuade and can pacify the Capelletti and the Montecchi together) (380). But, in the same breath, he adds a political motivation, namely a hope to please his ruler by putting an end to the infighting. There is no mention of blood or any concern for bodies coming together. Thus, these different versions of the original tale all attend to the social reality of the blood feud. Shakespeare's and Corneille's drama, on the other hand, underline the dangers of a materialist ideology that can support and justify such social infighting—the violent consequences that arise from investing blood with the literal power to influence bodies and persons.

Generally, if blood is present in these earlier source texts, then it is not the marker of kinship relations but rather the by-product of violence.[17] And even then, blood is less prevalent in the dénouement than in either of the early modern renditions of this premise. Bandello's Giulietta dies of "dolente" and da Porto's version involves leftover poison. The fact that Boaistuau's Julliette stabs herself is one of his important revisions: "Et ayant tiré la dague que Rhomeo avoit ceincte à son costé, se donna de la poincte plusieurs coups au travers du cueur" (And having pulled the dagger that Rhomeo had belted to his side, gave herself with the point many blows through the heart) (113). But surprisingly, Boaistuau's version does not emphasize the bloody by-product. Additionally, Shakespeare's tomb is a good deal bloodier because in Boaistuau's text there is no skirmish with Paris and thus no blood around the sepulcher prior to Julliette's awakening. Thus both Shakespeare's play and Corneille's *Le Cid* take the central premise from this story, a tension between romantic love and familial loyalties, and increase blood's dramatic presence.

There are, however, a few important exceptions to this general pattern, an occasional reference to blood that hints toward a deeper exploration of the blood discourses at work; and both Shakespeare and Corneille's adaptations of the Romeo and Juliet plot build upon these moments. During Bandello's balcony scene, for instance, Romeo insists that he would never allow

Giulietta's dishonor "perché io per conservarlo chiaro e famoso com'è mi ci affaticherei col sangue proprio" (because I, in order to preserve its clarity and honor, would strive as if it were my own blood) (378). His promise to labor for her honor as he would for his own kin insinuates the materialist assumptions of his early modern analogue that blood requires maintenance or honoring in a specific manner.[18] Boaistuau's version pushes these blood discourses further during Julliette's tirade against Rhomeo. In a lengthy monologue, she vilifies him for having broken their promise to one another. She asks, "Mais si vous estiez si affamé du sang des Capellets, pourquoy avez-vous espargné le mien?" (But if you were so hungry for Capulet blood, why did you spare mine?) (86); she then inquires why he would seek to crown his victory over her with her cousin's blood: "La victoire que vous aviez eue sur moy, ne vous sembloit-elle assez glorieuse, si pour la mieux solenniser elle n'estoit couronnée du sang du plus cher de tous mes cousins?" (The victory that you have had over me, did it not seem to you glorious enough, if to better solemnize it was not crowned by the blood of my dearest cousin?) (86).[19] Hence, though primarily in the sense of bloodthirsty, these references presuppose that Capulet is a blood type indicated by the possessive, the blood of the Capulets ("des Capellets"). Julliette's question might finally acknowledge a connection between herself and her family dictated by their consanguinity, which persists despite what she and Rhomeo promised: to renounce all familial honor or loyalty. Boaistuau's Julliette also suggests she may be in conflict with herself after Rhomeo has murdered Thibault. She protests that she cannot live without him and should accompany him to exile; there is no other option for her since "je me suis faicte ennemye de moy-mesme, me desdaigne et contemne" (I have been made enemy to myself, disdain myself and condemn myself) (90). This sense of turning against oneself is clear in Shakespeare's exposition of Juliet's dismay upon Tybalt's death ("I am not I if there be such and 'Ay'" (3.2.48)), and Corneille, too, will cite this moment with the potential conflict in Chimène's blood if she were to marry Rodrigue. Though one play did not directly influence the other, both Shakespeare and Corneille elected to expand upon brief hints at a physical or visceral interpretation of blood's ideology within this plot premise in order to underscore how it functioned: the belief that blood determines one's person and affinities had bloody consequences regardless of the intention of the person citing it. And as the devalued female voices demonstrate, it would be better to acknowledge this *as* a discourse, to acknowledge its basis in rhetoric rather than nature, and then to abandon it altogether. To speak of blood will only prompt more blood.

ROMEO AND JULIET: THE GREAT BLOOD PURGATION

Shakespeare sets his tragic tale in a literally blood-soaked world. The blood feud is not merely a footnote that prompts the central tragic action; rather, it is made explicitly violent and bloody. The conclusion of *Romeo and Juliet* is not typically remembered as a gory spectacle however. The chorus might prepare the audience for Juliet's death, but it does not obviously presage a bloodbath. Nonetheless, Juliet's blood should be prominent at the play's close. As theater historian Lucy Munro notes in her extensive catalog of stage blood, the stage direction "Fryer stoops and lookes on the blood and weapons" (5.3) in the quarto edition of 1597 indicates that stage blood was likely used ("Proceedings and Conclusions" 30). Blood is likewise mentioned several times in the dialogue in a compressed time span. "The ground is bloody," the Chief Watchman observes, just before he finds "Juliet bleeding, warm and newly dead" (5.3.172, 175). And most strikingly, Lord Capulet gasps at the sight of his daughter's body bleeding, "Oh Wife, look how our daughter bleeds!" (5.3.202).[20] Lord Capulet's exclamation calls the audience's attention to a shocking display; it asks them to remark *not* that she is dead but *that she bleeds.*

Yet the play's performance and scholarship often sterilize this aspect of the play and disregard the textual suggestions of a bloodbath reminiscent of those in *Titus Andronicus* and *The Spanish Tragedy*.[21] Even Jennifer Low's work, which importantly calls attention to the placement of Juliet's body onstage, or Ramie Targoff's piece discussing the inconvenient presence of Paris at the Capulet tomb do not attend to Juliet's bleeding. This critical omission is curious because, as noted above, the bloody language of Shakespeare's play text is likely a considered addition to the source material. I would, therefore, like to accentuate Capulet's exclamation: that there is something particularly horrible, noteworthy, and tragic about the spectacle of Juliet's bleeding body.[22] I argue that Juliet's blood onstage emphasizes the tragic consequences of not only the blood feud but also the logic of blood and kinship beneath it. Her blood onstage foregrounds the violent consequences of investing in blood's ideology; it suggests that the characters' literal investment in these beliefs begets a literal answer: a purging of the families' bloody enmity.

Prior *Romeo and Juliet* scholarship has treated the play's emphasis on blood mostly within the framework of the blood feud, which, thanks to historical studies like Edward Muir's examination of the vendetta in Friuli, can be glossed in socioeconomic terms. As Muir demonstrates, preserving aristocratic honor embodied in bloodlines was often an excuse for other, more practical disputes over property, feudal dues, taxes, political appointments, and so on—a truth often acknowledged by the *Romeo and Juliet* source material. Consequently, when *Romeo and Juliet* scholarship addresses the blood feud in the context of these larger societal structures or pressures, it often loses sight of the emphasis on blood's materiality in the play. Paul Kottman, for instance, recalls Hegel's influential proposal that the political setting subjects the lovers to violent chaos, and the lovers, either deservedly or undeservedly, reap the unfortunate consequences of rejecting social confines and familial duties.[23] Feminist critics have developed this line of interpretation, suggesting that the play highlights problems of masculine identity: violence and aggression is a necessary evil that allows men to prove themselves as men, and the lovers simply get caught in the crossfire of this social mandate.[24] Alternatively, the blood feud's violence has been interpreted in light of contemporary political riots and class conflicts or changing definitions of "civility" in the early modern period.[25] And Susan Snyder holds that *Romeo and Juliet* condemns the social system of the blood feud by noting its similarities to Althusser's ideology, maintaining that the feud-system, like other ideologies, "is not in fact predicated on any substantive difference between Montagues and Capulets" ("Ideology and the Feud" 88).[26]

But, while this treatment of the blood feud can explain a great deal about the play, it does not fully satisfy Romeo and Juliet's negotiations with blood, which not only pertain to familial enmity but also to love and their marriage. The lovers imagine that their marriage will end the blood feud due to its material component, namely mixing bloods upon becoming one flesh. Ultimately, I agree with Snyder that the play as a whole, using Juliet's voice, criticizes the blood feud by exposing it as discursive; but I maintain that of great importance to the play's drama is the fact that the characters nevertheless invest in a forceful, substantive difference between them. To understand the play's critique of this blood discourse, we must reinvest the characters' ideology with its literal sense: their conception that blood embodies "Montagueness" or "Capuletness," which must be honored.[27] As previously outlined, the ideology of blood assumes two things: first, that there *is* a material difference between bloodlines, and second, that blood has the agency to determine either enmity or affinity.[28] These underlying assumptions beneath the blood feud require further analytic attention.

The Montagues and Capulets hold that blood materially differentiates persons, substantiating both their identities and the perpetual enmity between their families. This investment, moreover, sustains violence by ascribing agency (and thus blame) to the blood that determines this enmity.[29] In Shakespeare's Verona, this discourse only creates more bloodthirsty Tybalts who genuinely believe that their bodies cannot do otherwise but hate and destroy another family's bloodline for the sake of their own.[30] Moreover, the belief that blood determines one's person and affinities has bloody consequences *regardless* of whether or not the person citing it has good, nonviolent intentions. Romeo and Friar Laurence hope to end the violence of the blood feud, but they do so within the existing ideological framework.[31] They embrace the belief that consanguinity determines affinity and simply hope that a new blood admixture (effected by the process of commixtio sanguinis in marriage) will forge peace between the two families by establishing new kinship ties.[32] And although their scheme "works"—it is the final bloody spectacle that prompts the two patriarchs to forge peace—its success is highly problematic.[33]

Hence, rather than attribute this tragedy to the inevitable workings of Fate,[34] Shakespeare's *Romeo and Juliet* alludes to the dramatic tradition of tragic sacrifice as catharsis; here, presented as the bloodletting of a young girl, this catharsis recalls, as Catherine Belling shows, the spectacular bloodletting of Lucrece in Shakespeare's narrative poem.[35] This conclusion illustrates that the persistence of this bloody logic requires violence and, in this case, blood sacrifice. For as with the classical example of Euripides's *Iphigenia*,[36] though the tragic hero (here the tragic couple) makes the "correct" choice given the ideological framework at hand, the nature of the sacrifice calls the larger value system into question. Shakespeare's *Romeo and Juliet* thus plays out the Galenic foundations of the blood feud in order to demonstrate the violent cost of this belief system wherein blood secures kinship and social structures. In short, the familial enmity that contaminates the bloodlines of these two houses must be violently purged, let out of the family body via the bloodletting of Juliet, sacrificed to this very belief system.[37]

And yet, in the famous balcony scene of Shakespeare's play, Juliet establishes herself as a tragic heroine capable of interrogating, at least for a time, the terms in which her family and society see the blood feud: she recasts it as a game of names. She speculates that blood is not a defining substance but rather a rhetorical tool, one used to perpetuate hostilities that continue to endanger the lives of both Capulets and Montagues. Juliet thus emerges as a figure who, by virtue of her youth and desire to love, can see beyond the assumptions of her society. Juliet's iconoclastic voice, however, is eventually overwhelmed by Verona's bloody discourse. At the end of Shakespeare's play, the spectacle of Juliet's stabbed, bleeding body in the Capulet tomb drives this point home viscerally; Juliet as well as the alternative viewpoint she represents are sacrificed to an ideology of blood that she so poignantly discredited. Shakespeare's *Romeo and Juliet* does not close with a celebration of social reintegration. Rather, evoking the tradition of tragic sacrifice with Juliet's body encourages its viewers to interrogate the means of the play's closure. This ideology, when played out to the letter, appears to demand either the blood feud's violence into perpetuity or this manner of sacrifice.

The play's opening chorus at first appears to utilize the sonnet structure to formally mirror a tidy closure, one that will redeem the deaths of the "star-crossed lovers" with the families' reconciliation.[38] Yet the chorus simultaneously hints at the bloody purge to come and the sacrificial nature of this tragic narrative:

> Two households, both alike in dignity,
> In fair Verona, where we lay our scene,
> From ancient grudge break to new mutiny,
> Where civil blood makes civil hands unclean.
> From forth the fatal loins of these two foes
> A pair of star-cross'd lovers take their life;
> Whole misadventured piteous overthrows
> Do with their death bury their parents' strife.
> The fearful passage of their death-mark'd love,
> And the continuance of their parents' rage,
> Which, but their children's end, nought could remove,
> Is now the two hours' traffic of our stage;
> The which if you with patient ears attend,
> What here shall miss, our toil shall strive to mend. (1–14)

Since both families are "alike in dignity" and thus have similarly worthy noble lineages, Romeo and Juliet should be deemed an appropriate match.[39] In fact, the existence of a feud itself speaks to this appropriateness, since feuds ordinarily stem from competition within a socioeconomic class. These two households, however, insist upon their differences, especially as embodied in their blood, which substantiate and perpetuate their feud.[40] The prologue swiftly establishes the literally bloody atmosphere by calling attention to the "unclean" hands covered in "civil blood."[41] The use of *civil* here, as defined by the *Oxford English Dictionary* as "of or relating to citizens or people who live together in a community; relating or belonging to members of a body politic," underlines that which unifies all of the familial bloodshed, namely that it belongs to the larger community of Verona (3a). Yet, immediately after this reminder of community, the prologue connects the physical blood spilt to the principle of inheritance along bloodlines, the "fatal loins" which here prove "fatal" to the lovers.[42] Moreover, rather than concluding with the notion that the families ended the blood feud solely on account of the lovers' deaths—"do with their death bury their parents' strife"—the prologue restates this plot point.[43] The break in the otherwise highly regular iambic pentameter with the spondee "which, but" further calls attention to this rephrasing.[44] The lovers' deaths are not a simple "wake-up call" that this blood feud has lasted too long but it is the *only* means by which the "rage" could be "removed," as if the death literally purges this familial hatred. "Nought [nothing] could remove" this rage except this particular circumstance, much like a leech might draw out a choleric humor. This figure of speech thus subtly figures the lovers' deaths as a humoral bloodletting, dispelling the familial violence once and for all by shedding blood that is *both* Capulet and Montague at once. Shakespeare's purgative end, determined from the play's outset, demonstrates that to take on these assumptions about blood—as that which determines characters, kinship, loyalties, and love—means to assume all of the material consequences of its logic.

This logic of blood can only result in a world of Tybalts: persons who feel licensed to engage in violence since they are supposedly provoked by their bodies. These people need not take responsibility for the community's violence; their blood has the agency and they are merely its servants. The addition of Tybalt's dialogue with Lord Capulet to the masque scene—for Tybalt does not make an appearance in the versions of Bandello, Boaistuau, or Brooke—highlights the force and impact of blood's ideology in Verona. While one could argue that his addition simply heightens the dramatic tension between Tybalt and Romeo, his conspicuous absence in prior versions and its emphasis on the role of the body in the family feud justifies closer consideration. Tybalt believes that his choler is not simply a feeling of anger, but it is the output of a material within his flesh that carries the same name. In accordance with a Galenic framework, Tybalt's anger toward the Montagues reinforces itself, further provoking his choleric innate heat that then prompts more anger; but his choler is likewise a by-product of his consanguinity with other Capulets, those who have suffered or bled due to the "ancient grudge" (chorus 3). Lord Capulet demands that Tybalt not quarrel with Romeo, but Tybalt insists that his body demands differently: "Patience perforce with willful choler meeting / Makes my flesh tremble in their diff'rent greeting" (1.4.206–7). Tybalt deems his choler "willful," having agency that will dictate his actions; his blood's enmity for that of Montague is insatiable and enduring, over which he supposedly has little control. For otherwise, Tybalt has no independent reason to dislike Romeo in particular because Romeo was not involved in the opening brawl (1.1).[45] The fact that those around him like Lord Capulet or Benvolio do not experience choler determined by their consanguineous connections does not appear to disturb his conviction.

The opening of the third act then explicitly connects the physiological experience of choler with material blood that purportedly provokes this anger and violence. Benvolio, looking to avoid a quarrel with the Capulets, asks Mercutio if they can leave since "if we meet we shall not scape a brawl, / for now these hot days is the mad blood stirring" (3.1.3–4).[46] Benvolio's speech invokes the Galenic precept that the physical environment will only further aggravate the "mad blood," inevitably provoking fighting between the two antagonistic bloodlines. This moment contains a seemingly innocuous references to Galenic humoralism, used so casually that it is practically evacuated of any genuine, literal significance. Yet this passing remark reinforces an ideology that absolves the individual actors of personal responsibility, invoked to excuse acting upon this blood feud. Throughout the play, therefore, this Galenic notion that blood has agency paves the way for the bloodletting of Juliet's body, wherein the blood that is supposedly responsible for familial enmity must be purged.

The staging of bloodshed thus emphasizes the intensely visceral way in which the blood feud operates or what happens when blood assumes the blame for perpetuating the cycle of violence as opposed to individual actors. From the play's outset, Prince Escalus calls for the end of Verona's blood feud by threatening, "On pain of torture, from those bloody hands / Throw your mistemper's weapons to the ground" (1.1.89–90). Of course, the weapons themselves are not "mistemper'd," but rather, this adjective extends metonymically to the blood within the bloody hands, which, contaminated by familial enmity, theoretically has a choleric character that instigates violent action. The characters ceaselessly appeal to blood and its demands especially after the "bloody fray" resulting in Tybalt's death. As the call of the Capulet family "blood for blood" illustrates, blood is here the agent of these demands not the individual persons or actors (3.1.152–53, 155). Moreover, since no individual agent need take responsibility for this demand, bloodlines will continue to demand violence indiscriminately. Prince Escalus laments this feature of the blood feud in his response to Lady Capulet, inquiring, "Romeo slew him [Tybalt], he slew Mercutio. / Who now the price of his [Mercutio's] dear blood doth owe?" (3.1.186–87).[47] His rhetorical question reveals the fact that the blood feud obscures personal responsibility; and once the individual agents responsible are reinserted in the discussion, the question remains about who should pay. This question has no clear answer, however, for clearly no individual can be held entirely responsible when they conceive of themselves as vessels for a collective identity. To take on this responsibility, as Juliet does in the conclusion, has a clearly tragic result.

Romeo and Juliet accentuates the degree to which this belief system remains fundamental to Veronese society (and Shakespeare's), since it cannot be divorced from the basic assumption that kinship—and with it affinity and allegiance—is based in consanguinity. Moreover, the foundational nature of this assumption means that it remains extremely difficult to extricate oneself from this ideological system, as Juliet will try, though fail, to do. Even the most vocal opponent of the blood feud's violence, Prince Escalus, struggles with relinquishing its underlying logic—the very personal, physical connection between those who share blood—when his own blood is at stake. He mourns for his kinsman Mercutio, "My blood for your rude brawls doth lie a-bleeding" (3.1.193). In addition to highlighting the blood of Mercutio's dead body onstage, *his* blood, marked by the possessive pronoun "my," is not simply metaphorical but physical due to the prince's consanguinity with the victim. The general obsession with possessives, which insists that a person shares blood with their entire bloodline, clearly permits this perpetual enmity. The prince's remark is structurally no different from the Capulet call for vengeance: "Prince as thou art true, / For blood of ours shed blood of

Montague" (3.1.152–53). In other words, an individual's blood requires the protection of their kinsmen since it is their own blood spilt and this logic functions across generations; as a result, individual actors need not take any measure of responsibility for the continuation of the blood feud. While Escalus's resolve to exile Romeo rather than kill him marks a superficial rejection of the blood feud, his language betrays the logic that undergirds this system. Furthermore, Mercutio's blood (the prince's blood) has already been avenged by Romeo; the leniency of exile could be read as an expression of appreciation, since Romeo absolved the prince of having to avenge the death himself. In sum, Prince Escalus's claim that *his* blood issues from Mercutio's body compromises his removed position above the bloody fray and renders it difficult for him to insist that the feuding families abandon their convictions. According to this system wherein consanguinity defines kin, a person's blood determines this course, thereby foreclosing the potential for change. This ideology of blood does have a mechanism for establishing affinity and likeness with others: mixing bloods when becoming one flesh in marriage. Yet, even here, the belief that love has a material foundation has the potential for unforeseen, gory consequences. Shakespeare's play suggests that this violence is not an abuse of this ideology but is rather an unavoidable by-product or consequence of it.

While typically the love and marriage of Romeo and Juliet has been read as wholly oppositional to the families' blood feud, Shakespeare's play demonstrates that their marriage as the blood feud's solution (termed a "holy physic") rests on the same logic. Namely, this solution concocted by Friar Laurence and Romeo equally depends on the precept that blood has agency, thus remaining squarely within the existing ideological framework. The only difference between the two camps is that Romeo and the friar hope that new consanguineous ties might determine affinity rather than enmity. And it is precisely because blood only operates discursively—though with unfortunately material consequences—that the sacrifice of Romeo and Juliet to this ideology is so tragic.

While Romeo calls marriage a "holy physic" to cure his wound of love (2.3.47–48), Friar Laurence takes up this vocabulary and endows it with a different literal sense, not as a cure for love but a cure for the families' hatred. "In one respect I'll thy assistant be," he promises Romeo, "for this alliance may so happy prove / To turn your households' rancor to pure love" (2.3.86–88). According to early modern thinking, in the act of marriage—or as Friar

Lawrence puts it, when "Holy Church incorporate two in one" (2.5.37)—husband and wife become "one flesh," though the wording in Q2 here makes a significant difference. In Q1, the friar pronounces that the two should not be alone "Till holy Church have joined ye both in one." The wording of "jointure" certainly appeals to the principle of *unitas carnis*, emphasizing the unification of the two individuals. However, Q2's chosen verb *to incorporate* denotes "to mix or blend thoroughly together" particularly "material substances" (*OED* 1a), a choice that more closely signals the process of commixtio sanguinis as the by-product of matrimony. The incorporation here is not merely of two people, but it is a mixture or blending of two bloodlines. Verona's friar assumes that the consummation of Romeo and Juliet's marriage will create a new Montague-Capulet bloodline, thereby forging new consanguineous connections and pacifying the households' rancor.[48] He holds to this belief even after the slaying of Tybalt and Romeo's banishment, assuring Romeo in his bout of despair that he need only "blaze [his] marriage" to "reconcile [his] friends" (3.3.151). And the friar's conviction or faith in this biological process might also explain why, after Tybalt's death, he insists that Romeo "Go get thee to thy love" and "comfort her" such that they might consummate their marriage (3.3.146–47). Caution would, of course, dictate that Romeo leave Verona immediately rather than be caught flouting the proclamation of Prince Escalus, but the consummation of the lovers' marriage might merit such a risk because it has an important role to play in the transformation of rancor to love.

Encouraged by Friar Laurence, Romeo accepts the premise of the blood feud: that he and Juliet are defined by their bloodlines, an embodied Montague or Capulet character, and that this blood has agency for either hatred or love.[49] Romeo has certainly always been susceptible to believing in a material basis for love. When expressing his love for Rosalind, Petrarchan tropes take on a humoral color: "Love is a smoke made with the fume of sighs, / Being purg'd a fire sparkling in lovers 'eyes, / Being vex'd a sea nourished with loving tears" (1.1.193–95). Embodied love becomes a bodily humor with several physical states; love likewise assumes agency in this formulation. Nothing much changes after his marriage to Juliet, except that love becomes material as blood instead of smoke. Yet what the Friar and Romeo do not anticipate is that by this logic the enmity of the blood feud cannot simply disappear by establishing new consanguineous ties; instead, the blood that embodies and determines the enmity must be somehow purged.

In this light, the conflict between Romeo and Tybalt is not simply another eruption of the blood feud; it is rather a conflict between competing opinions about blood's "agency": Does blood incite hatred or love? For Romeo, this newly forged mixture of Montague and Capulet blood should, in theory, incite Tybalt to abandon his quarrel due to their new consanguinity. Romeo pleads with him,

> Tybalt, the reason that I have to love thee
> Doth much excuse the appertaining rage
> To such a greeting. Villain I am none,
> Therefore farewell, I see thou know'st me not. (3.1.63–66)

Romeo here hints that his "reason" to love Tybalt is his new blood bond with Juliet; and his new Capulet kinship fights against the "rage" that he would otherwise feel. He concludes that Tybalt must not "know" him as anything other than a villain, certainly not as his new family member.[50] But just as Romeo and Juliet's bodies do not act as they should have during the masque—that is, by provoking immediate enmity as Tybalt would insist—neither does Tybalt's body recognize his newfound (spiritual if not yet material) consanguinity with Romeo. After all, blood does not actually possess the agency to evoke either emotional response.

Yet Romeo's persistent conviction in the ideology of blood further helps to explain his histrionics as he struggles with how to relate to Juliet after killing her cousin. Despite the clear failings of this blood system, Romeo's assumptions—both that blood determines sentiment and that his actions have inflected the material blood he shares with Juliet—are key to reading this scene. His appeal to violence to fix his current situation with Juliet, moreover, evidences the degree to which his logic resembles Tybalt's. As he sees it, now that he is responsible for the murder of her kinsman, Juliet's love for himself, solidified by their "one flesh," conflicts with her affinity for her kin, likewise determined by consanguinity. The two sentiments are seemingly unable to coexist in the same body and she might no longer love him, as he says, "Now I have stained the childhood of our joy / with blood remov'd but little from her own?" (3.3.94–95). Juliet should have lost blood in the form of her maidenhead upon the consummation of their marriage, but she has lost her kin instead. Juliet's blood has been removed by proxy of Tybalt's death. Romeo subsequently offers to purge his own blood in an attempt to remedy this conflict, and in retribution for his crime against her. On hearing that his name has hurt her, he exclaims,

> as that name's cursed hand
> Murder'd her kinsman. O tell me, Friar, tell me
> In what vile part of this anatomy
> Doth that name lodge? tell me, I may sack
> The hateful mansion. (3.3.101–7)

The stage direction immediately follows: "He offers to stab himself, and Nurse snatches the dagger away" (3.3). Romeo intends to stab himself, "the hateful mansion," because he believes that his name lodges *somewhere*. The answer to his rhetorical question about the part of his anatomy that houses his name is undoubtedly blood; thus he immediately undertakes to spill this blood as reparation since he has caused his blood to act against hers. It is precisely this manner of thinking that then governs the logic behind Juliet's bleeding body at the play's close: the enmity is in the blood that cannot be assuaged unless removed. In Romeo's struggle, the play illustrates the violent consequences of viewing one's blood as the material that contains hatred and sins committed. One cannot bestow upon blood this kind of agency, even in the hopes to foster love, since this logic permits the blood feud in the first place. In other words, giving blood agency means that blood can call for blood. The play thus encourages the audience to interrogate blood's ideology as the foundation for kinship and existing social structures when it can only foster and apparently justify violence.

Although Juliet's actions inevitably come to accept Romeo's blood-inflected terms, she is the first critic of blood discourses in the play. Certainly, to call Romeo something other than a Montague would solve the problem of her family's resistance to her desired match, but her speech likewise suggests an alternative, non-naturalized conception of identity and sentiments, one *not* determined by the blood running through one's veins or anything other than names. Juliet first wrestles with the existing vocabulary of the blood feud, one of innate sentiments, upon first discovering Romeo's identity: "My only love sprung from my only hate! / Too early seen unknown, and known too late!" (1.4.255–56). One might easily gloss this moment as typical Petrarchan fare, but her grief likewise makes sense in the context of the traditional assumption that like should like its like, reflecting the conundrum of her love for Romeo. If blood indeed worked according to this assumed affinity mechanism, then she should not have fallen in love with Romeo at first sight. Her Capulet blood would have negatively reacted to his Montague blood, signaling an enemy even if it could not precisely report his identity. In short, it should not have been "unknown." Juliet should have intuitively hated Romeo like her cousin Tybalt, but she did not. While Romeo seeks to accommodate this love that defies natural philosophy within the existing system, Juliet instead flirts with an unconventional outlook that abandons these terms altogether.

Shakespeare's play uses Juliet's voice to offer the iconoclastic alternative to, and potential escape from, the existing violent system of blood. This is not altogether surprising because Juliet is young and not yet fully indoctrinated. Moreover, as a woman, she remains outside of the predominantly masculine realm of the blood feud. From this vantage point then, Juliet can suggest that the problem might merely be one of naming, an arbitrariness of signifiers. As she imagines herself speaking to Romeo, she casts family belonging not in terms of consanguinity, which cannot be changed, but rather in terms of the choice to either assume or deny a name: "Deny thy father and refuse thy name; / Or if thou wilt not be but sworn my love / And I'll no longer be a Capulet" (2.1.76–78). This suggestion contradicts the tenets of kinship based in consanguinity, according to which it is impossible for her to be anything other than a Capulet. But if one can deny the responsibilities of kinship by refusing a name, then this, as the play suggests, is the only means to refuse the call of blood for blood: refuse membership and thus responsibility on the basis of consanguinity. Juliet makes this point even more explicitly:

> Tis but thy name that is my enemy,
> Thou art thyself, though not a Montague.
> What's Montague? it is not hand nor foot
> Nor arm nor face, O be some other name
> Belonging to man. (2.1.80–84)[51]

According to Juliet, the young lovers have no reason to be separated by this familial rancor: it is not Romeo's blood, but only his name, "Montague," that is her enemy. Juliet's speech aims to reject the premise that bodies, or their parts (with blood implied), dictate identity, sentiment, and thus violent action. While Peter Holbrook sees in Juliet's balcony speech the ability to divorce one's familial identity from oneself (216–17), Juliet's iconoclasm consists in not only the suggestion that one can discard one's social identity but in the very notion that this identity is social, is a *name* in the first place. Unlike Romeo who appeals to his "vile anatomy," Juliet proposes that the name does not lodge in any bodily substance. She thus begins to fashion a belief system wherein a person can act according to their individual will and desires, rather than be destined to act in accordance with consanguineous demands.

Yet developing an alternative picture nevertheless faces an issue that Juliet herself addresses: the difficulty to know how *else* to define Romeo. Juliet says simply, "Thou art thyself," a circular definition. This "thou" could take on some other name, but who or what is "thou"? Who is Romeo if not a Montague? She may not know. Juliet's speech at her balcony thus illustrates the struggle to establish a new belief system when inundated and restricted by the vocabulary of a dominant ideology. After all, there does not appear to be space in the play for alternative ways of thinking or speaking beyond the level of suggestion. Certainly, when Juliet attempts to use language creatively in order to reject her father's proposal to marry Paris, she meets with sound and even violent punishment from him.[52] And Juliet's inability to fully complete an alternative picture for what constitutes identity, lacking both the resources and vocabulary, results in the sacrifice of Juliet's possible world to blood's ideology, and her body with it.

Unsurprisingly, there is no receptive audience for her alternative suggestion of names; Romeo does not take up Juliet's suggestion to doff his name but rather will marry her *as* a Montague in an attempt to quell the blood feud. Juliet, it seems, has embraced the friar and Romeo's solution when she presumes an affective connection determined by consanguinity with both her kinsman and her new husband. Juliet's exchange with her Nurse after the altercation between Tybalt and Romeo exposes her investment in her newly shared blood. The Nurse cries about the blood from Tybalt's body, "I saw the wound . . . / A piteous corse, a bloody piteous corse, / Pale, pale as ashes, all bedaub'd in blood, / All in gore blood, I sounded at the sight" (3.2.52–56). Although this moment should call Juliet's attention to her cousin's blood spilt, Juliet misattributes the wound to Romeo. She proclaims that if Romeo has died, "I am not I if there be such an 'Ay'" (3.2.48). The proclamation "I am not I" sounds like nonsense, but it acutely identifies her imagined conundrum since she would have lost half of her newly married, fused self upon Romeo's death. The Nurse corrects her that the "bloody, piteous corse" belongs to Tybalt, and hereafter, Juliet struggles between which is the greater sorrow: her cousin's death or her husband's banishment. Juliet wrestles as Romeo did within blood's conceptual framework. Her loyalties are quite literally split down the middle because she now shares blood with both the perpetrator and the victim. If "Romeo's hand shed Tybalt's blood," then her blood has turned against itself—a conundrum not altogether different from her response, the litany of oxymorons "Beautiful tyrant, fiend angelical" and so on (3.2.71, 75).[53] However, viewing her own struggle as a bodily one has violent implications for it suggests a bodily solution, that is, purgation. Tybalt's blood, his piteous *corpse,* determines the bloody and piteous *course* of the plot to follow: Juliet's bloody body sacrificed to this bloody discourse at the Capulet tomb.

Romeo and Juliet's acceptance of the existing terms of blood, therefore, explains why the lovers continue to reference blood throughout the play though they oppose the blood feud. And with the lovers' bloody language, the play presages the bloodletting to come. Juliet foresees calamity and death before Romeo leaves Verona: "Methinks I see thee, now thou art so low, / As one dead in the bottom of a tomb. / Either my eyesight fails or thou look'st pale" (3.5.55–57). Romeo responds with a physiological explanation, that she too looks pale because "dry sorrow drinks our blood" (3.5.59). Highlighting this pallor, it turns out, prefigures her bloodletting, tragically sacrificed in order to purge the choler of their families. The blood being consumed or "drunk," moreover, alludes to Juliet's status as a kind of sacrificial animal, a scapegoat to be consumed in this ritual purging. Later, Juliet even offers to spill her blood as a solution to her impending marriage to Paris, proclaiming,

> this bloody knife
> Shall play the umpeer, arbitrating that
> Which the commission of thy years and art
> Could to no issue of true honor bring. (4.1.62–65)

Juliet's threat of suicide could indicate her desperation, but Juliet here invokes a long tradition of honor suicide beginning with Lucrece, cleansing one's family honor of sexual shame or, as is the case with Juliet, their enmity in a tragic sacrifice. In addition, Juliet's restaging of Romeo's prior threat to stab himself reflects her final adoption of Verona's bloody logic. She has finally succumbed to (or become overwhelmed by) this ideology to which she—and the possibility for an alternative, non-naturalized conception of identity and social structures—must be slaughtered. For, as the play dramatizes, if Veronese society maintains the tenets of blood's ideology, especially that blood materially determines the families' "ancient grudge," then only by spilling the newly mixed blood of Capulet and Montague could the enmity of the two bloodlines be cleansed.

The death and blood of Paris onstage at the Capulet tomb, whose presence is otherwise inconvenient, renders visible the necessity of this bloody purge due to the omnipresent and inescapable nature of this ideology. Rather than mourn Paris as a mere innocent bystander, the altercation in front of the Capulet tomb occurs as a result of Paris's mistaken belief that Romeo has come to do "some villainous shame" to the bodies, pursuing vengeance beyond death (5.3.52–55). Accordingly, Paris's blood marks just how embroiled Verona is in this logic of blood such that Juliet's alternative is untenable. Although Paris is neither a Montague nor a Capulet, he sincerely believes in the strength of these bloody ties such that Romeo would continue to pursue vengeance on the bodies of the dead. His belief is so entrenched that he will not listen to Romeo's protests—"By heaven, I love thee better than myself; / For I come hither arm'd against myself" (5.3.64–65)—and Paris rather responds, "I do defy thy conjurations, / And apprehend thee for a felon here" (5.3.68–69). Paris thus violently pays for this limited lens. According to Friar Laurence, this is no small amount of bloodshed given its prominence in his language:

> Alack, what blood is this which stains
> The stony entrance of this sepulcher?
> What means these masterless and gory swords
> To lie discolor'd by this place of peace?
> Romeo, O pale? What else? what! Paris too?
> All steep'd in blood? (5.3.140–45)[54]

Paris's blood here stains the stony entrance to the sepulcher, but this blood merely mirrors the figural blood that has already stained the Capulet tomb: the collection of this family's perpetuation of these bloody assumptions.

Thus, to return to Capulet's otherwise odd lament, "Oh Wife, look how our daughter bleeds!" denotes the gruesome consequence of this bloody framework.[55] The scheme of the friar and the lovers has worked, albeit tragically so, since the purging of Juliet's blood upon her suicide ends the familial enmity. The disappearance of blood from the language of the last scene after Capulet's exclamation further supports this reading. After all, the omission of the word blood from the remainder of the play is otherwise remarkable because Friar Laurence must summarize the entire plot for the grieving families, and Prince Escalus proclaims once and for all that this event must end the blood feud, both of which could easily mention blood. But Juliet's bleeding apparently removes all of the play's bad blood as the vessel of the new Montague-Capulet admixture or "incorporation." Ironically, she does "die with a restorative" as she says (5.3.166).[56]

All told, the final scene of purgation effectively realizes Juliet's nightmarish vision of the Capulet tomb, symbolizing Veronese society and the oppressive force of bloodlines. She envisions this tomb as follows:

> As in a vault, an ancient receptacle,
> Where for this many hundred years the bones
> Of all my bury'd ancestors are pack'd,
> Where bloody Tybalt yet but green in earth
> Lies festering in his shroud. (4.3.39–43)

This "ancient receptacle" is packed with the bones of her ancestors and the blood of Tybalt, who in life wholeheartedly demanded that Capulet blood be honored by vengeance. This image of Juliet suffocated by bloody Tybalt's "festering" aptly figures a Juliet overtaken by the demands of her bloodline and her blood-obsessed society. For, although Juliet's bloody sacrifice satisfactorily purges the familial enmity, the visual impression of the bloody, tangled mess of the Capulet tomb thwarts any clean sense of closure for the audience.[57] Shakespeare's *Romeo and Juliet*, rather than encourage the audience to celebrate the peace-making patriarchs and social reintegration, prompts them to question the means of this closure. Endowing bodies and blood with this kind of agency and power means refusing to recognize the role that human actors play in constructing this dangerously pervasive discourse and, therefore, to recognize their capacity to change it. These beliefs, after all, apply to Shakespeare's England as much as Romeo and Juliet's Verona.

Romeo and Juliet thus stages Juliet's bloody nightmare, but, in so doing, it calls on the audience to consider instead Juliet's imagined alternative dream, wherein one might call a rose by another name.

LE CID: A TRAGEDY OF TALKING BLOOD

In one respect, to compare *Le Cid* to *Romeo and Juliet* is to examine a similar narrative premise with a different and, according to some critics, potentially romantic end. Set in eleventh-century Seville, the play begins with the besotted Chimène hoping that her father agrees to betroth her to her beloved, Rodrigue, over the rival suitor, Don Sanche. Meanwhile, the Infante of Castille, also loves Rodrigue; but as he is below her station, she vows to help him marry his beloved Chimène. However, trouble erupts when the King passes over Chimène's father, Don Gomes, for the position of the prince's tutor, given instead to Rodrigue's father, Don Diègue. The two patriarchs meet in an attempt to smooth over the situation, but Gomes ends up slapping Diègue, requiring retaliation from Rodrigue to avenge his father's honor. This confrontation inevitably results in the murder of his beloved Chimène's father. After the conflict, Don Sanche offers to avenge the death of Chimène's father by dueling Rodrigue. She declines the offer and, after he leaves, professes her persistent love for Rodrigue, who happens to be hidden in a closet and overhears. Rodrigue reveals himself, imploring her to kill him in order to avenge her father, but she cannot bring herself to do so, instead resigning herself to ask the King for justice. After leaving Chimène, however, Rodrigue receives news that he must leave to fight the Moors. Rodrigue returns from battle a war hero, earning the new honorific title Le Cid, which gives the Infante renewed hope that his elevated reputation might permit their marriage after all. All convene in court, where the King pretends that Rodrigue has been killed in battle in order to trick Chimène, trusting that she will reveal her love for Rodrigue. Once she has, he can insist that she relinquish the blood feud to marry her beloved. However, once Rodrigue is found alive, Chimène refuses to fulfill her desired match and instead accepts Don Sanche's offer to duel Rodrigue, promising that she will marry the victor. After Don Sanche returns with a bloodied sword, Chimène once again implores the King to allow her to renege on her promise; she begs him that she join a convent instead so that she need not marry Don Sanche. But, as it turns out, Rodrigue disarmed Don Sanche but chose not to kill him. The

King once again insists that Chimène marry Rodrigue, arguing that she is released from her duty to her family's bloodline, and declares that the marriage will take place in one year's time, during which time Rodrigue will continue to fight in the war against the Moors.

Curiously at odds with the action's tragic bloodshed, the play's conclusion seeks to foreclose blood's demands. The King promises that the blood feud will cease, that the two lovers can wed and live together happily thereafter.[58] And yet, despite the attractiveness of this promise, Chimène's "point of honor" described in graphic detail—"d'avoir trempé mes mains dans le sang paternel" ("to have soaked my hands in paternal blood") (5.7.1838)[59]—appears to be neither an arbitrary point of contention to hide her true feelings nor a mechanism for Corneille to appease the critics of the French Academy. Rather, her complaint carries enormous weight if the theories of blood are taken seriously, as she clearly does. If one's blood determines one's character, and blood is mixed in marriage, then husband and wife ultimately come to share both the traits and the crimes of one another. In this case, if Chimène's blood were to mix with Rodrigue's, then she would be guilty of patricide.[60] Chimène clings to these reservations because for her, though the King can pardon Rodrigue's crime in a legal context, his blood cannot be changed by royal sanction. She thus clings to this ultimately destructive discourse in order to save both herself and her entire bloodline from "reproche éternel" ("eternal reproach") (5.7.1837).[61] If we recognize in this moment the seriousness of the characters' ideological framework—namely that blood is a substance that literally determines identities, duties, and affinities and that has the agency to make explicit demands—then Chimène's final speech and the uncomfortable nature of the dénouement becomes clearer.

Viewing the talk of blood in this play as not merely metaphorical but as embodied or fleshy gives new weight to Chimène's refusal; in this framework, her marriage to Rodrigue could not be morally permissible since she will assume his faults physiologically. Moreover, attending to the materiality of blood's call also prompts us to reexamine the play's standard gloss as a conflict between two antithetical ideologies: that of the nobility and that of the would-be absolutist monarch. Chimène's "point d'honneur" highlights that these assumptions about blood's nature, if earnestly believed, are more powerful than either the aristocracy or the monarchy that cite them to substantiate their authority. Much like the fate of Shakespeare's Romeo and Juliet, "it is the characters' fate to be caught up and trapped in their own code" (Bornedal 168).[62]

As noted, the classic political reading of *Le Cid* characterizes the play's struggle as one between two value systems: the emergent absolutist monarchy and the older feudal structures of the aristocracy. Corneille scholars have understood *Le Cid* as part of this historical story of noble decline, with the language of blood, characterized by Chimène's point of honor, squarely on the side of the aristocracy.[63] This gloss is understandable given the role that blood played in the debates around nobility in this moment as outlined in this book's introduction. Paul Bénichou, for instance, understands Corneille's dramatic corpus as embodying "l'esprit aristocratique" (the aristocratic spirit), which includes rebelling against the aristocracy's humiliation by royal authorities. For Bénichou, although *Le Cid* reinstates the King's power at the close, it nonetheless reflects a certain nostalgia for this older feudal political system.[64] Bénichou and other scholars generally speak of these competing value systems in a disembodied way. Elliott Forsyth, for example, underlines the "puissance trancendante" (transcendent power) of the feudal honor code in *Le Cid*, the "devoir se venger" (duty to avenge), though he views this discourse as a disembodied one of ideals, especially glory (393). Peter Bornedal's reading depends on his basic assumption that the code of honor is fundamentally about debts paid and maintaining a "name" (synonymous with honor), a value that outweighs all else. These systems, however, are very much attached to bodies; and when examined in this light, the two opposing sides are not so antithetical as often assumed.[65]

Rather than understanding this phenomenon as a story of the nobility's decline,[66] one could view this appeal to blood as part of their attempt to negotiate new institutions by using this vocabulary in novel ways; and in this negotiation the King might likewise make use of the same vocabulary.[67] Marcus Keller notes in the play that the King usurps the nobles' blood discourse (of lineage and race) to construct a new absolutist state and a national identity defined by a common ethnicity.[68] Yet Keller's analysis treats blood metonymically, connected to kinship and ethnicity with little reference to its literal purchase in both the play's action and in the characters' conceptions of their respective identities. In short, he fails to note that these blood discourses actually demand and involve literal blood through blood spillage and violence. Corneille's persistent emphasis on blood—the word *sang* (French for blood) appears no fewer than forty times in the space of the play—emphasizes the violence necessary to perform blood's ideology on *both* sides of the political divide. And even though blood does not cover the stage as it does for Shakespeare, Corneille's play, like its English cousin, enlarges blood's presence in order to emphasize this violence.[69] While, as Rebecca Zorach notes, the early modern French court also used blood to symbolize power and vitality, in *Le Cid* the blood that embodies honor and identity demands blood spillage to preserve these characteristics—whether in the context of the aristocratic blood feud or in the King's new body politic. Otherwise stated, even though the aristocracy and the monarch cite blood discourses toward contradictory ends, they invest in similar assumptions with similarly violent consequences.

Corneille's play reveals that blood discourses, when taken seriously, can only produce more violence as the physiological material dictates these actions regardless of individual will or affinities: Rodrigue must salvage the honor in his father's veins; blood likewise demands violence from Chimène, mandating death to the man responsible for spilling her family's blood;[70] and King Fernand is equally dependent on blood spillage for his own identity and authority as (absolute) monarch. He can attempt to argue that his royal blood (the blood of the nation, as he puts it) matters *more* than that of the nobility. But if he holds the same premise, that blood matters, then *all* blood matters; and he cannot convincingly apply this principle to his body alone. So long as these characters invest in these discourses, regardless of their side in this political divide, blood will demand more blood, and future generations will be determined by the same violent requirements and demands. The alternative to this dialectic, to have a future without blood, necessitates rejecting this bloody ideology entirely; as *Le Cid* intimates, defining others by their bodies here only yields violence, whether in a blood feud or in a holy war against the Moors. In this case, *Le Cid* offers another woman, one similarly disadvantaged though due to her station rather than her youth, to voice this alternative: the maidservant Elvire. Placing the voice of peace, previously reserved for the humanist heroes of Garnier's drama, in Elvire's mouth is both to place an enormous emphasis on this small moment and to render it inconspicuous.[71] Yet, like Juliet, her voice has no apparent influence on the play's conclusion, and thus, both endings appear equally bleak.

To briefly rehearse the violent consequences of this belief in blood, the nobility clearly accept blood's agency, which incites the blood feud. The two patriarchs, and Don Diègue especially, quickly invoke the belief that blood materially distinguishes families and dictates both characteristics and affinities, reminiscent of the possessives littered in Shakespeare's play.[72] With these affinities come violent duties to one's bloodline, despite the feelings of its particular individual members. Don Diègue rehearses several assumptions about how blood functions when he calls upon his son to avenge him. He importunes,

> Je reconnais mon sang à ce noble courroux,
> Ma jeunesse revit en cette ardeur si prompte,
> Viens, mon fils, viens mon sang, viens réparer ma honte;
> Viens me vanger. (1.5.264–67)
>
> (I recognize my blood in this noble rage; my youth revives in this ardor so prompt. Come, my son, come, my blood, come to retrieve my shame—come to avenge me!)

Don Diègue here not only equates his blood with his son, marked by the parallel construction "viens, mon fils, viens mon sang," but also because his son is a vessel of their shared bloodline he also must bear the same personality traits, his "noble courroux" and "ardeur." And as Rodrigue's blood has dictated his identity and personality, it must also dictate his actions; it is, after all, ambiguous as to which entity is "coming to avenge him" ("Viens me venger"), is it "mon fils" or "mon sang"? Revenge, moreover, demands spilling blood to wash away dishonor as Diègue spells out, "Ce n'est que dans le sang qu'on lave un tel outrage, / Meurs, ou tue" ("It is only in blood that one can wash away such an insult; die or slay") (1.5.274–75). As Kalervo Räisänen notes, we should understand this conundrum through the stain metaphor; one has to get rid of the stain by plunging it in the same material so that the stain disappears. Diègue's argument relies on the assumption that Rodrigue inherits the stain of his offense due to their consanguinity, and therefore, to avenge his father is to avenge himself: "Enfin tu sais l'affront, et tu tiens la vengeance, / Je ne te dis plus rien. Venge-moi, venge-toi" ("In short, thou knowest the insult, and thou holdest [in thy grasp the means of] vengeance. I say no more to thee. Avenge me, avenge thyself!") (1.5.286–87). Don Diègue does not frame his request as merely something that a son should do out of duty to his father, but rather Rodrigue and he are materially connected and his son likewise requires vengeance.

After Don Gomes's death, Don Diègue further foregrounds his own body that has been cleansed by his son's action, telling his son, "Touche ces cheveux blancs à qui tu rends l'honneur" ("Touch these white hairs, to which thou restorest honor!") (3.6.1036). Here the grammatical construction emphasizes Diègue's body; his honor *belongs to* or *resides in* his white hairs since the white hairs are the vessels of adulation and not himself. Diègue's engagement with this discourse here, and as a whole, stresses his belief that blood operates literally. He praises his son for the act that confirms their blood/lineage "et ton illustre audace / Fait bien revivre en toi les héros de ma race" ("and thy brilliant prowess causes the heroes of my race to live again in thee!") (3.6.1029–30). Though a moment of praise, Don Diègue's compliment also functions as a hefty demand. The heroes of his entire lineage, "ma race," depend on Rodrigue; they live on only in him. His body, his blood is of utmost importance not only for himself and his father but for an entire family history. And it is this expectation that ultimately convinces and compels Rodrigue.

Both Rodrigue and Chimène assume blood's ideology, and though conflicted, when blood is understood as the physiological foundation of one's person, it must trump previous affinities. In other words, blood conceived in this way necessarily dictates and determines without concern for any other emotional considerations, the latter of which should prompt the characters (or at least the audience) to interrogate this principle. For instance, Rodrigue voices this reasoning after a lengthy internal debate: "Que je meure au combat, ou meure de tristesse, / Je rendrai mon sang pur comme je l'ai reçu" ("Whether I die in the combat or die of sadness, I shall yield up my blood pure as I have received it") (1.6.343–44). Rodrigue insists that he must put love aside because his blood will not change; his basic nature, his integrity as an individual endures no matter the manner of his death. Moreover, according to blood's logic, material purity of the family honor must be his primary concern since his blood belongs not simply to his individual person. He is the mere vessel of a force greater than himself, which cannot be changed by either his love for Chimène or the King's absolution. Rodrigue fully internalizes these assumptions regarding consanguinity and even (re) cites them back to the older generation. He berates Chimène's father for his actions, asserting that his blood is the source of the ardor that Gomes once admired, the same blood he shares with his father: "Cette ardeur que dans les yeux je porte, / Sais-tu que c'est son sang? le sais-tu?" ("This fire which

I carry in mine eyes, knowest thou that this is his blood? Dost thou know it?") (2.2.401–2). Repeating twice "sais-tu"—particularly brash since he addresses his elder using the more informal "tu"—Rodrigue's challenging tone highlights his internalization of this discourse.[73] And this discourse apparently justifies his future actions: killing the father of his beloved.

Once completed, Rodrigue's subscription to the blood feud incites Chimène's response in kind, for she likewise accepts its violent injunctions.[74] Chimène cannot reject familial duty that has been physiologically determined. Blood's ideology equally implicates her noble body even if she lacks the capacity to act on it as Rodrigue does. As a woman, it would be considered unseemly for Chimène to effect Rodrigue's death by her own hand; and lacking any other male family members, she must continue to call on the King for justice/vengeance (Forsyth 394). Moreover, Chimène cannot find a place in the King's discourse of blood: spilling Moorish blood for that of her nation. As Barbara Woshinsky notes, gender accounts for the protagonists' respective measure of success: "He [Rodrigue] can say almost anything and be believed. *He* not *she*. For the guarantee does not work for heroines. While men can both act and speak, women can only speak, a disparity which will ultimately deny Chimène heroic stature" (154).[75] Chimène is thus at the mercy of the king's justice to satisfy blood's demand, but this blood, an intractable part of her, cannot be denied or set aside. When asking the King for justice with Rodrigue's death, she forcefully underlines blood's power:

> Sire, mon père est mort; mes yeux ont vu son sang
> Couler à gros bouillons de son généreux flanc,
> Ce sang qui tant de fois garantit vos murailles,
> Ce sang qui tant de fois vous gagna des batailles,
> Ce sang qui tout sorti fume encor de courroux. (2.8.659–63)
>
> (Sire, my father is dead! My eyes have seen his blood gush forth from his noble breast—that blood which has so often secured your walls—that blood which has so often won your battles—that blood which, though all outpoured, still fumes with rage at seeing itself shed for any other than for you!)

The first feature to note is that, after proclaiming her father dead, his blood is her very first charge. Second, her description of this blood is extremely visceral, flowing in "gros bouillons" from his flesh. While one might argue that this gruesome description acts as mere rhetorical flourish to underscore the violent nature of Rodrigue's crime, the physical nature of the blood discourse throughout this play asks the audience to take this bloody description more seriously. Third, Chimène's use of anaphora repeats the word *sang* three times over. The blood here becomes the subject of her proclamations: It protected the kingdom's borders; it won battles; it still fumes with rage.

Blood calls for action and not metaphorically. She describes her father's blood speaking to her,

> Son sang sur la poussière écrivait mon devoir;
> Ou plutôt sa valeur en cet état réduite
> Me parlait par sa plaie, et hâtait ma poursuite;
> Et pour se faire entendre au plus juste des rois,
> Par cette triste bouche elle empruntait ma voix. (2.6.676–80)
>
> (His blood upon the [surrounding] dust dictated [lit. wrote] my duty; or rather his valor, reduced to this condition, spoke to me through his wound, and urged me to claim redress; and to make itself heard by the most just of kings, by these sad lips, it borrowed my voice.)

This could easily be glossed as a conventional symbol of her familial duty, but the heroine's account of blood in these lines, in keeping with the phenomenon of cruentation (spontaneous bleeding to disclose a murderer's identity) is more than mere poetic device.[76] Chimène's language not only emphasizes blood's material presence spilt in the dirt but also figures blood as agent that calls for action since "sang" is the speech's subject with its own "bouche," the wound. Her father's blood speaks to her on account of their consanguinity, but it then borrows ("empruntait") her voice to speak for it in a public forum. Blood has agency in this drama, an agency that Chimène recognizes when she calls on "le sang par le sang" ("blood for blood") (2.8.692). Her actions are not hers alone but are instead compelled by her bloodline, a force larger than herself. Blood hastens her pursuit, and she apparently lacks the ability to refuse its directive. The familial blood dictates her actions and commands her to speak for it, just as a god might speak through a possessed human host. Blood calls for blood seemingly in spite of human volition. When

Don Rodrigue asks Chimène to kill him with his own sword, she exclaims, "Quoi! du sang de mon père encor toute trempée!" ("What! still imbrued with the blood of my father!") (3.4.858). And only a few lines later, she insists that this is likewise her blood, "Il est teint de *mon* sang" ("It is dyed with *my* blood!") (italics mine) (3.4.863). Chimène thus makes no distinction between her blood and her father's. Like Rodrigue's acceptance of this discourse outlined by his father, if Chimène assumes this discourse, then she believes she has no choice but to suppress any love for her enemy and plead for his blood to cleanse her own.

These lovers thus understand this mandate as a matter of personal integrity because it comes from within themselves, the very substance/essence of their noble beings. As other scholars have noted, the young lovers accept this "devoir du sang" ("mandate of blood") in part due to a double bind (5.2.1592). For in order to be worthy of the beloved's love, the lover must maintain their honor, even if doing so entails violence toward the beloved or their family.[77] Rodrigue begging for Chimène to kill him, insisting that only her hand can do it, aligns with this reasoning. In order for her to be worthy of his love, she must kill him to remove the stain of dishonor from her own. Rodrigue pleads, "Plonge-le dans le mien, / Et fais-lui perdre ainsi la teinture du tien" ("Plunge it in mine, and cause it thus to lose the death-stain of thine own") (3.4.863–64). The stain of her family's blood and thus hers, "la teinture du tien," can be washed away by spilling his own blood. Like the language's parallelism, the system is one of perfect parity, and cruelly, what is best both for her and their love is that he be repaid in kind. Rodrigue later explicitly reframes this demand in terms of honor, "Immole avec courage au sang qu'il a perdu / Celui qui met sa gloire à l'avoir répandu" ("Sacrifice with courage to the blood he has lost he who constitutes his glory in having shed it") (3.4.903–4). A sacrifice must be made to her father's blood lost, and with this premise she agrees. But the command "immole" takes on an even more disturbing tenor in this scene. The blood of the father assumes a quasi-divine status, which requires that one sacrifice *to* it, threatening some measure of retribution if she were to refuse. But only Rodrigue appears to have a means out of this bind: he can kill the Moors and subscribe to the King's new economy of blood in order to gain his favor, but Chimène lacks this recourse.

The Infante on the other hand, unlike the protagonists, sees the love between Rodrigue and Chimène as a possible means to end the feud and its violence. Much like Friar Laurence's reasoning in *Romeo and Juliet*, she entertains this marriage as a means to cleanse the bad blood between them. The marriage would establish new ties of affinity between the families and consequently would "snuff out" the discord. She tries to comfort Chimène,

> Le saint nœud qui joindra don Rodrigue et Chimène,
> Des pères ennemis dissipera la haine,
> Et nous verrons bientôt votre amour le plus fort
> Par un heureux hymen étouffer ce discord. (2.3.473–76)
>
> (The sacred tie which will unite Don Rodrigo and Chimène will dispel the hatred of their hostile sires, and we shall soon see the stronger [feeling], love, by a happy bridal, extinguish this discord.)

The Infante's hope is quite explicit; it is the marriage tie, and the power of Hymen, the goddess of marriage, that would dispel the hatred between the two families. If the reference to this "sacred tie" is at all ambiguous, then the allusion to Hymen clearly implies a sexual bond with its mixing of flesh and blood (especially in the form of the maidenhead) upon entering married life. As with the Montagues and Capulets, by mixing bloods in marriage (commixtio sanguinis), the two feuding families would now be members of the same extended kinship group, thereby determining new affine relations that could overturn the stains to family honor. Her appeal sounds very much like the friar's hope for ending the Veronese blood feud. Yet Chimène knows that a marriage would not absolve her duty, for to incorporate his blood into her own would mean to incorporate his character and liabilities, which was not a problem, of course, until he killed her father. Chimène frankly acknowledges this problem when speaking of King Fernand's command that she marry the victor of the duel between Don Sanche and Rodrigue. Whoever wins, "De tous les deux côtés on me donne un mari / Encor tout teint du sang que j'ai le plus chéri" ("In either case [lit. on both sides] they give me a husband, still [all] stained with the blood that I cherished most") (5.4.1659–60). Regardless of the outcome, when married, she would be guilty for, and thus tarnished by, the blood ("tout teint du sang") of the one that she most cherishes, either her father or Rodrigue. Her strain against the King's decree, therefore, reflects the seriousness of her investment in blood's nature and agency.

Thus, as the play concludes, Chimène worries that she will be *forced* to take on the horrifying consequences of marrying Rodrigue. She asks King Fernand if she *must* marry him,

> C'est trop d'intelligence avec son homicide,
> Vers ses Manes sacrés c'est me rendre perfide,
> Et souiller mon honneur d'un reproche éternel,
> D'avoir trempé mes mains dans le sang paternel? (5.7.1835–38)
>
> (It is too much to know with his homicide,
> Toward the sacred Manes it renders me treacherous,
> And to sully my honor with endless reproach,
> From having soaked my hands in my father's blood?)[78]

Chimène's language reflects that this marriage would be no small compromise because it would defile her honor with an "éternel" reproach. This contamination of her blood has implications both for the status of her own soul ("vers ses Manes sacrés") and for all of her future kin. Needless to say, she does not conceive of this stain on her honor as a metaphorical one. Her hands would be soaked in "sang paternel" ("paternal blood"); though not visible to the audience, she nonetheless believes that her body would still be compromised in a very visceral sense. If two become one in the sanctity of marriage and their bloods mix, then she is guilty of patricide, *as though* she herself had stabbed her father. And if this principle of contamination is true, if blood in fact determines and signifies in this way, then the King does not have the authority much less the capacity to absolve it. King Fernand's final pronouncement at the play's close, however, utterly ignores these implications. He effectively turns a deaf ear to Chimène altogether and instead calls on Rodrigue to hope:

> Espère en ton courage, espère en ma promesse;
> Et possédant déjà le cœur de ta maîtresse,
> Pour vaincre un point d'honneur qui combat contre toi,
> Laisse faire le temps, ta vaillance, et ton roi. (5.7.1837–40)
>
> (Hope in thy manly resolution; hope in my promise, and already possessing the heart of thy mistress, let time, thy

> valor, and thy king exert themselves [lit. do or act], to overcome a scrupulous feeling of honor which is contending against thee.)

As Cordell Black notes, despite the King's trick that forces Chimène to expose her feelings publicly, "to acknowledge [Chimène's] love for 'le Cid' is not at all tantamount to making the ultimate concession: marriage" (79). Her "point d'honneur" holds far more weight than the King will acknowledge; for her, it is not some small point of pride. Despite the King's appeal to time to assuage her concerns, blood's material will persist from one generation to the next. If the material of her body, the demigod blood, has a power and force outside of her reach, then it is likewise beyond King Fernand's purview. Namely, the King's sanction cannot change the physiological character of Rodrigue's blood. This belief is a serious problem for his authority. Yet he cannot deny the power of blood to the aristocracy because he likewise depends on the assumption that blood embodies power and authority, especially his own blood.

Although King Fernand promotes a romantic conclusion, his command that Chimène marry and relinquish her convictions betrays a double standard rather than a uniform critique of blood's ideology. His command conflicts with his own investment in blood, his royal blood and that of the nation, which both depend upon the same assumption as Chimène's point of honor, namely that blood signifies. The King calls upon his own power and authority to mandate that others relinquish this assumption when it pertains to the blood feud—"un sanglant procédé" ("a bloody practice") (4.5.1452) as he calls it—simply because it suits his purposes. Yet if his blood determines his power, authority, and identity, then there is an incongruity to his insistence that the nobility's blood does not.[79] His stance, therefore, does not fully accord with his promise of a comedic conclusion, and, as it turns out, shoring up his own authority and identity likewise proves to be a "bloody practice." To illustrate, the Infante insists upon maintaining her blood's purity allied with her father's investment in his blood's authority. Her status as a princess impedes her own love for Rodrigue since her union with a noble would debase her royal blood. Despite some inner turmoil at the beginning of the play, she resolves, "Oui, oui, je m'en souviens, et j'épandrai mon sang / Plustôt que de rien faire indigne de mon rang" ("I remember it so well, that I would shed my blood rather than degrade my rank") (1.3.85–86). The

Infante would rather *spill* her blood than do anything unbecoming of her blood's quality (rank), which is the supposed key to the power of the monarchy.[80] However, the Infante wavers once Rodrigue has returned victorious, once he has been named Le Cid and thus has theoretically obtained worthier blood by violence. She muses to herself, "Et ce grand nom de Cid que tu viens de gagner / Ne fait-il pas trop voir sur qui tu dois régner?" ("And this great name of Cid, which thou hast just now won—does it not show too clearly over whom thou art destined to reign?") (5.2.1587–88). As she conceives it, this new name is a function of a change in person, and one that is founded in blood. Admitting to her maidservant, "Si j'aime, c'est l'auteur de tant de beaux exploits, / C'est le valeureux Cid, le maître de deux rois" ("If I love him, it is [as] the author of so many brilliant deeds; it is [as] the valiant Cid, the master of two kings") (5.3.1635–36). According to the Infante's formulation, she loves an altogether different man, the author of these exploits, a new blood-borne identity.

King Fernand adamantly expresses his own investment in Le Cid and other heroes who can spill Moorish blood and themselves bleed for the safety of the body politic. For this blood is central to his royal identity. King Fernand first exposes this reliance on their blood after news of Don Gomes's death; King Fernand laments the loss of his life but especially his loyal career measured in blood spilt for his country, "Après un long service à mon État rendu, / Après son sang pour moi mille fois répandu" ("After long service rendered to my state, after his blood has been shed for me a thousand times") (2.7.643–44).[81] The diction here emphasizes his personal investment; the blood was spilt not for his nation or crown but their equivalent, "pour moi" ("for me"). Earlier in the play, the Infante echoes this attitude toward blood when daydreaming of a Rodrigue who would be worthy of her, "Porter delà les mers ses hautes destinées, / Du sang des Africains arroser ses lauriers" ("advancing his proud destinies beyond the seas, laving his laurels with the blood of Africans!") (2.5.542–43). The very visceral "watered" or even "soaked" of the verb "arroser" marks the somewhat grotesque and violent nature of the action needed to transform Rodrigue's nature; it is literally dipped in blood and dripping from the laurel crowns on his head. This repulsive image highlights the destructive center of even blood's positive symbolism. Moorish blood will be spilt for the preservation of the nation's lifeblood and that of its king.

King Fernand further paints his subjects' blood as his primary concern, although only as part in parcel of his own political body.[82] He explains to his adviser,

> Un roi dont la prudence a de meilleurs objets
> Est meilleur ménager du sang de ses subjets,
> Je veille pour les miens, mes soucis les conservent,
> Comme le chef a soin des membres qui le servent. (2.6.595–98)
>
> (A king, whose prudence has better objects in view [than such quarrels], is more sparing of the blood of his subjects. I watch over mine; my [watchful] care protects them, as the head takes care of the limbs which serve it.)

Caring for his subjects' blood and well-being is the primary aim, the *best* ("meilleur") object emphasized twice over, of the prudent ruler. And via metaphorical extension, their blood becomes his blood as "le chef" of the body politic. But as he admits earlier, his concern for his subjects' blood does not pertain to their sense of honor or individual blood purity but rather the blood shed in violent combat in service to his nation. Rodrigue, too, seemingly invests in the King's economy of blood in this moment, crediting the King and his empire for his own blood:

> Je sais trop que je dois au bien de vostre Empire
> Et le sang qui m'anime et l'air que je respire,
> Et quand je les perdrai pour si digne objet,
> Je ferai seulement le devoir d'un sujet. (4.3.1243–46)
>
> (I know too well that I must for the good of your Empire, and the blood that animates me and the air that I breathe, and when I would lose them for so dignified an object, I would do only the duty of a subject.)[83]

Rodrigue recognizes that his blood, and more specifically his blood that can or will be lost for his king, effectively belongs to the king. Moreover, the third scene of act 4 exhaustively renders this bloodshed in Rodrigue's narrative as "des champs de carnage où triomphe la mort" ("the fields of carnage where death triumphs") (4.3.1310). Further describing the surprise ambush, Rodrigue recounts, "Nous les pressons sur l'eau, nous les pressons sur terre / Et nous faisons courir des ruisseaux de leur sang / Avant qu'aucun resiste, ou reprenne son rang" ("We press them hard on the water, we press them hard on the land, and we cause rivulets of their blood to run before any [of them] can resist or regain his position") (4.3.1300–1302).[84] The streams of blood recounted here, after all, are no different from those streaming from the body of Chimène's father. Thus, this play does not simply expose the tragedy of some persons caught in an older value system during the transition from one social or political structure to another.[85] Despite the notable political change, both the feudal and monarchical structures invest in a common ideology. Both sides of this power struggle depend upon the premise of blood's signification, and in both cases, this investment is destructive.

Yet, by way of conclusion, Corneille's play does make a small space for hope, although not in the form of a prospective marriage between Chimène and Rodrigue. Instead, Chimène's servant, Elvire, provides a small opportunity for the audience to recognize that because blood does not literally operate in the way that the characters imagine, there is a way out of its violent consequences. Apart from the play's conflict, she suggests that rejecting this framework entirely might be the only escape from its violence and the only potential for a peaceful conclusion. One cannot merely insist that the blood feud cease (as King Fernand declares in the interest of his own power), but one must explicitly *recognize* the violent by-products of defining others, or oneself, by blood. Elvire might look like merely a stock character, playing the role of the confidante and sometimes chorus; and in fact, most critics generally ignore the cast of supporting characters since they do not garner much sympathetic attention from the audience. Mohammed Kowsar even goes so far as to say that "to be sure there are the two confidantes (Elvira and Leonor), the respective governesses of Chimène and the Infanta, but their personalities reflect the expediency of convention so faithfully that not a single perception of any consequence is emitted by either character" (291). But upon closer examination, Elvire voices the only bloodless option. Elvire's question to Chimène articulates the ultimate critique of the blood feud's nonsensical logic:

> Que prétend ce devoir? et qu'est-ce qu'il espère?
> La mort de votre amant vous rendra-t-elle un père?
> Est-ce trop peu pour vous que d'un coup de malheur?
> Faut-il perte sur perte, et douleur sur douleur? (5.4.1689–92)

> (What means this duty, and what does it hope for? Will the death of your lover restore to you a father? Is one [fatal] stroke of misfortune insufficient for you? Is there need of loss upon loss, and sorrow upon sorrow?)

Rather than look to bodies and who or what they determine, she instead belabors the physical consequences of maintaining these discourses, loss on loss and woe on woe. As Clifton Cherpack notes in his work on the trope of the cri du sang, Elvire's skepticism and critique of this discourse is depicted more clearly in other tragedies by Corneille. For the protagonists of *Le Cid*, the power of blood lies squarely on the side of reason, duty, and honor, which they take very seriously, but "for many Cornelian characters, while they seem to recognize the privileges and responsibilities attendant upon blood kinship, are quite cynical about their sanctity" (34–35). Elvire's concern should also resonate for Corneille's audience on account of recent historical events: the enormous bloodshed caused by the civil wars of religion, a conflict of faith but also of noble factions, as well as King Charles IX's probable role in the Saint Bartholomew's Day Massacre.[86] While blood's discourse functions as a means for the nobility and monarch to maintain and stabilize power, providing a material foundation for their virtues and privileges, it also, as Elvire notes, produces more hurt.

Elvire's moment of resistance, though brief, resembles speeches of humanist heroes, for instance, Agamemnon's call for peace in Robert Garnier's *La Troade* (1579).[87] Placing the voice of peace, previously reserved for such humanist heroes, in Elvire's mouth is both to place an enormous emphasis on this small moment and to render it less conspicuous. Giving Elvire the critique of blood's ideology, in one sense, is most fitting; for unlike her mistress, Chimène, Elvire is in no way implicated in the power struggles of the play supported by this ideology. However, on account of her social position, her voice is likewise muted and ignored, preventing her words from having any immediate impact on the plot. Blood's voice overpowers her own. Nevertheless, Elvire's critique underscores that defining others by their bodies can only yield violence, whether that be violence to the bodies of an opposing family (as mandated by the nobility) or excessive violence upon Moorish bodies (charged by the monarchy). As Corneille's *Le Cid* seems to suggest, if both the aristocracy and the new monarchy invest in blood, then only blood can follow.

CONCLUSION: DISCOUNTED VOICES

Corneille's and Shakespeare's similarly bloody renditions of their shared (or kin) Romeo and Juliet plot, a conflict between romantic love and familial loyalties, explore the violent necessities of performing discourses that not only demand blood for blood but that also insist that consanguinity dictates affinities. In making their sources bloodier, these plays speak to a shared cultural concern on both sides of the channel; they remind the audience that a literal investment in blood's discourse, though it may materialize and thus solidify social duties, authority, and power, has immensely troubling consequences. Both plays illustrate the conviction that honoring consanguinity, filial love, and affection by privileging blood over persons requires more blood spilt of one's enemy, only perpetuating violence. Moreover, staging these characters so heavily invested in these claims that do not obviously hold—namely, blood has not determined the affections of all of its members—calls upon the audience to view these appeals as rhetorical rather than factual. In both cases, blood's ideology has tragic ends; even those who look to cite blood discourses for positive ends fail. The mixing of bloods in marriage is not a way out of a tragic framework for Shakespeare's Romeo and Juliet, and neither is Corneille's King's insistence on marriage, since his own interests likewise appeal to blood and violence. Only a rejection of these logics outright appears to promise a space for romantic love rather than the blood feud's violent and tragic propulsion.

Yet, that these dissenting voices are women, both doubly discounted by society due to age or station, makes change very difficult to effect.[88] Shakespeare's heroine Juliet is both a woman and exceedingly young, a mere thirteen years old. Her voice, one not yet fully indoctrinated, can see an alternative to society's present structure. Juliet, albeit briefly, presents a different future, but here a future that has been tragically sacrificed to the status quo. For *Le Cid*, a serving woman who is disadvantaged by the system remains untainted by it. This disenfranchisement encourages this *thinking otherwise*, but it likewise diminishes the capacity for her words to have any influence in the confines of the play. But maybe, the audience could hear them, looking to possibilities beyond the existing, flawed, and violent system—the dramatic space rendering their voices audible even if their society does not.

2

When Blood Cries

INCEST AND BLOODY AFFINITY

Say that we had one father, say one womb
(Curse to my joys!) gave both us life and birth;
Are we not therefore each to other bound
So much more by nature? By the links
Of blood, of reason? Nay, if you will have't
Even of religion, to be ever one,
One soul, one flesh, one love, one heart, one all?

—GIOVANNI TO FRIAR BONAVENTURA
'Tis Pity She's a Whore (1.1.29–34)

In the opening of John Ford's *'Tis Pity*, Giovanni's speech begins as a seemingly customary expression of love for one's blood relations; but a potentially conventional expression of sibling affection quickly turns into an unconventional, high-flown declaration of an all-encompassing erotic love for his sister, whom he deems his perfect beloved given her biological similarity to himself. Most scholars gloss this moment, as well as Giovanni's arguments for incest in general, as either a perversion or some misunderstanding of Neoplatonic language.[1] But what if Giovanni is not mistaken?[2] What if incest were not a perversion so much as an extension of these "links of blood"? If, as outlined in the introduction, kin love one another on account of consanguinity, similarity in blood or "kind" was presumed to be the foundation of erotic love matches,[3] and that love only increased upon mingling bloods in marriage, then what (other than a taboo of course) could prevent the conclusion of incest? When reexamined in the larger context of contemporary assumptions about blood, incest (especially sibling incest) provides the limit case for the assumption that consanguinity determines affinity. If love is based in likeness, then a person would, theoretically, most like the person who is most like themselves; in which case, one would *expect* siblings, having exactly the same blood mixture, to potentially harbor romantic love.[4] After all, if married spouses love one another as a result of their shared blood postcoitus, effectively becoming kin, then the love between siblings is theoretically no different from a physiological perspective, except that they do not require sexual relations to achieve it. In short, incest is the result of affinity's logic when taken to its natural conclusion—if, as Giovanni says, the "links of blood" are extended to those "of reason."

Critics have glossed the profusion of incest plots in the early modern period in a variety of ways, but as Richard McCabe remarks, these explications have something in common, namely "a genuine and sometimes courageous willingness to think the previously unthinkable, to question prevailing theories of family and state, to lend voice to suppressed misgivings and challenge the normative doctrine of nature itself" (141). For instance, incest could be demonized as the mark of the consummate sinner (Bueler); the ultimate Faustian dream for the person who believes themselves to be above the law; or the fulfillment of a narcissistic desire (Smith; Forker). Alternatively, it might reflect an anxiety about knowing one's kin in a period when children were routinely shipped off to wet nurses or caretakers; or, it could be a brand of wish-fulfillment for emotional intimacy when kinship relations were vexed given the practice of primogeniture and the unequal distribution of economic resources (Forker). Incest might likewise be the natural outcome of the ideal of Christian "universal siblinghood" (Shell); or it might even provide a space for female agency as the sibling relationship best approximates perfect equity between partners (Quilligan). Yet most criticism interprets incest as the desire to be exempt from society's mandate of exogamy as famously outlined by Claude Lévi-Strauss. As Bruce Boehrer notes, shoring up social, political, or economic resources would have been particularly appealing in this historical moment because "the Renaissance literature of incest develops during a period (roughly from the first quarter of the sixteenth century to the middle of the seventeenth) in which the established ranks of English gentry and nobility come under increasing pressure from an upwardly mobile underclass, and in which the official equipment of English government is progressively stripped of its investiture in the family" (*Monarchy and Incest* 13). Yet, to build upon Boehrer's observation, incest is not simply the perfection of endogamy, it is the natural conclusion of the aristocracy's *already* endogamous practices and the logics that support their hegemony, namely a biological principle wherein blood prescribes affinity.[5] Or put another way, in this framework incest is not properly monstrous since it mimics acceptable, and even ideal, conceptions of familial and spousal love.

This chapter thus takes up incest drama, and particularly sibling incest, as yet another limit case for blood's ideology. Rather than separate these plays based upon genre or the incest-realized versus incest-averted distinction, tracing kin plays that share this plot premise within a particular genealogy reveals a common project shared among these dramatists, namely, to expose the folly of the elite's intense desire to maintain endogamous practices. Sibling incest is, after all, the most extreme instance of the dictum that "like likes like." I argue that these plays critique the elite's endogamous establishment, one wherein noble persons claim that they cannot help but love their likes, thereby abdicating any responsibility for the reproduction of this system and their hegemony. In sum, this social system reproduces itself via reproduction, and the nobility attributes the status quo to the biological, and thereby inescapable, workings of blood. Sibling incest drama thus not only demonstrates the lunacy of this logic but also the ultimately rhetorical workings of this appeal to bloodlines. These plays highlight the power of this rhetoric to transform lives by fashioning social station, though it should rather provoke disgust and shock at its violent outcomes.

Comedies averting incest—like Giambattista della Porta's *La Sorella* (c. 1590) and Thomas Middleton's *No Wit, No Help Like A Woman's* (1612)—appear to endorse the status quo by affirming bodies' importance: both in establishing affinity among kin (affirming the cri du sang) and avoiding the taboo by ensuring that the love matches are *not* consanguineous siblings after all (with birthmarks as proof). And while these romantic tropes ordinarily bring about a satisfactory comedic conclusion, their employment in these plays opens up a problem for these methods rather than closing them down; they, in fact, suggest that bodily distinctions mean very little, exposing the fundamentally discursive nature of blood that purportedly solidifies existing kinship groups and power structures. For in these plays, most of the cris du sang—the instinctive recognition of one's kin accompanied by feelings of affection—turn out to be incorrect, and, in fact, characters invoke blood as it suits them. Della Porta's play successfully averts the incest threat, primarily, by swapping names. And Middleton's *No Wit* takes this suggestion even further, lacking any mention of blood movements or verifying birthmarks intended to distinguish the two women, proffering a more pragmatic attitude toward identity that amounts to "name and place" (5.1.463). This alternative account of identity formation undermines blood's ideology that serves as the biological foundation for the elite's nobility. Moreover, *No Wit* further suggests that blood's discourse obscures the actual motivations beneath this account of kinship structures and love, namely wealth, which by virtue of its ability to circulate, can easily be divorced from noble bloodlines. Hence, even in the case of incest-averted and appropriate matches happily concluded, the entire system supporting the elite's endogamy and hegemony begins to fall apart, made explicit by the crib-switching that can change the course of a person's life by elevating their station.

John Ford's tragedy of sibling incest, in a sense, turns the incest-averted plot inside out, asking what might happen if there were no deus ex machina to forge a comedy. *'Tis Pity She's A Whore* (1630) considers the assumptions about consanguineous affinity from another vantage point, illustrating the tragic repercussions of taking this principle to its natural, uncomfortable conclusion. Giovanni appeals to logics of blood such that shared blood naturally pulls siblings together in body and soul within the Neoplatonic framework—an instinctual attraction beyond their control. And though the people of Parma, and Friar Bonaventura especially, reject Giovanni's interpretation and application of blood's logic, they fail to provide a genuine alternative. Blood's ideology is so powerful and capacious that appealing to any component—for instance, that blood determines both class and character—hinders one's ability to propose an alternative to even its most extreme application. Reminiscent of the workings of tragedy in both *Romeo and Juliet* and *Le Cid*, Ford stages the violent consequences that result from this investment in blood. In this case, Giovanni cannot only blame his blood for his attraction to his sister, but it can also (apparently) justify the violence that he enacts upon her, even excavating her heart. *'Tis Pity* thus calls upon the audience to direct its moral shock not merely at Giovanni as an individual actor but at the system that produced him.

This particular selection of plays is not motivated solely by the sibling incest premise, for there is a probable kinship relation between these three plays. As first noted by Renaissance scholar D. J. Gordon, Thomas Middleton's *No Wit* appropriates Della Porta's plot. This connection then presents a possible point of transmission from Della Porta's play to Ford's.[6] Although not a perfect comparison since it compounds the types of incest, mother-son, father-daughter, and sibling, nouvelle 30 from Margueritte de Navarre's *L'Heptaméron* (1558) might also provide a common point of departure for all three.[7] Yet this is not to suggest a purely literary lineage or teleology, for the connective threads provoke a common consideration: that incest might be the by-product of prevailing blood logics. In De Navarre's story, the mother figure commits herself to a life of celibacy after her husband dies. When her son has grown up and becomes lascivious (making designs on one of her chambermaids), she asks the chambermaid to lure him to bed and plans to get in herself in order to chastise him. Yet, after allowing him to embrace his desires to see just how far he will go, she herself becomes overwhelmed by desire due to her "nature si fragille" (fragile nature) and then beds him. Conceiving a child in that union, she sends her son away to war and her daughter to another household to be raised. Then, of course, as luck would have it, her son returning from the war meets his daughter/sister, falls in love, and brings her home. De Navarre's narrative underlines the central assumption that consanguinity determines affinity when articulating the happy marriage between the two siblings: "Son filz et sa belle fille, lesquelz s'entre-aymoient si fort que jamais mary ny femme n'eurent plus d'amitié et semblance, car elle estoit sa fille, sa seur et sa femme, et luy à elle, son pere, frere et mary. Ilz continuerent tousjours en ceste grande amityé" (Her son and her beautiful daughter, who loved each other so strongly that never husband nor wife had more friendship and semblance/similarity because she was his daughter, his sister, and his wife and him to her, her father, brother, and husband. They continued always in great friendship/amity) (233).[8] The fact that their amity or affinity for one another surpasses all others is explicitly due to this great similarity or semblance between them, the degree of their relatedness, indicated by the preposition "car" or "because" they were all of these relations at once (brother/sister, father/daughter, husband/wife). Hence, while the natural or biological foundation for this logic is not rendered perfectly explicit, it can be safely assumed.[9] Incest does provide theatrical appeal—an opportunity for shock and awe by breaching a primitive taboo—but what these plays highlight is that the monstrosity onstage (either real or potential) is not the incest per se but rather the ways in which blood's ideology leads to society's (re)production of its oppressive structures, over and over again.

LA SORELLA (*THE SISTER*), OR IS SHE?

Giambattista Della Porta's *La Sorella,* at first glance, appears to be a fairly straightforward play that depends on incest's horror as well as the truth of bodies' testimony (both to avoid incest and to establish its romantic conclusion). Yet the play's baroque quality of extremes, which no doubt contributes to the play's unpopularity among scholars, might likewise encourage further investigation of the sibling incest premise. Della Porta has been studied primarily in reference to his works in natural philosophy, especially *Magia Naturalis* (1558) and *De humana physiognomonia* (1586), but he was an equally prolific playwright and wildly popular in his time. Italian scholar Giorgio Pullini's comments might explain the relative dearth of criticism on Della Porta's drama.[10] He says, "I soggetti non presentano alcuna originalità di concezione e, oltre ad essere pesantemente ripresi dalla tradizione classica e dalla novellistica medioevale, si ripetono con insopportabile monotonia. Non c'è opera che si salvi per novità di invenzione e non c'è, quindi, opera che conservi un interesse per la scena a noi contemporanea, all'infuori di esperimenti intellettualistici o di elevata e ristretta esplorazione culturale" (299). (The subjects do not present any originality of conception and, other than being heavily taken from the classical tradition and that of the medieval novella, repeat themselves with insupportable monotony. There is no work that saves itself by novelty of invention, and there is not, therefore, a work that preserves an interest in the plot for a contemporary audience, apart from intellectual experiments or elevated and narrow cultural exploration).[11]

Potentially resulting from this type of dismissal, most studies of Della Porta's theater have been limited to *Gli Duoi Fratelli Rivali,* a potential source for Shakespeare's *Much Ado about Nothing.*[12] And *La Sorella* in particular has received little scholarship, though it is occasionally referenced as a source for Middleton's *No Wit.* Pullini's comments, while undoubtedly harsh, are not unfounded. Della Porta's plots and conceits are not novel, but I would maintain that his composition is. Della Porta deliberately plays with these tropes not purely for the sake of melodrama, but for the sake of the plot's greater questions, in this case, those about the importance of blood and bodies, hardly a "narrow cultural exploration."

Yet, since this play will be unfamiliar to most readers, I will briefly rehearse its plot here. Our protagonist, the young Attilio, is sent abroad by his father, Pardo, to find and ransom his mother and sister who were lost at sea and captured by Ottomans many years ago. En route to Constantinople, however, Attilio is distracted by a beautiful young slave named Sofia. After falling madly in love with her, he uses the ransom money that his father gave him to buy her freedom instead. Attilio then brings Sofia home with him to Nola and lies to his father, telling him that *this* is his sister, Cleria, and that his mother, sadly, has died. All goes well for the lovers as they live under the same roof as "brother and sister," until, that is, Pardo decides to marry Attilio to Sulpizia, who is the beloved of Attilio's best friend, Erotico. The drama begins with Attilio begging his crafty servant, Trinca, for help out of this conundrum. With a few misunderstandings along the way, ultimately all parties decide that they can convince Pardo to marry "Cleria" to Erotico in addition to the planned marriage. Then after a wedding ceremony conducted by an actor, all four lovers can live under the same roof, swapping brides at night until the old man kicks the bucket, at which point, each can marry his true beloved. This scheme only requires that Trinca thwart the marriage designs of the braggart soldier Trasimaco, who desires Cleria, by harming the credibility of Trasimaco's marriage broker, the parasite Gulone. Everything goes according to plan until Pardo's old friend from Turkey shows up proclaiming that Pardo's wife, Constanza, *is* alive, and that Attilio never made it to Constantinople. Constanza herself then returns home. Yet, after she hears about Attilio's love for Cleria/Sofia, she agrees to participate in the scheme and pretend that this woman is her daughter for Attilio's sake. The two women meet and (bum bum bum) it is the real Cleria! Attilio is, of course, completely distraught, but insisting that he cannot relinquish his love for his sister, he decides to exile himself instead. Only by happenstance, a nursemaid quarrels with her master, and she reveals the secret that Cleria/Sofia is *actually* Sulpizia. The two were switched at birth so that "Cleria" (in fact Sulpizia) would have a better life. Moreover, her uncle Orgio decided to go forward with the actual incestuous coupling of Attilio and "Sulpizia" (in fact Cleria) for his own monetary gain. All true identities are revealed by verifying birthmarks, each man can have the woman that he loves, and (hooray!) a double wedding.

Commenting on Della Porta's work as a whole, Sergius Kodera suggests that both his drama and natural philosophy "became such a success because

[they] naturalized (and thereby legitimized) the social realities of many other members of educated elites in the various absolutist cultures of Europe" (35). *La Sorella's* final insistence on the body's import certainly speaks in favor of this interpretation since birthmarked bodies provide evidence that the two lovers are not consanguineous siblings. Other scholarship on *La Sorella* generally falls into two camps. The first strand of criticism views the theme of incest as the opportunity for a formal experiment, exploring how the models of Roman New Comedy and the *commedia erudita* might be inflected by tragic elements and pathos within a new genre that borders on the baroque (sometimes called *tragedia con fin lieto*).[13] A second group of scholars view Della Porta's incest plot as an opportunity for cultural critique. For Tessa Gurney, *La Sorella* is one of many plays in the period that illustrates the lasting psychological impacts of the recent wars with the Ottoman Empire (1570–71), wherein "the negative Turkish stereotype reaches an all-time high as time spent in Constantinople among the Ottomans and their customs is credited as the force that drives the two siblings to commit incest" (161).[14] And for Raymond LePage, it is a "satire directed against Church morality" by staging the murky lines of incest (343). In this view, the drama becomes a venue for Della Porta to voice his criticisms of the Roman Catholic Church because the pope had already formally censured his scientific writings (Poulsen 188).

But upon closer examination in the context of blood, the play seems to use this trope of the romance genre to critique *both* incest and the logic of consanguineous affinity. Although the plot's saving grace (the reveal of the crib swap) sanctions the central love plot as natural rather than monstrous, it feels contrived since, by thwarting generic expectations, the play persistently undermines incest's gravity. Incest is the play's central question, but because *La Sorella* consistently draws our attention away from the act and its "horror," it invites the audience to question its status as sin as well as blood's assumed import that undergirds it, namely that blood meaningfully determines either identity or affinity. In general, the characters treat kinship connections quite casually. Even when they appeal to these principles in earnest, be it reading birthmarks or the cri du sang—the "alarm bell" that depends on the body's natural affinity for its kin—the play quickly exposes the failure of these processes and, subsequently, their status as essentially rhetorical. Ultimately the characters, and Attilio especially, cite blood connections and affinity as it suits them—in this case, a performance to cover up other sins such as abandoning his family members (in favor of bringing home a new girlfriend) or explaining his supposedly incestuous attraction (for his girlfriend-turned-sister). As the clowning subplot of *La Sorella* likewise intimates, bodies and blood mean very little since they are the products of signification rather than biological determination. Finally, since Cleria and Sulpizia were initially switched at birth because a lower-class family wanted to provide their biological daughter with greater opportunities, the exposure of blood's fundamentally discursive character evacuates its argumentative force in the context of social class. Specifically, bodies can no longer take responsibility for creating the existing social structures and maintaining the status quo. The characters' continued reliance on bodies, however, likely results from their belief that in a world of deception bodies are more reliable than persons.

To begin, for a play entitled *The Sister*, its structure—including its manipulation of generic conventions and the nature of its agon or central problem—greatly downplays incest. This curious shift in attention probes the gravity of incest's larger ideological framework, namely that blood determines both one's identity and affinity for kin. From the opening of the play, the audience would likely not anticipate a story of incest, but rather a comedy in which "Cleria," her false identity as long-lost sister to Attilio, is finally revealed as Sofia, the serving girl with whom Attilio fell in love. In this imagined, more conventional romance plot, once Sofia's "true" identity is revealed, and Attilio's father, Pardo, discovers that she is not kin, Pardo could both *pardon* his son (after the safe return of his real wife and child, of course) and bless the love match. This is the standard Roman New Comedy plot that the audience expects because although the play begins with Sofia playing the sister, the issue at hand is not properly incest but scheming lovers trying to overcome social obstacles. In short, the dramatic focus (agon) is the fact that this "feigned" sibling relationship blocks their union because Attilio has been promised to Sulpizia. Despite the hint from the title, the actual incest threat crops up quite late in the play, only with Constanza's reveal (4.2). Rachel Poulsen observes that, given the delay of this revelation, "viewers witness all but the incestuous act itself, and that they are encouraged to imagine. By withholding the fact of the lovers' consanguinity until after the vicariously imagined consummation takes place, Della Porta makes the audience complicit in their deeds. Viewers are thus forced to examine their own responses, judgements and relationship to the contested category of 'natural law' in a particularly vexed and immediate way" (189).

Then, once the incest *is* revealed, the play does not abide by the conventions of incest-averted comedy or cavalier drama, wherein upon the discovery of the unwitting incestuous desire, "brother and sister respond to moral law by giving up their sexual love for each other, they promptly bestow themselves on others" (Bueler 125). In Della Porta's play, Attilio does not relinquish his love for his "sister" Cleria. His "incestuous" love persists even though he promises to flee so that he cannot further act on it. Attilio laments just prior to hearing the reversal of his fortune, "Niuna buona nuova può esser per me, se non che Cleria fusse mia moglie; ma ciò non potendo essere, dunque non è buona per me" (No news is good unless Cleria becomes my wife, but this cannot be, therefore no news is good for me) (5.6).[15] He has insisted that his love cannot be arbitrary and refuses to embrace a socially acceptable alternative. As he says, nothing else will ever be good. He thereby casts incest in a romantic light and rejects the disgust reaction, just as Ford's Giovanni will do. Other characters disregard the "horror" of incest for pragmatic reasons. Balia's disclosure of the secret, the deus ex machina that allows for the nonincestuous coupling of Attilio and Cleria, reveals that Uncle Orgio would have permitted the actual incest for his own profit. The lack of concern over the incest taboo exhibited by both Orgio and Balia should encourage us to reconsider the status of incest as abomination.[16] Richard McCabe notes, with regards to this moment, "Society is seen to connive for purely mercenary reasons at the very relationships it professes to condemn. Once the marriage is financially exogamous—joining together two previously unrelated fortunes—little else matters. Attilio's moral 'horror' is hereby undercut by society's cynical pragmatism" (136). Though the horror and threat of incest is seemingly the play's central concern, especially at its close, the play's plot structure gives it little weight or prominence. It thus encourages the audience to question not only whether incest is a sin but also whether bodies or blood determine much of anything, much less justify the endogamous practices of the elite.

And while the incest-averted plot routinely depends upon the belief that bodies testify to the truth of persons' natures—that faces, hearts, and birthmarks can be read as reflections of genuine differences to allay the threat—*La Sorella* repeatedly highlights the moments when these assumptions fail.[17] The play should establish that bodies accurately report the truth in order to lay the foundation for the play's conclusion, but it does not. Pardo, for example, assumes that he can read faces accurately when searching for a reason to disregard his friend's report that his son has been lying. He rationalizes, "Certo che dovea star ubbriaco; e già lo tengo per tale, che stava rosso nel volto" (Of course he must have been drunk; and indeed I maintain it as such, that he was red in the face) (3.4). But Pardo has reconstructed this evidence after the fact to suit his own desires. And even if his friend Pedolitro were somewhat red in the face, it would not necessarily signify that he was lying or unreliable; the audience already knows that he was not. Later, Pardo claims "che con più d'una guardatura ho confrontato l'imagine tua con quella che nel cuor impressa mi lasciasti" (that with more than one look I compared your image with the one that I left engraved on my heart), which confirms his wife Constanza's identity (4.3). In this instance, the image imprinted on his heart testifies correctly, but theoretically, this principle should have applied to his daughter as well—either Cleria, whom he raised, or Sulpizia, his biological daughter—and he remains unable to identify either of them.

Issues and ambiguities that arise from reading bodies reach their peak in the moment that *should* allay the incest threat: the reading of birthmarks to verify Balia's testimony. Without an accurate reading of marks, there is no way to distinguish between the girls. But since the play has already rendered the act of reading bodies suspect so too is the play's conclusion. Upon revealing the crib-switching plot, Balia references Sulpizia's features as proof of her true parentage to convince Pardo that she is his biological daughter: "Sulpizia vostra è di pel rosso, come voi sete; gli occhi azurri, come i vostri; e il volto simile al vostro" (Your Sulpizia has red hair, like you; and blue eyes, like yours; and a face similar to yours) (5.2). Sulpizia's appearance is a miniature copy of Pardo. But apparently, this is not enough, so Balia cites a very particular birthmark on Sulpizia's body as the most reliable identifier: "E se ben vi ricordate, ha una macchia rossa nel braccio sinistro, come goccia di vin rosso" (And if you well remember, she has a red mark on her left arm, like a red wine stain) (5.2). Pardo does not comment on the fact that he has missed Sulpizia's remarkable resemblance for all these years. Instead, this strange mark satisfies his desire for evidence. He exclaims, "O Dio, veramente mi ricordo di quella macchia rossa; . . . e nella vostra Cleria mai più ve l'ho vista" ("Oh God, truly I do recall this red mark; . . . I have never seen such a mark on Cleria") (5.2). Sulpizia's new biological mother, Constanza, then verifies that these birthmarks testify to their relationship. One might find it suspicious, however, that Constanza is entrusted with verifying the "true" daughter on the basis of identifying birthmarks that neither the audience nor the other characters see. Constanza, after all, has already demonstrated that she is willing to lie in the name of her son's happiness. Trinca, too, references these birthmarks when he promises Attilio that his crisis has been averted. Trinca affirms, "A signali Constanza ha trovato vero quanto ha detto" (The signs Constanza found true just how she [Balia] said) (5.6). But what if Sulpizia had not been graced with this wine stain? The fact that the deus ex machina rests upon a birthmark is not only troubling but it also highlights the fact that blood's supposed capacity to dictate affinity for one's kin has failed to correctly inform the characters of their consanguineous relations. To put it bluntly, the cri du sang is a sham.

Just as faces and hearts do not reliably indicate fact in *La Sorella,* the cri du sang—the notion that one's body can intuitively know or sense kin because one's blood communicates love—is more rhetorical than biological, thus cracking the physiological foundation of, and justification for, maintaining existing sociopolitical structures managed via kinship. After all, the play's beginning depicts a world in which kinship has little ethical sway. And if blood ties determine any affinity at all, it is minimal at best. Attilio has coarsely forsaken the quest for his mother and sister in favor of using his father's ransom money to free his beloved Sofia from slavery. When Attilio relates the story to his friend Erotico, he describes his reaction upon falling in love, exclaiming "Che madre? che sorella? che viaggio?" (What mother? What sister? What journey?) (1.3). As previous scholars have noted, one *should* feel horror at this disloyal and wrongful act; Attilio has utterly disregarded the importance of blood ties. But the play's other characters do not appear fazed. Erotico, for instance, does not voice any moral objection to his friend's behavior. Only a short moment later, Trinca's plan to overcome the lovers' obstacles contains an equally crass treatment of kin. To assuage Attilio, who is concerned that he and Erotico will be trapped in pretend marriages according to the scheme, Trinca promises, "E bisogna scambiar le mogli fin che vive il vecchio, il qual non potrá viver molto" (It is necessary to switch the wives only as long as the old man is alive; certainly, he will not likely live much longer) (1.3). Or in other words, "don't worry, your father will likely die soon, and then you and Erotico can marry the right women without any issue." Attilio, evidently satisfied, has little concern for his brethren. The principle that blood determines some kind of overwhelming love for one's kin, much less one that would potentially result in an overflow of erotic desire and incest, is highly suspect in this context.[18]

Attilio invokes the cri du sang to explain his unwitting incestuous desire but only *after* discovering Cleria's "true identity." Constanza reveals the sibling incest using extremely visceral language, pointing to her body and their joint conception. She sadly reports to him that they were "conceputi d'un istesso seme, portati nove mesi e partoriti dal medesimo ventre mio" (conceived by the same sperm, carried nine months and birthed from the same womb of mine) (4.5). Trinca also emphasizes bodily connection when revealing the "incest" to Erotico, "Cleria è sua vera sorella carnale" (Cleria is his true flesh sister) (4.6). Only after this persistent emphasis upon their consanguinity does Attilio subscribe to this belief. He laments, "Ahi, che tanto movimento di sangue, che mi occupò il core nella prima vista, stimava che fosse dalla tua bellezza; ma era dalla forza del sangue, perché eravamo nati di un medesimo sangue" (Ay, that such a movement of the blood that overcame my heart at first sight, I gathered was from your beauty; but it was from the force of blood because we were born from the same blood) (4.5). Born of "un medesimo sangue" (the same blood), Attilio either believes that he mistook the cri du sang for erotic love, or, as his language suggests, his natural affinity for his sister transmuted into erotic love since his body registered it as such. Attilio further supports this premise by insisting that love cannot be transferred and that love objects are not arbitrary. After Erotico asks Attilio if he could love another, Attilio proclaims simply, "Amor non vuol cambio" (Love does not want change) (4.5). According to the logic of blood affinity, the specific object of love is extremely important; one's blood naturally responds to its like, which, though it possesses the possibility for incest, cannot be changed. Attilio bucks against the incest taboo, but since he firmly believes in the concurrence of blood similarity and love, he insists that he cannot love otherwise. As the audience soon discovers, however, this is not true! Attilio and Cleria are not siblings, and Attilio merely cites this logic to explain what he *thinks* he knows. The cri du sang invoked to explain incest as a natural extension of blood's ideology is hereby called into question since his body has *not* determined his love for Cleria after all.

Attilio's love is not the cri du sang's only failure; blood repeatedly fails to operate as advertised in disclosing kin. The cri du sang that should allow Constanza and Attilio, mother and son, to recognize one another fails (4.2), and Pardo likewise appeals to blood affinity only after the discovery of his "true" daughter, Sulpizia. Following the report of the telltale birthmark, he marvels, "Ma io non conseguisca mai desiderio di mia vita, se, sempre che ho visto Sulpizia, non mi sentiva un certo movimento di sangue per la persona, tra carne e pelle, e non potea imaginarmene la cagione. La natural veramente facea; l'ufficio suo" (But I never deduced the longing of my life, if, always when I saw Sulpizia, I did not feel a certain movement of blood for her person, between flesh and skin, and I could not imagine the cause. Nature truly does her office) (5.2). Pardo claims to have felt this cri du sang for his daughter, Sulpizia, all along, although his final words on this subject expose just how ridiculous this sounds.[19] He exclaims, "Nature was truly doing its office," but Nature clearly did not achieve this office. The girls' true identities went unknown and unacknowledged for years. If Pardo's blood was supposed to "move" in a way to get his attention and signal kinship, then it did a very poor job of it. Blood clearly does not have the agency that the characters take it to have, and thus kinship, if dependent upon naturalized affective feelings, may not amount to much, certainly not enough to justify existing sociopolitical structures that depend upon the force of these connections.

This principle of blood affinity becomes especially suspicious because the one theory that would explain, and by virtue of assigning blood agency potentially excuse, incest falls flat; one's body does not reliably communicate love for one's kin as the cri du sang would suggest. Long before the real incest threat appears, Trinca appeals to blood's affinity in order to explain Attilio and Cleria's amatory behavior, which bothers Pardo as overly affectionate and thus inappropriate. Trinca seeks to appease Pardo by insisting that "son sorelle e fratelli carnali al fine, e il sangue tira e fa l'ufficio suo" (they are sisters and brothers of flesh in the end, and blood pulls and does its office) (1.5). In this case, Trinca believes that he is lying, for he is still under the impression that the two are not siblings, but his use of the adjective "carnali" is purposefully ambiguous. It could mean of the same flesh or "lawful issue" as John Florio defines in *The World of Wordes*, but it could also signify carnal, as in a sin of the flesh. Trinca here cites the ubiquitous logic that blood acts as an agent of attraction; it quite literally "tira" (pulls) thereby simply performing "l'ufficio suo" (its office). In response, Pardo voices his concern about this natural affection and anticipates concerns about incest, suggesting that sibling affection can bleed into erotic attachment. He grants, "Io non voglio che non trattino insieme con molta amorevolezza, ma in fin ad un certo termine onesto e di creanza, e non con modi cosí disonesti e di scandalo a chi li vede" (I do not want that they not treat each other with great affection, but there is in the end a certain limit of honesty and decency and not with ways so dishonest and of scandal to whomever sees them) (1.5). As Pardo says, sibling affinity *should* have a certain end point, one that is both "honest" and "decent." But notably neither adjective connotes a natural barrier, only a social one; it is ultimately about who might *see* them.

Hence the comedic subplot, featuring petty fighting between the parasite Gulone and the braggart soldier Trasimaco, may seem to provide purely comedic relief as part of the lazzi tradition. But the central position of this subplot should draw our critical attention since it occupies much of the play's dramatic space—the majority of the second and third acts. This clowning plot provides a foil to the central incest by further demonstrating that bodies themselves are subject to signification. Bodies are not impervious to rhetoric and manipulation; it matters how they are read, but they do not determine their own signification. For one, the fools' argument about their respective preoccupations—food for Gulone and women for Trasimaco—implies that there is little difference among bodies. Gulone insists, "Non ci è differenza tra l'amor mio e il tuo: io fo l'amor con vitelle mongare, tu con vacche: carne ami tu, carne ancho io: tu cruda e io cotta" (There is no difference between my love and yours: I make love with tender calves and you love cows [also sluts]: you love meat, I love meat too: you have it raw and I have it cooked) (2.2). This is a misogynistic joke, of course, but it likewise establishes a spectrum of being on which there appears little difference between species, much less persons. Later, Trasimaco calls the value of blood into question by boasting of his supposedly superior qualities and lineage. He insists, "Son cavaliero da tutti i quarti: cerchesi nel mio parentado" (I am a knight from all the quarters: look in my pedigree) (3.6). While Trasimaco can assert his lineage or "parentado" all he likes, he, true to the braggart soldier type, is all show, just like his many letters from princes worldwide, which, the audience can assume, are forgeries.

Lastly, at the play's structural center, Gulone and Trasimaco's battle reveals the folly of appealing to the body as testimony since it is subject to arbitrary, and self-interested, signification. Trasimaco, preparing to fight with Gulone, dubs his body the proof of both his physical prowess and his bravery, bound to send adversaries running after only a glance. He even gives himself a blazon of sorts: "Come egli si vedrá intorno questa statuaccia del mio corpo, queste spallaccie di Atlante, con questi torreggianti gamboni, con queste nerborute braccia fulminar la mia taglianasi, troncabraccia e mietigambe, tu vedrai i motivi che fará. Considera se son bravo, vedi che viso sfreggiato" (How he will see around this great statue of my body, these shoulders of Atlas, these towering large legs, with these sinewy arms stunning, my nose cutting, my arm dicing, my leg mowing; you will see what he does. Consider if he is brave, look at this [my] gashed face) (3.7).

Trasimaco's description of himself, however, is promptly undermined by his decisive failure in an indecorous fight with Gulone, who leaves him with an abundance of bruises. But in an attempt to save face, Trasimaco concocts a story of an ambush. Attilio's servant Trinca rejects this story having witnessed the scuffle firsthand: "L'ho visto con questi occhi!" (I saw it with these eyes!) (3.9). Yet, Trasimaco counters, "Ch'i lo può saper meglio di me, che ho patito le maladette bòtte su la braccia, sul collo e su le spalle" (Who could know it better than me, who suffered the wretched bumps/bruises on my arms, my neck, my shoulders) (3.9). Eyewitness testimony, in theory, *should* verify the fact of the matter, but Trasimaco knows that it is simply a war of words. Although bodies apparently testify to the truth, Trasimaco has shown how these bodies can be easily resignified as one pleases.

In light of this comedic foil, *La Sorella* insinuates that incest might likewise be a social proscription rather than a natural barrier. Sulpizia's wine-stained birthmark might be no different from Trasimaco's bruise, conveniently remembered to signify her "true" identity and bring about the comedic close. The characters' language similarly hints toward this possibility that signification matters more than bodies, that the threat of incestuous coupling is solved by swapping names and not by swapping babies. For instance, although Constanza believes in the gravity of Attilio's consummated incest, she nevertheless describes it as "la tua infamia e vitupèro" (your disgrace and shame) (4.5). Rather than calling it a sin, these words emphasize an anxiety about reputation or social shame rather than a spiritual issue. The role of society, and not nature, likewise figures prominently in Attilio's complaint about the incest taboo. He bemoans, "Per esser tu troppo congionta meco, è forza che da te mi disgiunga. O leggi, o costumi umani a me contrari!" (For you being too joined with me, it is forced that I separate myself from you. O laws, o human customs set against me!) (4.5). These are human or societal dictates, "customs" or "conventions" even, and not those of God or nature, at least according to his account. And despite all previous talk of blood, one might ask how much it matters in comparison to a name. Attilio laments about the incest in this way: "Quel nome di Cleria che fu prima lo spirito della mia vita, or è morte della mia vita" (The very name "Cleria" that once gave the spirit of my life, now is the death of my life) (4.6). While Attilio previously had to call "Sofia" (her name as a servant girl) by his sister's name "Cleria," it is still curious that Attilio loves her *as* Cleria. And finally, Attilio's account of the miraculous discovery highlights the fact that the transformation is one of signifiers: "O stupori, o meraviglie grandi, che da moglie mi diventi sorella e da sorella moglie!" (O wonders, o great marvels, that from my wife she has become my sister and from my sister a wife!) (5.6). Relationships are a matter of appellations, and with Sulpizia now Cleria and Cleria now Sulpizia, everyone can have their cake and eat it too.

In closing, despite the characters' continual reference to bodies' determination of persons and their affinities—logics that both give the incest taboo meaning and reconfigure incest as the natural extension of kinship—Della Porta's *La Sorella* time and again picks away at the facade of these beliefs. It exposes a different principle of identity underneath: one based in signifiers and names. Signifiers, after all, successfully transformed the girls' identities for years so that an impoverished girl might have a better upbringing. And if based merely on signifiers, then the hegemony of the nobility loses its persuasive power. The elite cannot simply point to their bodies as responsible for the reproduction of the status quo: their affinity for a closed system to the point that it flirts with incest. But then, the play likewise provides a brief explanation for one of the motivations behind blood's ideology. The characters appear to hold onto blood logics—that blood "cries" to alert familial love and that a birthmark can finally sort out any confusions about identity—because bodies *seem* to be more reliable than persons.

In a moment of misunderstanding and miscommunication, Balia, the nursemaid and a minor character, expresses her anxiety. Alone on stage she gives this monologue: "O mondo immondo, o mondo tutto piano di fallacie e d'inganni, or chi può vivere in te, che sia sicuro delle tue insidie? O etá maladetta, o credeltá, o barbarie, che a pena può adeguarsi col pensiero!" (O dirty world, o world so full of lies and deceits, now who can live in you, who is safe from your tricks? O wicked state, o cruelty, o barbarity, who has pain can adapt himself to this thought?) (2.5). Who *can* adapt themselves to this horrible, deceptive world? It proves extremely difficult to adapt, which is seemingly why early modern persons search for ways to compensate, seeking infallible methods to manage this insidious world. In this case, they hold onto the belief that bodies could reliably indicate persons. Moreover, they hold that bodies and the blood within themselves could accurately report kin and/or similar persons with whom they can ally themselves so as not to compromise their power and position. And what in Della Porta remains at the level of suggestion becomes far more explicit in Middleton's translation. In *No Wit* identity resides entirely in "name and place," divorced from bloodlines. And Middleton's play further intimates that blood's discourse hides the actual motives beneath these fictions of kinship structures and love, namely wealth. And wealth, by virtue of its flexibility, can easily pass hands to become detached from the bloodlines of the elite.

ENGLISHING DELLA PORTA'S PLOT: THOMAS MIDDLETON'S *NO WIT, NO HELP LIKE A WOMAN'S*

Thomas Middleton's *No Wit, No Help Like a Woman's* appropriates Della Porta's plot.[20] And like Della Porta's comedy, the scholarship on this play, especially regarding the incest-averted or Twilight family plot, has been limited. As Kenneth Muir explains, "The qualities of the play are theatrical, rather than poetical, and this has made literary critics undervalue it; and one may suspect that they are shocked by the callous behavior and by the doting imbecility of his mother. Nevertheless, it is a crowded, lively and varied comedy which does not deserve its neglect" ("Two Plays" 158–59).[21] Thus, a more sustained analysis of this secondary plot reveals that as Middleton imports Della Porta's sibling incest premise, he further emphasizes the feeling of ambivalence toward the incest threat from the Italian original.[22] D. J. Gordon remarks on this point, "We do not feel in these speeches that Middleton is lingering over the fact of incest, as della Porta certainly does. On the contrary, we have the impression that everyone treats it in a rather cavalier way. Lady Twilight is even more summary in dismissing it than Constanza was" (411).

Gordon, to my mind, misconstrues Della Porta's stance on incest, but it is fair to say that Middleton's attitude more explicitly dismisses incest's "horror."[23] This shift in tone to a more obvious, comedic register is due, in part, to the nature of the incest's reveal. Della Porta reserves it for the very end; we have no hint of the crib-switching until the concluding deus ex machina. In *No Wit*, on the other hand, while we may not know of the possible incest threat or that the secret involves swapping the two girls at birth, Lady Goldenfleece gives the audience an indication that there *is* a secret about their identities (1.1). The audience can then anticipate the secret's reveal, which will resolve whatever plot complications follow.

Yet there is a satirical point to Middleton's dismissal of incest's horror, one that builds on Della Porta's social critique. *No Wit*, by experimenting with this limit of blood's ideology, further demonstrates how blood and bodies provide little means for distinguishing or determining persons;[24] and in so doing, the play undermines the basis for the elite's hegemony and its perpetual reproduction via endogamy. As John Jowett's introduction observes about the nature of identity in the play, "It is important here that in both plots a character is available to double with one of the marriage partners. Philip's supposed and real sisters are dangerously interchangeable (Middleton gives the women's fathers the virtually synonymous names Sunset and Twilight), and the evasive device of the stand-in marital partner links this resolution with that of the Low-water plot" (780). Building on Jowett's observation, people are interchangeable not merely as love objects in this play but as people in general. Even though the incest premise assumes that consanguinity matters (and the cri du sang depends upon it), Middleton's play exposes this notion of blood-determined love to be an utter farce. *No Wit* exaggerates the point from Della Porta's source text that language has a greater role to play than bodies in identity formation. Consequently, the play similarly undercuts the physiological basis for kinship and affinity, which serves as the justification for endogamy ("like likes its like") and ensures the elite's reproduction of their hegemony. This critique then inflects the wooing widow plot, which also interrogates this sociopolitical system based in blood. The drama around the second marriage of the wealthy Lady Goldenfleece exposes not only that avarice is the primary motivation in making marriage matches, rather than a concern for pairing like persons, but also that *wealth*, not bodies, dictates "worthy" persons. And ultimately, wealth is fluid and fungible. After all, the figure of the wealthy widow is so threatening to society, in part, because *if* she has no heir from her first husband, familial wealth can be separated from its "rightful" bloodline. Just as Kate Low-water's cross-dressing demonstrates the flexibility of gender, so power and privilege can be divorced from blood lineage.

Middleton imports several aspects from *La Sorella* in the name of expanding Della Porta's social critique. The first is the casual, bordering on crass, attitude toward kin, which undermines the belief that familial love based in consanguinity could be confused with, or extended to, erotic love and incest. Additionally, this apathy toward kin utterly contradicts the nobility's justification for their closed kinship networks. Both of these logics are out of place in a world wherein Philip can forego rescuing his mother and sister in favor of his beloved. When Philip reports to his friend Sandfield how he was able to bring Grace home, he says, "Pish easily" since he simply told his father that his mother was reported dead (1.1.94). And this account does not bother Sandfield one bit; the two of them are far more concerned about maintaining their friendship than their familial relationships. As George Rowe remarks, "No one acknowledges the seriousness of what has been done; no one seems to care about 'injuries' to family which make Philip's difficulties with Sandfield insignificant in comparison" (117). Thus, it is difficult in this context to understand how bodies dictate, and then communicate, affinities for kin when these affinities receive little consideration.

Middleton then translates problems with the cri du sang from the Italian source, namely that blood fails to communicate kinship relations accurately, though Middleton makes these concerns even more obvious. As with Pardo, Sir Oliver Twilight makes explicit reference to this trope upon seeing his wife for the first time: "Oh, my reviving joy! The quickening presence / Makes the sad night of threescore and ten years / Sit like a youthful spring upon my blood" (4.1.1–3). But just as in *La Sorella,* blood fails to incite recognition between separated mother and son; instead, it requires verbal testimony (2.2). Moreover, Middleton further complicates the mother-daughter recognition scene. What is an immediate recognition of her daughter in the Italian original is here delayed. Lady Twilight says after seeing Grace, "Methinks the more I look upon her, son, / The more thy sister's face runs in my mind" (4.1.174–75). Having already faked a recognition in front of her husband for her son's benefit, she recognizes her daughter only upon hearing Grace's story: "Oh, I begin to feel her in my blood! / My heart leaps to be at her" (4.1.189–90). But if the body could truly identify its progeny as a feeling in one's blood (cri du sang), then Lady Twilight should have known her daughter almost immediately from instinct or bodily stirring. And of course, the play calls this exclamation into question because, as the audience will later discover, Lady Twilight and Grace do not actually share blood at all.

To further complicate matters, Philip, incensed by his mother's words, wants verification of the body instead. He exclaims,

> Her words? Shall bare words overthrow a soul?
> A body is not cast away so lightly.

> How can you know 'tis she? Let sense decide it;
> She then so young, and both so long divided. (4.1.230–33)

His argument, of course, is undercut on two fronts. First, the cri du sang has already been exposed as unreliable. And second, what he later accepts as the *actual* proof of Grace's identity is the ring that his mother gave to Grace, which Grace then gave to Philip in marriage and thus is *not* a confirmation of Grace's body after all (4.1). Having to abandon the notion that shared blood would cry out, the characters ultimately rely on other means of identification. Middleton suggests that these kinship ties are the products of signification and, in this case, the signification of an object (the ring) rather than biology. After all, just prior to this scene when Philip plots with his own mother about continuing the charade, he urges her, "Say but she is my sister, and all's well" (2.2.163). His entreaty not only devalues his true blood relations but it also brings forth language's role in establishing kin via performative utterance.

Middleton's *No Wit*, therefore, makes the critique of Della Porta's play—that blood fails to prescribe identities—more explicit. The characters, in the end, privilege language over and above bodies, suggesting an alternative principle of identity formation that undercuts the elite's reliance on blood's ideology for their hegemony. This argument can be seen most clearly in the conclusion of the Twilight plot. Early in the play, when Lady Goldenfleece debates whether or not to divulge the secret of the switched girls, she betrays that she basically sees very little difference between them; she favors pragmatic signification over any concern for biological truth. She resolves that the difference between Jane and Grace "'tis now no more / Than like mistaking of one hand for t'other" (1.1.188–89). Her use of metaphor describes the two girls as though they were both attached to the same body and, consequently, without any material difference between them. Although what material difference supposedly exists between the two girls is even further confused by the fact that they are consanguineous to a degree, for "both were nursed together" (4.1.139).[25] Nonetheless, Lady Goldenfleece uses a wildly inappropriate analogy in the context of an incest threat, wherein one's material body and bodily difference are immensely important to classification and appropriate mating as made obvious in the period's proliferation of family trees.[26] Yet, as Goldenfleece suggests, what material difference there is between the two girls should not, or does not, matter. Once Lady Goldenfleece finally reveals the secret in the final scene, she says simply "now each has her right name and place" (5.1.463). Even the language that marks them compromises their very particularity. Just as Sir Oliver Twilight goes to complete the "appropriate" marriage matches, he says pointing at Jane, "This? For my son," to which Lady Goldenfleece responds, "Oh back, sir, back! This is no way for him!" (5.1.414–15). Not even names but rather empty signifiers indicate the girls—"this," a pronoun that does not even disambiguate persons from things. Middleton's conclusion is thus even more ambivalent and anticlimactic than Della Porta's since it lacks any mention of blood movements or birthmarks meant to verify the girls' identities. This significant departure from the original suggests that identity amounts to "name and place," though that "name and place" has great consequences for one's life and station.

In the context of Middleton's play at large then, the sibling incest threat—one limit case for the principle of blood affinity—demonstrates that if identity is not determined by bodies but by name and place, then wealth, not bloodlines, dictates privilege. Just as with the original Della Porta text, the children were swapped for the purposes of sociopolitical advancement. Mistress Sunset "changed the children, kept your [Sir Twilight's] daughter with her, / And sent her own to you for better fortunes" (5.1.446–47).[27] This switch not only exposes that a child's biology dictates neither kinship nor affinity—for the Twilights and Sunsets loved their respective children no less when ignorant of the switch—but it also reveals the ideology of blood as powerful yet foolish. Namely, it underscores the lengths that one would go to in the name of providing a better life for one's offspring. The incest plot in this frame becomes especially strange considering that clearly wealth, as reflected in "name and place," dictates the girls' respective worths and not lineage. In fact, as the switch evinces, the entire system of blood that preserves the elite's hegemony can no longer be tied to bodies at all. Turning to the wooing widow plot, the same principle holds. The widow is a prime object to be wooed precisely because she "owns" the wealth, privilege, and power of her former husband, and, without an heir, this wealth can now be divorced from its former bloodline. Furthermore, this separation is possible *because* the bloodline had no intrinsic privilege or worth to begin with. It is no wonder, then, that in Middleton's *No Wit,* avarice is the central motivation in marriage and not a "proper" matching of like persons. Wealth (not blood) dictates love, and property (not bodies) is joined and exchanged in the process of becoming one flesh.

Avarice is the guiding principle behind what is deemed an appropriate marriage match in the world of *No Wit*. This principle is a notable departure from the conventional doctrine of matchmaking touted in the marriage treatises of the period: pairing like characters, that is, those similar in station, lineage, bloodline. For instance, Lady Goldenfleece wishes for both Jane and Grace, "A fair contented fortune in your choices" (1.1.181). And though *fortune* could simply indicate well-wishes, *fortune* is a loaded word for a woman whose former husband made his fortune by "fleecing" others.[28] Sir Gilbert, the villainous suitor, likewise discloses his real motivations for pursuing the widow: "For my love, / In matching with the widow, is but policy / To strengthen my estate" (1.2.125–27).[29] Of course, the audience would know that this admission characterizes nearly all marriages between elite persons, involving a lusty widow or no, especially in the changing economic market that requires liquid assets, as opposed to property, to maintain social position and "strengthen estates."[30] Sir Oliver Twilight's motivation for matching his daughter, Grace, is the most extreme example of an avarice that ignores all other principles of matchmaking, for he promises Grace to Weatherwise solely because the fool requires no dowry. The trickster Savourwit outlines the situation,

> And yet, because the fool demands no portion
> But the bare down of her smock, the old fellow,
> Worn to the bone with a dry covetous itch
> To save his purse and yet bestow his child,
> Consents to waste [her on] lumps of almanac stuff. (1.1.115–19)

Savourwit, in an attempt to carry out the plot to pair Grace with Sandfield, tries to speak reason to Sir Oliver. He begins by giving generally well-regarded advice to pair like persons in marriage: "There's no proportion, height, or evenness / Betwixt the equinoctial and your daughter" (1.1.201–2). Sir Oliver, however, responds with the single-mindedness of his avarice, saying, "I can abide no word that ends in portion; / I'll give her nothing" (1.1.205–6). While Sir Oliver is clearly ridiculed for his view that marriage does not pertain to likeness, proportion, or fitness, his "covetous itch" marks nearly all of the players. And this itch motivates the vocabulary of fitness in station and bloodline in the first place.

Thus, rather than likeness in blood dictating love and affinity—which is, after all, the entire premise of the incest plot—wealth inspires love in Middleton's play, just as, Middleton would likely contend, it does in the world outside the playhouse. Although Jane assures her cousin Kate that "love not by wealth but by desert is prized," Kate knows better that the two are not clearly distinguishable (1.2.30).[31] The conversion of wealth into love is most obvious in a comedic context, wherein Sir Gilbert bribes Pickadille:

> GILBERT. [. . .] Honest Pickadille, thou wast wont to love me.
> PICKADILLE. I'd good cause, sir, then.
> GILBERT. Thou shalt have the same still; take that.
> [*Gives him money*]
> PICKADILLE. Will you believe me now; I ne'er loved you better in
> my life than I do at this present. (3.1.155–60)

Here the "cause" for loving is explicitly money, and love grows in direct correlation with the amount of money involved. This same assumption then applies to the more serious dealings of matchmaking. For instance, Sandfield, still hoping to receive some kind of dowry for marrying Grace, asks of Sir Oliver, "Faith, something sir; / Be't but t'express your love"; but Sir Oliver replies, "I have no desire, sir, / To express my love in that way, and so rest satisfied" (1.3.10–12). Hence, though Sir Oliver refuses to transmute his love into money, he nonetheless holds that Sandfield's love can be equated with a monetary value, namely the dowry that he will forego.

The slippage between wealth and love then takes center stage in the conclusion of the wooing widow plot. Kate reveals herself and Lady Goldenfleece is now aware of the trick; with their brief marriage, Kate regains the wealth that was stolen by Sir Avarice Goldenfleece and transfers the remaining wealth to her own family by orchestrating the marriage between the lady and Kate's brother, Beveril. Lady Goldenfleece is somewhat perturbed by this trick, but Kate reproaches,

> What matter is't? 'Tis the same man was chose still,
> No worse now than he was. I'm bound to love you;
> Y'have exercised in this a double charity,
> Advanced my brother and restored mine own.
> Nay, somewhat for my wrongs, like a good sister,
> For well you know the tedious suit did cost

> Much pains and fees; I thank you, 'tis not lost;
> You wished for love, and, faith, I have bestowed you
> Upon a gentleman that does dearly love you;
> That recompense I've made you. (5.1. 366–76)

Kate thus reminds her that the widow said she wanted love—as she swore after the revelation of Sir Gilbert's plot, "I'll never covet titles and more riches, / To fall into a gulf of hate and laughter. / I'll marry love hereafter, I've enough / And wanting that, I have nothing" (2.1.391–94)—well, now the widow has bought it. Wealth, therefore, dictates affinities between individuals not some similarity in character or "kind" marked by blood. This exposure hence undercuts the elite's appeal to bodily determinism to justify their worth and the closed social system, broaching incest, that it permits.

Moreover, Middleton recasts the purportedly physiological process of commixtio sanguinis in marriage as a purely social and economic exchange to further interrogate blood's ideology. For if wealth dictates affinities rather than likeness, then it follows that marriage is not about establishing kinship structures by exchanging blood in sexual relations. Rather, it concerns establishing social networks by combining and exchanging property. And this conception of marriage makes it possible for two women to marry. The legal, social marriage that takes place between Kate Low-water and Lady Goldenfleece is never consummated; there was no *unitas carnis,* which means that theoretically this marriage could be annulled, especially since Kate reveals herself as a woman. But it is not annulled on that account as one would expect it to be. Instead, the union must be dissolved by a divorce and a division of assets, which, as Kate rightly remarks, are legally hers: "Are not both mine already? You shall wrong me, / And then make satisfaction with mine own?" (5.1.224–25). Pepperton comments on the strangeness of this act, expressing his disbelief, "Parting of goods before the bodies join?" (5.1.264). But while this circumstance is strange in the context of blood's ideology, the wooing widow plot reveals that the process of two becoming one is never really about physical consummation but about the transfer of wealth or property. Wealth and not bodies undergirds marriages, both the motivation for matches and the act of marriage itself. In sum, the double plot of *No Wit* critiques the current sociopolitical system and the ideology of blood that supports it. First, the incest-averted plot undermines the role of blood in the formation of identity and kinship structures; it thereby subverts the assumption meant to ensure endogamy and thus reproduce hegemony based in and limited to certain, noble persons and bloodlines. The wooing widow plot then exposes that which the ideology of blood should hide: that privilege is based in the transmission of *wealth*, which provides a far more fluid and permeable system than that of bloodlines. Where Della Porta's *La Sorella* initiates the suggestion that signifiers dictate identity such that the hegemony of the noble classes loses its biological foundation, Middleton's *No Wit* furthers this argument by pointing to the repugnant motivation for this ideology as well as the actual matter involved, not blood but wealth.

By way of transition, and to further propose Middleton's *No Wit* as a point of departure for Ford's tragedy, I would like to signal one notable amendment to Della Porta's plot, which is the brief mention of a potential child of incest. Savourwit remarks in an aside that the bride-switching plot is time sensitive since they must hide Grace's pregnancy, her impending "round belly" (1.3.34). Savourwit later briefly reminds Philip: "The meat's too hard to be minced now; / She breeds young bones by this time" (2.2.93–94). Middleton's drama, however, does not dwell on this fact, which is especially surprising since Grace's child has the potential to be the "monstrous" product of incest when this threat appears. But just as the play dismisses incest's horror so too would a disgust response be misplaced upon the literal threat of a monster. One *might* direct disgust, however, toward this child as a symbolic monstrosity: as the product of a completely closed system of power that reproduces its dominion indefinitely.[32] Ford's play takes up this detail and fulfills the plot premise of sibling incest to show the tragic potential lurking within the belief that blood will know and love another most like itself. *'Tis Pity* dramatizes that this ideology produces a kind of terrifying claustrophobic system—one that is extremely difficult to circumvent if one holds any of its tenets.

SHARING BLOOD AND BLEEDING HEARTS IN *'TIS PITY SHE'S A WHORE*

In staging the consummation of sibling incest only threatened within the comedies, John Ford's tragedy considers the violent consequences of investing in the assumption that blood determines affinity. As Lisa Hopkins aptly remarks, *'Tis Pity* is easily the counterpoint to Shakespeare's *Romeo and Juliet*: "Both plays feature young lovers attended by a nurse and a friar, but this time the marriage is too much within the family rather than too far outside it" ("John Ford" 202).[33] I would further suggest that this plot turned inside out marks the two plays as limit cases on either side of the spectrum of blood's ideology. Giovanni invests in the vocabulary of blood affinity, highlighting incest's romantic potential;[34] but *'Tis Pity* subsequently illustrates that tragedy lurks within blood affinity's limit case: when the principle of "like likes like" reaches its natural, uncomfortable conclusion. According to this picture, the magnetism and love between family members threatens to bleed into an all-encompassing erotic love since the blood shared by siblings is functionally no different from that shared by spouses. Or, as Giovanni himself argues, it would not make sense that this principle has an arbitrary end point as it gets closer to fulfillment. And *'Tis Pity*—like both Shakespeare's tragedy and the comedies of Della Porta and Middleton—emphasizes that if blood has agency to determine characters and affections, it does so because the characters believe these principles to be true not because they have biological foundations. These logics seek to shore up existing kinship and social networks, maintaining power in the hands of the aristocracy due to their basis in an "unchangeable" biology. Yet, rather than avert the incest in a comedic space, Ford's tragedy shows that there is a monstrous, violent quality to the reproduction of exact likeness and its hegemony.

Many critics have glossed Giovanni's citation of blood's discourse as a misunderstanding (of icons as proffered by Laurel Amtower or of aristocratic discourses as proposed by Terri Clerico) or a perversion of this vocabulary (as a symptom of theatrical decadence suggested by Carla Dente, as the mechanism of Ford's satire held by Mark Stavig, or as a distortion of Lévi-Strauss's gift-giving principle proffered by Pompa Banerjee).[35] Clerico, in particular, views Giovanni's literal interpretation of the Neoplatonic discourse as evidence that, given his status as a merchant social climber, he apes a discourse that he does not fully comprehend in an effort to elevate his status. It is true that, as she says, "the political intricacies of sexual and social advancement find their way onstage in the guise of half-hidden fantasies of absolute power unremittingly drenched in the language of blood" (405). Blood's ideology, after all, *is* not only the sexual means for social advancement but also, better yet, for social stability since it promotes endogamy. Yet Giovanni does not misunderstand the discourse; his application of noble discourses merely highlights their literal underpinnings: shared blood physically pulls siblings together in body and soul as like should like its most like.

Incest is, as noted by Susan Wiseman, "the hidden precondition of platonic language" and, as I maintain, of blood's language too (186).[36] And though other characters, Friar Bonaventura especially, may fight against this particular application of their beliefs, for them blood still embodies a person's nature and the status of their soul. Thus still steeped in this vocabulary, these characters provide no real counterpoint to Giovanni's worldview.[37] The language of blood is so all encompassing, as Carol Rosen reminds us, that it applies "both to the violence and lust of Parma and to the blood-ties and rites of purification enacted by Giovanni and his sister" (359). As a result, the play seems utterly incapable of escaping it. As Michael Neill and others describe, "what is extravagant about *'Tis Pity* . . . is not so much its metaphors themselves as Giovanni's violent literalization of them"("'What Strange Riddle'" 165). But I maintain that the most terrifying aspect of Giovanni's limit case is that he highlights the *already* literal and violent character of this ideology, one that in the hands of the elite permits oppression and exploitation. Any disgust or moral condemnation of Giovanni's character, therefore, should be redirected toward the ideology that he merely applies in its fullest sense.

Just as Della Porta's Attilio laments, though mistakenly, that his cri du sang swelled into erotic feeling, Giovanni understands incest as the natural outcome of blood's attraction to or *liking* of its like. As his father, Florio, says regarding his match for Annabella, "My care is how to match her to her liking" (1.3.10). Of course, this choice of words disturbingly foreshadows his children's incest, for to match Annabella to her "liking" might involve pairing her with her "likeness," pairing those who already share blood. Giovanni illustrates that if Neoplatonic philosophy, which romanticizes lovers as perfect mirrors of one another, is combined with assumptions about blood—that blood holds or embodies the soul and similarly determines affinity for its like—then the incest taboo would appear to be an arbitrary limit to what theoretically should be the most perfect iteration of love. Or alternatively, one could say that the siblings are already materially joined, so what is to prevent their sexual relations? As I mention in this chapter's introduction, Giovanni opens the play considering the tension between cultural assumptions about blood and the incest taboo. He says to himself,

> Say that we had one father, say one womb
> (Curse to my joys!) gave both us life and birth;

> Are we not therefore each to other bound
> So much more by nature? By the links
> Of blood, of reason? Nay, if you will have't
> Even of religion, to be ever one,
> One soul, one flesh, one love, one heart, one all? (1.1.28–34)

The incest taboo, marked by parentheticals of an aside or pause, illustrates the intrusion of this limit to Giovanni's desires, literally severing the sentence in two. Saying that they have the same parents is to "curse" his "joys." But the "links of blood" likewise play into the fantasy of complete union with the love object, to be "one all." Giovanni, in a sense, has merely taken the romantic model, and especially that of the commixtio sanguinis, one step closer to perfection. If blood-mixing is part and parcel of sexual consort that already involves each partner acquiring a bit of the other, then to share the same familial blood from the outset would further dissolve any and all individual distinction. This complete dissolution of difference to reproduce sameness, after all, seems to be the ultimate aim of blood's ideology, which is to replicate the status quo.

Giovanni further elaborates on the fulfillment of the Neoplatonic ideal in incest when he seduces Annabella, insisting that Nature predetermined their love: "Wise Nature first in your creation meant / To make you mine; else't had been sin and foul / To share one beauty to a double soul" (1.2.236–38). From "one beauty," we must infer that, though not identical twins, these siblings are so alike that they physically mirror one another. Giovanni's language draws upon physiognomic assumptions, whereby a single appearance necessarily corresponds to a single soul. In his mind, Nature in her wisdom could not have created two things so alike, sharing one beauty, if she had not meant for them to ultimately share one soul in sexual consort.[38] He extends this reasoning to blood's discourse: "Nearness in birth or blood doth but persuade / A nearer nearness in affection" (1.2.239–40). If blood has a magnetic pull, then greater similarity in blood would only effect greater affection, being not only near but *nearest*. Or, as he deduces, "If hers to me, then so is mine to her; / Since in like causes are effects alike," both insisting that the affection is mutual and effectively summarizing the "like likes like" principle (2.5.25–26). They *are* like one another so they *will* like one another.

In light of this logic, Giovanni understandably characterizes the prohibition of incest as a social convention, an arbitrary limit inconsistent with

Nature's design. He asks,

> Shall a peevish sound,
> A customary form from man to man
> Of brother and sister, be a bar
> 'Twixt my perpetual happiness and me? (1.1.24–27)

He alludes to the appellations of "brother and sister" not as products of Nature but rather of custom, forms from "man to man" and mere "peevish sounds" that bar the fulfillment of his desires. In keeping with this conviction after having sex with Annabella, Giovanni proclaims that she is "no more sister now" (2.1.1). Yet his desire to discard sibling labels as a way to reject the incest taboo stands in tension with the logic of their love. That is, the terms *brother* and *sister* pick out a particular blood relation that supposedly determines their erotic love to begin with, such that they could not *help* but love each other. He thus apparently contradicts himself only a few lines later, telling Annabella that her heart is "a heart whose tribute is thy brother's life" (2.1.5).

While Giovanni's application of Neoplatonism to incest is novel, his attitude toward incest as a flexible concept is not exactly unprecedented. After all, the church routinely made exceptions to incest prohibitions for the purposes of royal marriages: King Henry VIII's marriage to Catherine of Aragon being one prominent example.[39] Giovanni takes advantage of this precedent to convince Annabella to disregard the taboo, promising, "I have askt counsel of the holy church, / Who tells me I may love you; and just / That since I may, I should; and will, yes, will!" (1.2.241–43). Not only has he lied, of course, but also just as his logic in favor of loving Annabella functions based upon an ambiguity between familial and romantic love—one that serves his society in the perpetuation of appropriate kinship structures—so too does his interpretation of the holy church's (pretend) permission rest upon ambivalence of the phrase that he "may love." Nevertheless, Giovanni's logic is not some misunderstanding or misapplication, for incest naturally follows as love's perfection rather than its limit, rendering the taboo wildly inconsistent.

So although Giovanni is undoubtedly invested in assumptions about blood, a closer look shows that this ideology is open to rhetorical manipulation. Calling the audience to witness Giovanni's rhetorical usage of this ideology then encourages the realization that the love of sameness and the reproduction of the status quo is not the inevitable outcome of Nature; rather, it is the means for those in power to maintain their power. Giovanni's love is no more determined by his body than is the aristocracy's nobility determined by theirs. To emphasize blood's rhetoricity, Giovanni dismisses aspects of blood's ideology that do not immediately suit his purposes, such as the importance of Annabella's maidenhead:

> I marvel why the chaster of your sex
> Should think this pretty toy called maidenhead
> So strange a loss, when, being lost, 'tis nothing,
> And you are still the same. (2.1.9–12)

His condescension is surprising first because he has so forcefully emphasized the pull of Annabella's blood in his reasoning and second because the maidenhead is incredibly important in the context of this discourse—as both the material embodiment of honor and the first bit of blood to be exchanged in the commixtio sanguinis of marriage's consummation. But since, in this case, concern for the maidenhead conflicts with his goal, he proclaims it a "pretty toy" that when lost "is nothing." Giovanni will, however, make a great deal of the maidenhead later when it suits him, pronouncing that it signifies his privilege to Annabella based in physiology, which takes precedence over the legal claim of her new husband. He insists,

> She is still one to me, and every kiss
> As sweet and as delicious as the first
> I reap't, when yet the privilege of youth
> Intitled her a virgin. (5.3.8–11)

His proclamation implies that by virtue of having taken her hymen, Annabella is still effectively his ("one to me"), although this assertion blatantly contradicts his prior repudiation of her "pretty toy." Giovanni clearly cites blood's discourse as it suits his desires in a particular moment, and the dramatization of this blatant manipulation should call attention to it as such. Ford's play shows that blood is subject to tampering, for it is decidedly not a force that has dictated Giovanni's erotic love for his sister.

But despite the fact that Giovanni's citation of this ideology signals its discursive status, Ford's Parma fails to provide an alternative to Giovanni's worldview and its ultimately violent outcome. Moreover, whatever distaste one might hold for Giovanni's character (which to be perfectly candid, I do), this aversion *should* be applied to the ideology itself. Undoubtedly, Parma is corrupt as other scholars have underlined;[40] yet I want to frame this issue not in terms of corruption but of faulty logics and assumptions that plague both Giovanni and his society at large. In Parma, blood marks both class distinctions and individual character,[41] a person's "true nature."[42] These assumptions support Giovanni's claim that Annabella's blood marks her as his perfect mate rather than contradict it. Soranzo, for instance, clearly invests in the notion that Annabella's blood embodies her (im)moral character. Upon discovering Annabella's pregnancy, he drags her onstage proclaiming her a whore and promising to revenge himself on "every drop / Of blood that runs in thy adulterous veins" (4.3.1–2). In Soranzo's mind, her actions have tainted her character/body as well as his, for as he laments to Vasques, "in this piece of flesh, / This faithless face of hers, had I laid up / The treasure of my heart!" (4.3.106–8). Even though he does not explicitly reference their blood mixing upon marriage, he clearly envisions his connection with Annabella as bodily, laying his heart in her face. He further imagines his own blood at stake on account of her former indiscretions: "All my blood, / Is fired in swift revenge" (4.3.149–50). This revenge, moreover, can only be satisfied with her blood: "Blood shall quench that flame" (5.4.27). Blood's vocabulary and logics have saturated this society, and within it, Giovanni's logic is equally as valid as Soranzo's.

Moreover, just as Giovanni claims that shared blood is the agent of his erotic attraction, alleviating both Annabella and himself of any responsibility for their incestuous desires, so too do the other characters believe that blood has agency and, consequently, culpability. Florio characterizes the duel between Grimaldi and Vasques as spurred by "your disordered bloods" (1.2.25). Hippolita likewise blames blood, both hers and Soranzo's, for their affair and the murder of her husband; she characterizes Soranzo's wrong as his "sensual rage of blood" and contends that she only agreed to conspire with him due to "the devil in my blood" (2.2.29, 74). Richardetto, albeit falsely, assumes that Annabella's "sickness" (in fact, her pregnancy) is attributed to "a fullness of her blood," an excess causing heightened sexual desire for which marriage is the only cure (3.4.8). And Soranzo blames his own blood—"my blood's on fire"—for his vengeful desires (5.2.26). In this society wherein blood is accountable for nearly everything, Giovanni's extension of this logic to incest is not out of place.

Friar Bonaventura then might contend with Giovanni's application, but he invokes the same vocabulary when describing how to repent, thus providing no genuine alternative to Giovanni's worldview. As Robert Ornstein notes, in general "the Friar's worldly 'realism' does not clash with his otherworldly piety; it is instead a direct consequence of it. His literalistic mind views morality wholly in terms of crime and punishment; he regards sin with the mentality of a criminal lawyer" (209). I would extend this point to say that the Friar's reasoning and method of rejecting incest does not escape the ideological framework that makes the incest enticing. After hearing about Giovanni's incestuous desire, he begs,

> Cry to thy heart; wash every word thou utter'st
> In tears, and if't possible, of blood:
> Beg heaven to cleanse the leprosy of lust
> That rots thy soul, acknowledge what thou art. (1.1.72–75)

Based upon the assumption of an intrinsic link between blood and the soul, the Friar calls for some manner of humoral bloodletting.[43] But plainly, Giovanni's love for Annabella, at least as he claims, is based upon this same conflation, for shared blood indicates a shared soul in the most perfect occasion of Neoplatonic love. Additionally, in accordance with Friar Bonaventura's vision of penitence, Annabella's blood serves as the ink for her letter of contrition—"this paper double-lined with tears and blood"—which the Friar conveys to Giovanni in order to encourage his repentance in kind (5.1.34). The Friar draws attention to the ink's medium: "The blood's yet seething hot that will anon / Be frozen harder then congeal'd corrall" (5.3.25–26). His remark draws an analogy between the newly spilt blood and fresh repentance that will become hardened and more resolute with time, especially with the coral (a symbol of marital loyalty and commitment) of her new marriage to Soranzo. Giovanni recognizes this blood, "'Tis her hand / I know't, and 'tis all written in her blood" (5.3.31–32). Yet the ink of Annabella's letter merely reminds Giovanni of his own investment in her blood. This bloody ink acts not as an inducement to repent, as the Friar intends, but as the agent of the siblings' attraction. Therefore, Friar Bonaventura's lack of an alternative vocabulary licenses Giovanni's final, violent application of this discourse.

Notwithstanding Florio's desire that the rivalries for his daughter's hand will not "cause the spilling of one drop of blood," the violence that dominates the play and especially Giovanni's conclusion exposes that these bloody logics have ultimately bloody outcomes (1.2.63). Violence in general is even termed "being bloody," as, for example, Vasques encourages the Banditti to "be bloody enough" (5.4.3). The play's bloodbath begins in earnest halfway through the action when Grimaldi kills Bergetto. Bergetto describes the event with some surprise, "I am sure I cannot piss forward and backward and yet I am wet before and behind," and "O, my belly seethes like a porridge pot!" (3.7.12–13, 20). Then, the moment before he dies, he asks, "Is all this mine own blood?" (3.7.32). Although this first death has a comedic tone, its visceral and overwhelmingly bloody spectacle marks the play's turn for the worse, foreshadowing the bloodiness of the play's remainder. Friar Bonaventura notes a similarly ominous moment with the death of Hippolita at Annabella's wedding banquet: "I fear the event; that marriage seldom's good / Where the bride-banquet so begins in blood" (4.1.108–10). Of course, his wording eerily foreshadows Annabella's own death, her bloody heart on a platter for a literal banquet of the bride.[44] Gazing upon Annabella's corpse, Giovanni underscores that his motivations for her murder are the result of his adherence to the logic of blood. He professes that he "killed a love for whose each drop of blood / I would have pawned my heart" (5.5.101–2). While his statement could express regret for what he has done, this interpretation would contradict the conviction with which Giovanni performs the final scene. As McCabe clarifies, Giovanni rejects Annabella's right to terminate the relationship, and he is, therefore, prepared to kill her to save her honor from her "adulterous" marriage (238). To take this observation one step further, Giovanni believes that her murder saves the honor carried in her blood by purifying it. Killing her and spilling this blood would purge her of said connection to Soranzo and, consequently, her infidelity to their own love.[45] According to Giovanni's gloss, Annabella's heart "trimmed in reeking blood" carries his own within it: "'Tis a heart, / A heart, my lords, in which is mine entombed. / Look well upon't" (5.6.10, 26–28). The blood they share, the love this blood demands, and their hearts subsequently imprinted with the proof of this love, all come to literal fruition.

Ford's *'Tis Pity* thus illustrates that incest is the realization of the combined logics of blood affinity and Neoplatonic love. Moreover, tragedy stems from the literal application of these principles since Parma's society apparently cannot provide a counterpoint to Giovanni's worldview. A vocabulary of blood remains the currency of character, love, and even salvation. Blood's ideology, as invoked by the elite, can further ensure that love, loyalty, and power stays in a limited number of hands by promoting endogamy for a class's preservation. And as Ford's play suggests, the tragic consequences of this ideology pushed to its limit should call our attention not to the monstrosity of the act of incest, or Giovanni's "perversion" of blood's vocabulary, but to the ideology itself. Only one brief moment in the play might question this state of affairs by rejecting the assumptions of this system. Interestingly, it is voiced by the titular character, or the character whose name literally translates to "whore." After Annabella confesses her incestuous love to her nursemaid, Putana, her confidante quickly replies that it is no great matter for "if a young wench feel the fit / upon her, let her take anybody, father or brother, all is one" (2.1.48–49). Putana holds that "the fit" of sexual desire makes all love objects arbitrary; consequently, incest does not matter particularly because neither does the sexual partner.[46] Putana's view utterly contradicts the very intentional choice of love object, based upon blood affinity, that Giovanni and Annabella believe justifies their incest. Her worldview may be crass, and it certainly lacks the romanticism of Giovanni's worldview. Yet this acceptance of any and all persons might nevertheless be preferable to society's obsession with the particularity of bodies that justifies the hardening of class boundaries, an obsession that likewise has Giovanni's disturbing application lurking beneath it.

PERPETUATING A SPECIES, REPRODUCING HEGEMONY

The trope of sibling incest found its way across Europe, from Navarre's novella to Della Porta's drama, then was translated both comically and tragically in England. And examining this grouping of kin texts together reveals a shared cultural critique across these traditions, namely pushing the logic of "like likes like" to its limit. The dramatized sibling incest plot speaks to the aristocracy's intense desire to maintain a closed system, which they achieve by supporting an ideology of blood that naturalizes endogamy, or that love is based in likeness. Comedies averting incest—though they apparently endorse the status quo by upholding bodies' importance in order to dissolve the incest threat—suggest that the distinctions in bloodlines mean very little. These comedies expose the rhetorical nature of these differences either by calling bodily markers into question (Della Porta) or by rejecting them all together in favor of "name and place," revealing wealth to be the true motivation and foundation of social identity and affine structures (Middleton). Ford's tragic culmination of this premise stages the violent consequences that result from this investment in blood, the monstrosity of not incest per se but of the ideology that permits this conclusion. Blood's ideology, especially its principle that "like likes like," has allowed hegemonic structures to literally reproduce themselves; and those that cite this ideology can disavow any responsibility for its oppressive consequences. One could envision the nobleman articulating his position as follows: we, the nobility, are in power because we are intrinsically worthy and that is not our fault but it is the consequence of our bodies; you, the commonfolk, cannot participate in this social network on account of your inferior bodies, and you cannot join this system through marriage because nobles are not attracted to commoners. As these plays suggest, however, incest exposes the monstrosity of this logic for what it is: an ultimately false, though powerful, mechanism of oppression.

Like the voices of Elvire and Juliet in our blood feud drama, voicing alternatives to this ideology come from below, though, unfortunately, none of the critiques sound particularly appealing in these plays. Della Porta's Trasimaco may assert the ability to signify his body as he pleases to protest his heroism, but, as a classic *braggadochio*, this is hardly the makings of a larger movement of social resistance. The nurse Balia—though she herself bemoans the "mondo immondo" (dirty world) where none are safe from deception—only reveals the truth that supplies the comedic conclusion when it suits her interests to do so, namely, because she is fighting with her employer. The suggestion of Middleton's fool, Pickadille, to trade money for love sounds repugnant; and Ford's Putana offers a picture of indiscriminate love that, while an iconoclastic alternative to consanguineous affinities that support hegemony, are all too fitting with her name. The alternatives to blood's ideology, though pragmatic, are not particularly pretty. They certainly do not convey the romanticism of the cri du sang, that one's blood miraculously cries out for its like in the world. And yet, as these plays suggest, the status quo is ugly enough that these alternatives are worth entertaining.

3

Bed Tricks and Bloodlines, or All's Well That Ends with a Baby

The trope of the bed trick, in one light, fundamentally questions the importance of a body's specificity, demonstrating that bodies and persons can be easily substituted for one another in the dark. But ironically, the biblical version of this trope presumes the opposite. In this literary lineage, a bed trick not only carries divine sanction but it also foregrounds the moral weightiness of blood. The worth of a child, one of a *specific bloodline* destined for greatness in a sacred genealogy, permits the otherwise morally dubious means of its conception.[1] The most germane example of this topos is the story of Tamar. The divinely appointed mother of the House of Israel, Tamar tricks Judah because she is without alternative recourse to fulfill the demands of a leviterate marriage (Gen. 38). Namely, Tamar should sleep with her late husband's relative, in this case his father, Judah, in order to provide an heir for her deceased spouse. The child here is first and foremost a vessel for preserving his bloodline, in this case an extremely important one. The lineage of the House of Israel *must* continue in a Judeo-Christian context, for this bloodline will produce King David and, ultimately, Christ. The value of the immediate child, the progeny of the bed trick, pales in comparison to the future-oriented value of the bloodline that will produce the son of God.

God's blessing or implicit sanction of Tamar's bed trick is not merely a question of moral justification; rather, it highlights the magnitude of bloodlines. This biblical story demonstrates that blood has intrinsic worth, producing worthy persons, which is apparently part of God's plan. In a sense, what I call the "biblical bed trick" is the greatest display of the assumption that blood has value; it implies that not only is blood's ideology built into God's design but the naturalized hierarchies supported by this ideology carry God's sanction. This chapter illustrates how staging a divinely sanctioned bed trick in drama—one orchestrated to preserve a specific bloodline—can question or frustrate assumptions about consanguinity, kinship, and affinity. For one, what is the value of perpetuating a particular bloodline in the first place? Is consanguinity a requisite foundation for the affective and social ties of kinship and marriage? And if so, could the commixtio sanguinis of marriage's consummation determine or prescribe affection? Or, alternatively, will merely the facade of consanguinity, a *name*, suffice for familial and societal organization?

Dramatic allusions to this biblical bed trick may assuage the audience's moral compunctions regarding the use of deception to arrive at a comedic conclusion. Yet both Shakespeare's *All's Well That Ends Well* and Machiavelli's *La mandragola (The Mandrake Root)* cite a very specific iteration of this topos, the divinely sanctioned, biblical bed trick by way of the novella from Boccaccio's *Decameron* (III.9), in order to interrogate the ideology of blood at its foundation. In following this literary lineage from the Bible through Boccaccio and into early modern Europe, we find another limit case for blood's ideology. Literary engagement with this scriptural trope asks whether it is really the case that a bloodline can justify an otherwise immoral act. Initially scrutinized by Boccaccio's novella, these plays build upon this critical project; in both Shakespeare and Machiavelli the bed trick does not emphatically affirm blood's value as it does in its original biblical context, but it rather accentuates the absence of intrinsic worth. Not only does blood not determine anyone of particular merit—after all, is there any real value or virtue in producing a child for the disagreeable Bertram or Nicia the fool?—but blood also fails to function as its ideology presumes: both in its alleged capacity to inspire affinity (between spouses in Shakespeare's *All's Well*) and in its function as the material basis for the family unit (as in Machiavelli's *La mandragola*). Moreover, the characters of both plays cite the biblical bed trick, and blood's logic behind it, to serve ultimately rhetorical, or pragmatic, ends: to ward off unwanted marriage partners, to entrap a spouse into affine bonds, or to conceive a child for the sake of social capital.

The divinely sanctioned bed trick in Shakespeare's *All's Well* troubles the status of blood's ideology since it fails to operate as the characters believe and as God supposedly designed. Blood's ideology is not merely some hobbyhorse of the elitist Bertram, but, rather, it deeply influences all of the characters in this play. It particularly affects Helena who appeals to this ideology and employs the bed trick in the hopes of forging a connection between herself and her beloved Bertram. Helena's desire for a child is a desire for proof of a newly mixed bloodline (commixtio sanguinis) and proof of her own subsequent bodily transformation upon the consummation of their marriage. This mixing would make her *like* him in blood, thereby overcoming the supposed biological incompatibility and forging an affinity between them. Shakespeare's *All's Well* invites us to see, however, that because Helena appeals to a fundamentally hollow system, her attempt to establish a romantic connection between her lowly self and the ennobled Bertram founders. Moreover, though the Countess and King recognize that blood's ideology cannot accommodate the incorporation of persons who are worthy of alliance and love, it proves extremely difficult to think outside of this system from a privileged position within it. Only the powerless Lavatch the clown presents a possible critique of this ideology, highlighting the ridiculous consequences of the assumption that similar blood dictates love. His humor likewise highlights that blood might appear to solidify class distinctions, but it should be a great equalizer among persons.

Due to their shared examination of Boccaccio's biblical bed trick, examining *All's Well* alongside *La mandragola* proves particularly fruitful, for what remains at the level of suggestion for Shakespeare becomes far more explicit for Machiavelli. *La mandragola* cites this biblical bed trick but without mentioning bloodlines at all. The comedy thereby exposes that blood has no intrinsic worth nor does it play a role beyond a rhetorical one in the foundation of kinship networks. Put another way, the omission of the bed trick's moral—that blood is worth something—demonstrates that this discourse is rhetorically convenient rather than biologically necessary. The friar, Fra Timoteo, cites God's sanction to convince Lucrezia to bed a stranger in return for a bribe, *not* to perpetuate Nicia's bloodline or with some other higher purpose. Machiavelli's play embraces a pragmatic, iconoclastic attitude toward bodies, blood, and families; in fact, the specificity of the child's body and consanguinity becomes insignificant. Lucrezia's husband, Nicia, simply needs a child to *name* as his own. The child's value is not inherent or biological but constructed and practical, providing a means to enhance reputation. In sum, the biblical bed trick trope emphasizes the extent to which God will supposedly venture for blood, and its divine sanction is rhetorically convenient for those persons whose blood embodies their power and privilege. Yet these playwrights place pressure on this limit case, either implicitly or explicitly, thus undermining its foundational assumption. If blood does not determine worth, then the elitist agendas that this ideology supports likewise fail.

AN ALL-IMPORTANT BLOODLINE, BOCCACCIO'S DIVINELY SANCTIONED BED TRICK

While both plays cite, albeit in different ways, this biblical, moralized strain of the bed trick, literary scholarship has yet to compare Machiavelli's *La mandragola* to Shakespeare's *All's Well* given the former's peculiar lack of influence upon English drama.[2] Despite the great mobility and influence of Machiavelli's political writings, this play seems to have escaped the purview of the European Renaissance community at large, at least until the mid-seventeenth century.[3] The two works nonetheless share a common source or literary "kin": Boccaccio's *Decameron*. Boccaccio's tale of Giletta di Narbona (III.9) is a well-documented source for *All's Well*,[4] and *La mandragola* likewise owes a debt to the *Decameron's* bed tricks.[5] While Boccaccio's tale of Giletta may not explicitly mention the necessity of perpetuating a specific lineage, God manifestly orchestrates and condones this bed trick like its biblical forebears. Moreover, the two children that result from the protagonists' coupling carry the visible imprint of their father, proof that they are products of his bloodline. Boccaccio's plot seemingly translates the divine bed trick in a straightforward manner, as framed and recounted by its moralizing narrator Neifile, but the novella's details and its frame likewise encourage the reader to question God's support for this action.

The summary of Boccaccio's tale is as follows. Giletta (a physician's daughter), raised with Beltramo (the young Count of Roussillon), falls hopelessly in love with him. After refusing all other suitors to her family's dismay, Giletta decides to take action to acquire her love match. Upon hearing of the King's terminal illness and knowing that she can cure him, she uses this skill to her advantage, thereby allowing her to ask the King for her choice of husband. Beltramo, upset that she is of lower rank, submits to the King's order but refuses to consummate the marriage. He then runs off to Florence and falls in love with a poor gentlewoman. Yet Beltramo promises to return to Giletta if she can fulfill a seemingly impossible task: procure his family ring and produce a male heir. Following Beltramo to Florence, Giletta convinces the mother of this poor gentlewoman to agree to a bed trick, substituting herself so that she may conceive a child for Beltramo as well as gain his ring. The scheme completed, Giletta brings home not one male heir, but two! Throwing herself at his feet, she says, "Ecco nelle mie braccia non un sol figliuol di te, ma due, ed ecco qui il tuo anello. Tempo è adunque che io debba da te, sí come moglie esser ricevuta secondo la tua promessa" ("Lo in my arms not one son only but twain, gotten of thee, and on my finger thy ring. Tis time then that I be received of thee as thy wife according to thy word").[6] The tale then concludes with complete assurance that Giletta has satisfied Beltramo's terms, verified by the physical similarity between himself and his children: "Il conte, udendo questo, tutto misvenne, e riconobbe l'anello e i figliuoli ancora, *sí simili* erano a lui" ("Whereat the Count was all dumbfounded, recognizing the ring and *his own lineaments* in the children, *so like* were they to him" [italics mine]). Beltramo subsequently welcomes back both Giletta and his twins, and all is well that ends well.

In this narrative, not only is Giletta's bed trick a licit consummation of their marriage, but the tale's narrator persistently alludes to God's sanction and support for Giletta's impregnation. The queen for storytelling day three, Neifile, looks to reconcile propriety with the heroine's wit and ingenuity by moralizing the tale and bringing attention to God's hand in the plot. She effectively transforms a novella—a genre that ordinarily emphasizes everyday or "real life" situations—into a Bible story. Giletta plainly conceives the twins by God's grace in this narrative. Although Giletta sleeps with Beltramo numerous times, the narrator insists that this repetition allowed Giletta to procure the ring. The twins, by contrast, were conceived in their first encounter: "Ne 'quali primi congiugnimenti affettuosissimamente dal conte cercati, come fu piacer di Dio, la donna ingravido in due figliuoli maschi, come il parto al suo tempo vent manifesto" ("In which first commingling, so ardently sought by the Count, it so pleased God that the lady was gotten, as in due time her delivery made manifest, with two sons"). But of course while the narrator Neifile can assert that Giletta's conception pleased God, the characters' language exposes that they have no such knowledge or certainty of God's sanction. For instance, as the King finally concedes to Giletta's cure, he muses, "Forse m'è costei mandata da Dio" ("Perchance she is sent me by God"). Giletta might have God's dispensation, but there is no assurance only a "forse" (maybe). Later, Giletta convinces the poor woman's mother by casting the trick as God's work but not without a similar sense of reservation and the same "maybe": "*Forse* mi farà Iddio grazia d'ingravidare; e cosí appresso, avendo il suo anello in dito e il figliuolo in braccio da lui generato, io il racquisterò e con lui dimorerò come moglie dee dimorar con marito, essendone voi stata cagione" ("*Perchance* by God's grace I shall conceive, and so, having his ring on my finger, and a son gotten of him on my arm, shall have him for my own again, and live with him even as a wife should live with her husband, and owe it all to you" [italics mine]). Thus, even God's emissary is uncertain about the moral status of her actions. Neifile works very hard to assure the audience that the act is sanctioned, but the characters, who are unaware of God's permission, enact the bed trick regardless. The tension between the certainty of the narrator and the uncertainty of her characters throws the "moral," the virtue and necessity of Beltramo's children, into relief. Although it matters greatly to Neifile, from the characters' perspective, God's warrant may not matter. The character Giletta invokes this tradition in order to serve her purposes nonetheless.

Like the characters' language, the placement of this novella in the *Decameron* similarly disturbs Neifile's moralistic enterprise. As Howard C. Cole convincingly demonstrates, Boccaccio tinges Neifile's gloss of the Giletta story with irony given its surrounding tales. While all stories of the third day end happily, it is "usually because the protagonist is rewarded for his cleverness, not his goodness" (28). Furthermore, Cole suggests that "if the queen was looking ahead to her heroine, whose designs would be blessed by God, many of her subjects look back to Alatiel, who used God as an excuse" (20). The tale's immediate context in the *Decameron* thus casts Neifile's religious exegesis into doubt. And both Shakespeare's and Machiavelli's plays translate Boccaccio's tongue-and-cheek depiction of Neifile's biblical bed trick into drama—a genre that removes the moralizing narrator altogether and further interrogates God's support for this bed trick, which would affirm blood's ideology. If blood ceases to operate as promised then, as Shakespeare's play suggests and Machiavelli's asserts, God did not in fact bestow blood with the power to determine either worth or affinity. *Maybe* the ideology is fashioned by men and is thus insubstantial rather than the product of divine design.

ALL'S WELL THAT ENDS WELL ENOUGH

Shakespeare's *All's Well* plays out Boccaccio's biblical bed trick to the letter as should be clear from the summary, though with an even greater attention to the supposed importance of bloodlines as the embodiment of worth. Blood indeed motivates the heroine to orchestrate a bed trick as it did for her biblical forebear Tamar. However, there are two notable differences. First, the child/bloodline is the motivation only insofar as sharing blood with her spouse promises to forge a romantic connection. And second, God is noticeably absent from this play, which calls the sanction of the bed trick, and the ideology of blood that justifies it, into question. In the context of *All's Well* criticism, these claims are contentious as many scholars view Helena as a Christ-like or Marian figure, alluding to the genre of the morality or miracle play.[7] Michele Osherow, in particular, maintains that Helena's bed trick, which plainly references the tradition of Biblical matriarchs and particularly the story of Tamar, divinely sanctions her deception. I agree, of course, with Osherow's identification of this allusion; but while for Osherow this allusion answers the problematic character of Helena's bed trick, I believe that Shakespeare alludes to Tamar in order to interrogate the biblical story's foundational assumption.

Like Boccaccio's novella, several moments in *All's Well* disturb our faith in an explicit divine sanction. For one, as Helen Wilcox notes, Helena "uses devotional vocabulary more frequently with reference to her very earthly love for Bertram than for the adoration of God" (144). Second, though the King's cure conspicuously references heaven's will, not only might this not characterize Helena's motivation but it also does not guarantee the extension of heaven's will to the bed trick.[8] Helena might frame the bed trick in terms of her God-given role as "dower" to Diana (4.4.19–20), and she might assure the widow to "doubt not," whereas Giletta allows room for doubt; but as the Boccaccian source text reminds us, it is God's will that Giletta *conceive* a child, not his will for her to provide the impoverished gentlewoman with a dowry (Wilcox 145). Third, it is clearly not the work of God that motivates Diana and the Widow to agree to this scheme, but, rather, as W. L. Godshalk puts it, "the money helps soothe the Widow's anxiety lest her reputation be stained by her complicity" (65). Lastly, and most importantly, the fact that Helena's initial, and likely primary, motivation for the bed trick is her desire for Bertram interrogates the preeminence of a bloodline's preservation, which, by contrast, Tamar's story assumes due to her status as the matriarch for the House of Israel. The child, at least for Helena, is the means and not the end; her true desire is the love that should arise from commingling her blood with Bertram's—the commixtio sanguinis of consummation. Yet this end depends upon her credence in blood's ideology as a part of God's design. Helena truly believes that she is different in "kind" from Bertram; she simply finds a mechanism, a loophole if you will, by which this could change: becoming like Bertram in becoming one flesh.

If, as some scholars speculate, the play's previous title were *Love's Labors Won* (written as the companion play to *Love's Labors Lost*),[9] then the play might be asking "could one *win* love?" For *All's Well's* Helena, the child's bloodline *matters* because given her knowledge of natural philosophy as a physician's daughter, she believes that the mixed bloodline will bring her a happy ending and the love she desires.[10] Bertram's mandate to produce a child shares in this common assumption, maintaining that for Helena to be a "true wife" and obtain his affection requires the marriage's consummation. Only sexual congress, mixing their bloods and producing a child from this new admixture, would *dictate* his affections and sympathies. In other words, Helena must share Bertram's blood to become like him in order for him to entertain loving her (even though he has no intention of doing so). While Bertram cites the blood discourse in order to maintain the hierarchy between himself and the lowly Helena, she hopes to establish a connection between them via consanguinity.

And with this bed trick now evacuated of God's explicit stamp of approval, the ideology of blood that undergirds it becomes subject to critique. For while the conclusion's discomfort may result from the formal differences between drama and Boccaccio's novella, the difficulties of fairy-tale tropes onstage,[11] and a profound dislike for one or both of the protagonists,[12] it may also stem from a purposeful staging of this ideology's failure. By failing to bring the play to a satisfactorily comedic close—in this case, any hint of future happiness for the protagonists—the play exposes that consanguinity fails as the foundation for affine relations.[13] And this failure then troubles blood's ideology more generally. God may neither support this ideology nor have designed a system in which bloods determine diverse kinds as elites would have it. As Helen Wilcox briefly entertains, the term *grace* in this play may in fact indicate "social preferment" rather than "the gift of God" (147).[14] Moreover, blood may not be a divinely ordained biological fact unless, as Lavatch suggests, it dictates humanity's universal sin rather than any individual's intrinsic value.

This reading diverges from previous scholarship on the play's attention to blood, which has presented Bertram's concern for his blood either as an excuse to conceal some fear of Helena's sexuality or as a metaphor for social class rather than its embodiment.[15] For instance, Robert Miola views blood as contributing to the depiction of Bertram as a snob who "values too highly external trappings and pays too little mind to that within which passeth show" ("New Comedy" 29).[16] The exception to these readings is the work of David Berkeley and Donald Keesee; they provide the most comprehensive analysis of blood consciousness in *All's Well* as it relates to historical social discourses and natural philosophy (especially as outlined in Galen's *On the Natural Faculties* and Aristotle's *Generation of Animals*). These scholars importantly highlight that Bertram represents conventional wisdom, and further demonstrate the characters' literal investment in these discourses throughout the play. Yet they finally reach a politically conservative view of Shakespeare, who, they claim, generally supports the ideology of blood.[17] But just because the characters believe in an ideology, it does not mean that the play endorses it. As John Love convincingly argues, blood or class consciousness functions as the play's source of darkness: "Class debases the characters of Bertram and Helena throughout the play, and in the final scene it determines their fates and that of Parolles, despite the measure of virtue and vice each character possesses" (520).[18] An examination of blood discourses thus reveals that all of the characters are tainted by the existing vocabulary for social identity and affinity, even those who search for alternatives. And since both protagonists invest in an ultimately unfavorable and hollow ideological system, nothing ends well at all. This is not a product of either character's individual flaws but the result of the particular discourse to which they both subscribe.

All's Well's characters grapple with the assumption that persons' identities accord with their bodies, and especially that a person's parents have translated virtues through their bloodline, which then dictates their affinities (i.e., loving others like themselves).[19] The young protagonists not only embrace this vocabulary but fetishize it. Although they have different motivations—and Helena, given her position, may feel that she has no other option but to play the game by the established precepts—they both mistakenly believe that the *right* heir of the right blood would establish affinity between them since their bloods will have mixed.[20] The *right* child is supposedly conceived, but the ending remains profoundly uncomfortable. Blood clearly fails to deliver on its promise of love and happiness. The Countess and the King begin to see the failures of this ideology, namely that it cannot accommodate love or favor for those who may merit it but are not of the right "kind." However, their proposals are restricted by their continued use of blood's vocabulary, likely prompted by their privileged position within the existing system. Once blood is emptied of its purported agency and deprived of its divine author, Lavatch's comedic foil furthers this critique, pointing toward the ridiculous implications of the logic that blood materially determines love. Lavatch proffers that blood might determine sin but probably not worth.

The play opens by citing the basic assumption that consanguinity determines a shared character and virtue—or at least it would ideally. As a result, some bloodlines might be worth more than others such that it would justify a bed trick, as is the case for Tamar's House of Israel. In the Countess's blessing to her son before his departure to Paris, she reminds Bertram of his envisioned, inherited personality. She bids farewell,

> Be thou blessed, Bertram, and succeed thy father
> In manners, as in shape. Thy blood and virtue
> Contend for empire in thee, and thy goodness
> Share with thy birthright. (1.1.61–64)

Her blessing assumes that Bertram's blood and virtue work together as a unit in order to succeed his father in both shape and manners. Though distinct, blood and virtue are inseparable, as the first carries the second. Of course, if his blood actually determines his moral character and actions, then it would not be "contending" for empire so much as firmly holding it.[21] But Bertram nonetheless accepts the belief that his blood and his "virtue" are intertwined, and it is no surprise that he becomes sensitive about his marriage to Helena, for he has been led to believe that to fail his blood is to fail in these other moral virtues. The King reiterates this ideal correspondence when referencing Bertram's inheritance, one of both body and personality. He addresses Bertram,

> Youth, thou bear'st thy father's face.
> Frank nature, rather curious than in haste,
> Hath well composed thee. Thy father's moral parts
> Mayst thou inherit too! (1.2.19–22)

Here, one inherits "moral parts" as one might inherit a nose or a chin; but the King's wish also hints that these moral parts may be distinct from the inheritance of physical beauty, that these two aspects are not inherited simultaneously, at least not as promised. If the biblical bed trick, therefore, depends upon the perpetuation of a virtuous bloodline, wherein the child must take after their father, then it is a problem for this story that Bertram does not fully resemble his own father in virtuous action.[22] Both Bertram and Helena nonetheless buy into these assumptions about blood determining both love and worth.

Unlike other glosses of this play, I maintain that Bertram and Helena have both absorbed blood's ideology that justifies the biblical bed trick, perpetuating this vocabulary and, consequently, the notion of naturalized and divinely ordained class hierarchies.[23] Bertram, of course, is the more adamant defender of this view because he has the most to lose from this prospective marriage.[24] As Claire McEachern notes in her introduction to the play, "While the social barriers to the union of Helena and Bertram can be removed by the powers—ruler and family—that sustain them, what these powers cannot do is change the way in which Bertram's character has been shaped by them. Bertram actively disdains her, not for who she is, but for what she is: beneath him" (566). Shaped by society's ideology, Bertram asks the King why he and his integrity, which he conceives of in physiological terms, should pay for the King's health:

> But follows it, my lord, to bring me down
> Must answer for your raising? I know her well;
> She had her breeding at my father's charge.
> A poor physician's daughter my wife? Disdain
> Rather corrupt me ever! (2.3.111–15)

Despite a likely degree of equity in their "breeding" at his "father's charge," marriage would not simply be a "disdain" or a social slight but a corruption of himself via his blood.[25] According to contemporary beliefs, Bertram "through repeated copulation with Helena would improve her blood" but to the detriment of his own (Berkeley and Keesee 252). To mix bloods would be a contamination for "ever." As further reason not to marry, Bertram insists, "I cannot love her, nor will strive to do't" (2.3.144). Lack of love, of course, is no reason not to marry, and his excuse sounds irrelevant. However, according to the logics he has imbibed, Bertram speaks to a kind of bodily incompatibility. He cannot "love" her, hold any sympathy or affection, on account of his superior body.[26] In his mind, like cannot like unlike. Moreover, "striv[ing] to do't" would mean compromising his own superiority; it would mean consummating the marriage and mixing bloods at the expense of his own bodily, noble virtue.

Due to the King's ordinance, however, Bertram ultimately relents,

> I find that she, which late
> Was in my nobler thoughts most base, is now
> The praised of the king; who, so ennobled,
> Is as 'twere born so. (2.3.169–72)

His words, however, are clearly empty and do not reflect a genuine change of heart. Bertram constructs the final simile to indicate that he accepts Helena as his wife as though she were born a noble, but his rhetoric also calls attention to the very fact that she was *not* born of noble blood but base. Even though Bertram must participate in the social rite of marriage, Bertram proclaims "I will not bed her" since he can at least prevent the consequences of its consummation (2.3.268).[27] Bertram upon leaving for Italy stipulates his return to Helena in the infamous letter, which reads,

> When thou canst get the ring upon my finger,
> which never shall come off, and show me a child begotten
> of thy body that I am father to, then call me husband
> but in such a "then" I write a "never." (3.2.56–59)

In order for him to be a true "husband" and have affection for Helena, he must consummate their marriage.[28] The sexual act would affect the mixture of their bloods that would, once shared, establish an affection between them as well as conceive a child of this new admixture—his heir. The ring merely represents this exchange symbolically—just as it does during a marriage ceremony when placed on the fourth finger of the left hand—but the important exchange for Bertram is really one of blood. Of course, he has no intention of entering into this exchange, and his appeal to blood speaks to his desire to maintain the perceived difference between them.

One might think that Helena's love for Bertram would incline her to reject the traditional discourses barring her from their union, but, surprisingly, Helena too believes that she is different in kind, which blocks their union.

She accepts the parameters of this discourse, namely the purported value of blood, agreeing to play Bertram's game by reenacting the bed trick of biblical matriarchs. In so doing, Helena takes advantage of the loophole outlined in Bertram's letter in order to force a connection between them. What she fails to realize, however, is that since mixing blood does not actually make her *like* Bertram, the scheme will not work to determine love from him as she intends. At the play's outset, Helena speaks of herself in relation to Bertram as if he belongs to a different species—in her version of the commonplace: A bird may love a fish but where would they live? In Helena's more violent rendition, she laments, "Th'ambition in my love thus plagues itself: / The hind that would be mated by the lion / Must die for love" (1.1.92–94). Helena reiterates the problems of her baser body when conversing with Parolles. He asks her why it is a pity that she only wish him well, and she replies it is a pity,

> That wishing well had not a body in't
> Which might be felt; that we the poorer born,
> Whose baser stars do shut us up in wishes,
> Might with effects of them follow our friends,
> And show what we alone must think, which never
> Returns us thanks. (1.1.180–85)

Unlike wishing, Helena desires to manifest her love physically, to take her love out of the realm of thought and put it into the material world, that of the "body." It is a problem with her body, being both "poorer born" and female, that means she cannot do anything other than wish since she lacks the authority to change the terms of the discourse. Moreover, one could easily read this moment as a veiled desire to sexually consummate her love for Bertram since with "effects" she might "follow our friends."[29] The "effects" of consummation, the commingling of blood with Bertram, would allow her to become something other than she is by obtaining a better blood.

Although Helena seeks an elevation in station, she expresses this hope as a natural one rather than a social one (unlike Parolles for example)—a hope for nature to bring together like things although they are of unlike fortunes. She pines, "The mightiest space in fortune nature brings / To join like likes, and kiss like native things" (1.1.220–21). Her wish operates according to the same logic as Bertram's: that like will like its like natively. Since nature only mates equals, she must believe that her desire reflects some native similarity though there would appear to be none. Or, at least, their mating can solidify a likeness between them and thereby establish a reciprocal love. That Helena puts faith in the same blood assumptions as Bertram is further evidenced by her commentary during her husband selection. She says to the fourth lord, "You are too young, too happy, and too good, / To make yourself a son out of my blood" (2.3.95–96). It remains unclear whether or not this remark should be read sarcastically because, after all, if she truly believed this, then she should not submit Bertram to this indignity either. Yet Helena has nonetheless accepted that the creation of mixed-blood progeny would ultimately benefit her, making her (more) "like" the man of her liking. In her mind, to forge a connection to her beloved requires a metamorphosis of sorts into a different type of being or species.

And, though the important exchange for both Bertram and Helena is one of blood, *All's Well* stages this exchange as one of fetishized rings, a symbolic token established in Bertram's lettered ultimatum, pulled from the Boccaccian source.[30] Nevertheless, this act of displacement illuminates something important about blood, namely that it is a matter of signification anyhow since Bertram is not *actually* attracted to his like. For instance, when Helena schemes with the Widow about how to obtain Bertram's ring, she says,

> Now his important [importunate] blood will nought deny
> That she'll demand. A ring the county wears
> That downward hath succeeded in his house
> From son to son some four or five descents
> Since the first father wore it. (3.7.21–25)

The ring clearly stands in for Bertram's blood or lineage, and it is actually the *blood* she needs as the means of both her self-transformation and the determination of his affinity. She simply cannot mention her real motivation in what would be a cruder description of the sexual act.[31] Yet Helena's scheme also exposes a difficulty for the multivalent nature of the blood discourse. Lust inspires Bertram's persistent "importunate" blood, and it dictates Bertram's attraction to Diana, which creates the opportunity for the bed trick. But Diana is *not* of noble birth; she lacks even a dowry. Bertram's affinity for someone unlike himself, inspired by his lusty, sinful blood, reveals the falsity of the "like likes like" logic that supposedly motivated his objection to marrying Helena in the first place.[32] Bertram's elitist logic, which maintains class distinctions through endogamy, quickly unravels, even if it is apparently the only means to effect the "ending well."

Just as the biblical bed trick demonstrates God's sanction of blood's ideology, *All's Well's* bed trick appears to support the importance of bloodlines with the protagonists' endorsement. And just as the biblical bed trick has a blessed conclusion so should *All's Well's* bed trick effect a happy, romantic one. This "happy" ending's uncomfortable tenor, though, throws these assumptions about blood into relief. The logics of blood's ideology have been played out to the letter, but the conclusion stages this ideology's failure to deliver on its promises; blood is neither inherently worthy nor has it determined affinity. At the beginning of this final scene, upon the prospect of being permitted to marry a woman more worthy of his station (Lafew's daughter, Maudlin), Bertram claims that it was his contempt, his "scornful perspective," that turned Helena into a "hideous object," and since having lost her, he "ha[s] loved" (5.3.47–55). Yet, given his lukewarm reaction to the surprise that Helena is not dead after all, Bertram must be feigning love in this moment in order to curry favor with the King. Furthermore, as Gerard Gross notes, we should not take this expression of love seriously because Bertram has lied about everything else up until this point and we have no indication that he has changed (268).[33] Bertram's reticence later in this scene likewise disproves his previously cited justification about affinity based in blood. If consanguinity determines affinity, Bertram, having already bedded Helena unbeknownst to him, *would* love her. Her blood mixed with his would dictate an affinity for her, which he could not otherwise explain. His assumptions about blood, if true, would justify a rather sudden change in heart and "perspective," but this proves to be mere lip service to the King. The bed trick's status as a limit case has not proven blood's power but debunked it.

Fittingly then, just when Helena should be reiterating her body and blood—which she now shares with Bertram both by having consummated their marriage and having conceived new flesh in her womb—her reveal rather suggests disillusionment. She first calls herself a shadow: "'Tis but the shadow of a wife you see" (5.3.304). Although her pregnant body onstage carries a measure of visual proof in many productions,[34] technically only her words alongside Diana's testimony guarantee her conception of a child for Bertram, a fact that Helena emphasizes. Theoretically, she would want to highlight her changed physique, but, instead, she presents Bertram with his letter, his word, and says, "This is done" (5.3.310). This strange reversal of expectations might intimate that Helena is already disenchanted, already aware that their consanguinity will not have the desired effect. Bertram then doubles down and requires more as part of a *condition* for loving her, demarcated by the all-important "if clause." He concedes, "If she, my liege, can make me know this clearly, / I'll love her dearly—ever, ever dearly" (5.3.312–13).[35] Bertram does not specify what it means to "know this clearly," yet one may assume (like Helena's Boccaccian predecessor Giletta) it would be a child that looks unquestionably like himself, one that is indisputably of his own blood. Although just as Bertram does not satisfactorily resemble his father's "moral parts" so too might this heir disappoint. Bertram's words are an excuse, not a profession of love.

Finally, and maybe the greatest disappointment for the audience looking for a romantic close, is that Helena *accepts* these conditions, even allowing stipulations for divorce in a new *verbal* contract. The marriage contract, supposedly concluded and finalized by its physical consummation during the bed trick, is clearly not binding. Helena establishes the new terms of their arrangement in her response, "If it appear not plain, and prove untrue, / Deadly divorce step between me and you" (5.3.314–15). This is hardly the beginning of marital bliss. The protagonists' attitude toward the fulfillment of their marriage and Bertram's contract seems purposefully unpleasant, a legal arrangement rather than a physical, intimate one. For Helena, the promises of blood's logic—that she in marriage would become like him, that the fish would sprout wings and fly and that her transformation would inspire love—have gone unfulfilled.

Throughout the play, the Countess and the King begin to interrogate these beliefs, recognizing that the logic of "like likes like" in blood's ideology prevents the incorporation of (and affection for) Helena. However, these characters remain entrenched in, and thus limited by, blood's vocabulary, which prevents any genuine change.[36] This vocabulary, after all, is the basis for their own positions and privileges. It is the explanation of and justification for their entitlements. Incorporating Helena into this existing system such that the Countess and the King do not imperil their own positions requires that they probe the existing discourse for possible exceptions, not dismantle it. The Countess, for one, aims to uncouple love from consanguinity in her description of Helena's "adoption."[37] The Countess tells Helena that she has a "mother's care" even though "You ne'er oppressed me with a mother's groan" (1.3.144, 143). Helena did not come from her body, but the

Countess can nonetheless pronounce, "I say I am your mother" (1.3.150). Her performative declaration suggests that one can become a mother simply by proclaiming it so and that the label has little to do with bodies. The same principle applies when the Countess denies that the absence of consanguinity affects her love for Helena. She announces, "If she had partaken of my / flesh and cost me the dearest groans of a mother, I / could not have owed her a more rooted love" (4.5.10–12). Love need not be "rooted" in blood or flesh.

Yet, as in the case above, this vocabulary persists even when the Countess aims to deny blood's determinations and implications. Denouncing her son after he has abandoned Helena, she uses blood's terminology: "He was my son, / But I do wash his name out of my blood, / And thou art all my child" (3.2.65–67). The Countess figures the process of disownment as a material one. It somehow involves a change to her blood in order to divorce mother from son, thereby dissolving any prior affinity between them determined by their consanguinity. Moreover, her pledge to wash away Bertram's name/identity presumes that this name is somehow embodied in the blood that they share. Whether the Countess understands this process of blood purification literally or merely as a metaphor, it is a metaphor that must be expressed in the corporeal terms of consanguinity whereby sharing blood *does* dictate affinity for kin. The Countess hopes to split love from blood and embrace Helena instead, but she has not escaped its terms. The society from which she directly benefits organizes people according to bloodlines, and this society seems to mandate the Countess's persistent emphasis upon the bodily difference between herself and Helena, despite her affection that clearly transcends these limitations.

The King's attitude toward blood possesses a similar degree of equivocation. He searches for a means to reject Bertram's resistance to wedding Helena based upon their supposed material difference, but he has difficulty expressing an alternative worldview confined as he is by blood's ideology. After all, Helena's request to select a husband reveals the reason for the King's ambiguity, namely a concern for the continuity and preservation of royal blood. She dictates that as repayment for her cure,

> Thou shalt give me with thy kingly hand
> What husband in thy power I will command.
> Exempted be from me the arrogance

> To choose from forth the royal blood of France,
> My low and humble name to propagate
> With any branch or image of thy state;
> But such a one, thy vassal, whom I know
> Is free for me to ask, thee to bestow. (2.1.194–201)

When establishing her terms, she specifies that she can choose anyone except royal blood. In fact, she spends twice as much time outlining what she will *not* ask of him: compromising his own royal lineage.

The King thus does not seek to dismantle this all-important ideology so much as find a bit of wiggle room to appropriately recompense his healer. Bertram's obstinance simply stands in the way of the King exerting his power or prerogative. In response to Bertram's refusal, the King insists that

> 'Tis only title thou disdain'st in her, the which
> I can build up. Strange is it that our bloods,
> Of color, weight, and heat, poured all together,
> Would quite confound distinction, yet stands off
> In differences so mighty. (2.3.116–20)

As M. C. Bradbrook notes, since the King's speech occupies the structural center, it should likewise be the center of our attention as the play's moral. His reflection on blood's nature underlines the inability to distinguish differing bloods visually from one another and further insists that these differences might be based in name only. However, the King has not denied that there might *be* differences, only that those differences in "color, weight, and heat" would confound distinction if all mixed together. In fact, the natural philosophy of the period would support the belief that Helena's base blood would have been perceived as darker, less sweet, and thicker (Berkeley and Keesee 251–52). But for the King, if there are differences, one cannot identify them though they be "so mighty." By juxtaposing the title, which he can "build up," with the "confounding" nature of blood, the King implies that these differences are not physical but social.[38] Yet the enumerated physical distinctions remain, and the King's language has not explicitly dispelled the belief that blood determines them, just cast a bit of doubt for rhetorical purposes. The ambiguity of the King's speech means that he need not dismantle blood's ideology from which he too derives power.

The ambiguity persists as the King perseveres in his quarrel with Bertram, maintaining

> That is honor's scorn
> Which challenges itself as honor's born
> And is not like the sire. Honors thrive
> When rather from our acts we them derive
> Than our foregoers. (2.3.132–36)

But his account of honor has not denied that one *can* inherit qualities such as honor. It has only established that honor fares better when obtained by action. Similarly, the King insists on his ability to bestow his honor where he chooses—"It is in us to plant thine honor where / We please to have it grow" (2.3.155–56).[39] Yet, by using a botanical metaphor, honor still sounds like a trait based in material, as though the King could create honor via mixed-blood cross-pollination. Both the Countess and the King begin to search for alternatives to the language of consanguinity to describe affine relations and worthiness, for this language cannot accommodate Helena. But they ultimately remain limited by this vocabulary due to their dependence upon it for their own positions, and thus, they fail to devise a fully satisfactory alternative.

The comedic interludes of the clown Lavatch, on the other hand, provide one mechanism for critiquing this system from the outside. While they are seemingly unrelated asides, these act as a foil for the central action by poking holes in the prevailing assumption that blood (pre)determines both a person's character or "kind" and their affinities. As I note throughout this book, the character most likely to be disregarded or ignored is often the only character to suggest a genuine critique of the existing system. In this case, Lavatch's dark comedy should shed light on the citation of blood discourses throughout the play, especially as it pertains to social class. Lavatch first critiques the belief that consanguinity inspires love effected when becoming "one flesh and blood"—the very assumption that prompts Helena's resort to the bed trick. He illustrates its folly due to the fact that it can have counterintuitive results. He reasons for the Countess,

> He
> that comforts my wife is the cherisher of my flesh and
> blood; he that cherishes my flesh and blood loves my

> flesh and blood; he that loves my flesh and blood is my
> friend: ergo he that kisses my wife is my friend. (1.3.45–49)

Lavatch succinctly shows that according to this logic, one should love the man who cuckolds him. A worldview that produces this result is clearly flawed; but his fellow characters seemingly ignore the larger implications of his comedy.

Not only is the logic that consanguinity dictates love unsound, if character or worth is to be determined by blood, as the biblical bed trick implies, then one might look to another part of the Bible to find that blood and flesh are far more likely to determine sin than virtue. Arriving onstage to ask the Countess for permission to marry, he explains, "I have been, madam, a wicked creature, as you / and all flesh and blood are, and indeed I do marry that / I may repent" (1.3.36–38). Lavatch's comment highlights that flesh does not distinguish people from one another but rather functions as a great equalizer. In a religious context, all are descendants of Adam and consequently of original sin. Lavatch here evokes the rough equivalent of the common rhyme critiquing inherited nobility: "When Adam dug and Eve span, who then was the gentleman?" Though couched in a comedic scene, and slightly less threatening as a result, Lavatch's request for marriage should call attention to Bertram's class sensitivity later in the play.[40] Moreover, as we witness with Bertram's susceptibility to the bed trick, it is his "importunate" blood that defines him; and his lustful nature is, as Lavatch jokes, common to all men. Lavatch's comedy proposes that blood neither establishes affinities nor biologically determines kinds as Bertram would have it. The play thus intimates that if blood serves these functions then it does so due to rhetorical efficacy not truth value. The appeal to the biblical bed trick to support and justify blood's ideology is, moreover, at odds with the more obvious reading of God's opinion regarding human flesh: that it is all equally meritorious given that it is all equally sinful. Consequently, *All's Well* suggests that there is no such natural hierarchy of persons built into God's design as blood's ideology would claim.

Shakespeare's allusion to the divinely sanctioned bed trick, rather than stress blood's significance and power, has used this test case to expose that blood's substance has no bearing upon either affection or intrinsic worth. Both Bertram and Helena look to blood's ideology to understand their prospective coupling and with good reason, for, as the biblical bed trick stresses, bloodlines (with children as their vessels) are extremely important. As Helena says, God allegedly designed a world "to join like likes," and those of different classes, having different bloods, are akin to different species. But this play intimates that blood's ideology with supposedly divine support is instead a convenient apparatus for elitist agendas, especially since, in a biblical context, flesh is just as likely to prescribe original sin as it is to designate intrinsic worth (as with the House of Israel). Yet, without an alternative vocabulary for conceptualizing relations between persons and when fully imbricated in this system—as the Countess and the King are—it remains difficult to establish a new state of affairs that does not separate persons according to purported material differences. Machiavelli's play then offers a radical course of action: one might simply choose to ignore this vocabulary altogether. Unlike Helena, for whom the bloodline mattered far more than the baby, for Nicia and Lucrezia, the baby matters, not the bloodline.

BLOODLESS BABIES AND MAGICAL MANDRAKE ROOTS

In the case of Machiavelli's *La mandragola* (c. 1518),[41] alluding to the biblical bed trick inserts a necessary "ends justifies the means" framework for conceiving a child. Lucrezia's husband, her mother, and a morally questionable priest are all convinced to put Lucrezia in bed with another man for the all-important pursuit of creating an heir. This plot, not only reminiscent of Boccaccio's Giletta and Tamar of the Old Testament, approaches other hagiographic episodes: Elizabeth's miraculous conception of John the Baptist and even the Annunciation.[42] For a divinely sanctioned bed trick to function properly in accordance with this tradition, blood should matter. And even in its secular application as with the Boccaccian source material, blood is the marker of legitimacy—a father's assurance that his legacy, a literal piece of himself, will survive. Yet Machiavelli's play illustrates that this heir is necessary *not* by virtue of his blood but by virtue of his name, his function as placeholder to inherit wealth and enhance reputation. The heir's value is constructed and social not physiological. Even though a child (a body) matters, the bloodline does not, and the import of consanguinity to the family's social function is exposed as a great hoax. For aside from needing a male for the mandates of patriarchy, so long as the son is *signified* as Nicia's, it does not matter whether there is any biological truth to this claim. Nicia can pass down both his name and his wealth to, as he calls him, the *naccherino* ("little brat") all the same (5.2, 112).[43] The citation of the biblical bed trick, therefore, is entirely rhetorical. And in omitting blood from this allusion, Machiavelli's play evacuates the worthy "end" of the biblical bed trick, namely, blood's ideology, of its worth. Machiavelli's translation of the biblical bed trick via Boccaccio iconoclastically suggests that a body can be made to signify whatever or whomever one likes. And absent its divine design or endorsement, blood's ideology crumbles. Lastly, just as Helena must feel a great deal of disappointment at the failure of blood's ideology so too is Lucrezia deprived of a locus for personal integrity or agency, in the form of bodily chastity, if the particularity of bloodlines and their purity no longer matter.

Existing scholarship on Machiavelli's *La mandragola* aims to integrate the play with his political philosophy as outlined in *Il Principe* and *I Discorsi.* Unsurprisingly, Machiavelli's political theory offers an attractive method for unpacking a rather perplexing play. Although sometimes interpreted as a specific political allegory,[44] most of these political readings, conducted by scholars of political philosophy and literary critics alike, ultimately amount to "who is the *savio*?" (the savvy political leader) or even who might be the stand-in for Machiavelli himself?[45] These readings provide competing arguments for the protagonist lover Callimaco, the trickster-parasite Ligurio, the exceptional Lucrezia, and even the duped lawyer Nicia.[46] The elaborate ploy to impregnate Lucrezia is subsequently glossed as a tale of renewal, a comedy that celebrates rebirth out of degeneration by establishing a new society ruled by this savio. Ezio Raimondi's version of this reading is the most well-known, though others echo it, including Salvatore Di Maria's "Ethical Premises," wherein Nicia's infertility symbolizes cultural stagnation until Callimaco rejuvenates the community, and Theodore Sumberg, who argues for an intellectual rebirth wherein "Callimaco, the new Prince [is an aristocrat] by intellect who will use the new science to overthrow the existing nobility and their ancient ideas" (330–31).[47]

Alternatively, scholars cast the play as a satire of this Florentine "rebirth." This more pessimistic "establishment of dystopia" reading can be found in Ronald Martinez's "Pharmacy of Machiavelli" that compares Lucrezia to Livy's Lucretia.[48] Jack D'Amico similarly acknowledges that "Lucrezia, unlike a young man, cannot escape the city and its customs, no matter how corrupt they become" ("Virtù of Women" 270).[49] Yet presenting *La mandragola* as a play about a great cultural shift has at least one wrinkle. Namely, the child from this union will be Nicia's *in name*, which means that this scheme, in fact, perpetuates the existing social order. In this light, the play's end does not establish some fundamentally new society. As Di Maria notes, "The play's society, too, in its resolve to survive, does not hesitate to adjust its values and to avail itself of any means at its disposal, including the foolish *messere* and the ignoble friar" ("Machiavelli on Stage" 186).[50] Nicia's heir, then, simply maintains the status quo that requires offspring for the perpetuation of the ruling classes. The satire, if you will, exposes the discursive quality of these ideologies that sustain the status quo. It might be dystopic, but it always was; Machiavelli simply brings out the practical or functional attitude toward talk of kinship in Florentine society.

Thus, while I do not want to suggest that Nicia is the Machiavellian savio secretly pulling strings under a facade of stupidity,[51] I would grant his character an awareness of his social system's actual requirements: that appearances matter more than facts. As Di Maria highlights, one of Nicia's defining flaws is his excessive concern for appearances and disregard for truth ("Ethical Premises" 29). The play might imply, however, that this is neither a fault nor a symptom of Nicia's folly because appearances might be all that matter anyway. If appearances matter more than truth—since, as Jon Snyder suggests, ultimately appearances (dis)simulated dictate a person's identity—then Nicia may not be so foolish to insist upon them.[52] There is no question that Nicia is a dupe, but it may nevertheless be the case that he is, as his name suggests "the Victorious,"[53] for to be *recognized* as a father is the relevant consideration regardless of the child's true parentage. Nicia needs progeny, but consanguinity with his progeny does not matter if this child's identity is the product of signification. As opposed to trying to wash a name from blood as does *All's Well's* Countess, *La mandragola*'s characters pragmatically take for granted that blood is a material without a name attached, whose signification is arbitrary. And as a result, they do not bother to mention it, even though it is the moral foundation upon which the biblical bed trick rests. Sharing blood with a son, guaranteeing the perpetuation of his biological lineage, does not affect the child's ability to inherit Nicia's wealth and assume a place *as* his son in society, which would bestow honor on his family, his primary desire. Nicia needs a "chunk of flesh," a body to signify and nothing more (3.4, 93).

Though it is true that much of Nicia's concern for appearances in the play generates comedic relief—for instance, the social caché that comes with speaking Latin (even if it is nonsense) or perpetually name-dropping—his insistence upon titles should prompt the audience to appreciate that public performance and social recognition are, in fact, the key features of identity and not bloodlines. For example, Nicia lectures Ligurio that he should address "doctor" Callimaco by his proper title "maestro," saying, "Non dir così, fa'il tuo debito. E, se egli si offende, si tolga pure la cinghia" ("Give his due title, and if he doesn't like it, he can ignore it") (2.1, 82). The title is a person's due, and, effectively, all that they have to distinguish themselves from others; to ignore it is to deny any such distinction. Ligurio's failure to address the "maestro" correctly means that he ignores or denies his "debito" (due or debt), his obligation to acknowledge the social distinction between himself, a parasite, and an illustrious physician (and, by extension, the distinction between himself and the illustrious lawyer Nicia!). As Nicia seems to recognize, status is based upon social distinctions. And what are these social distinctions but titles?

One could easily ignore Nicia and gloss this moment as a bit of foolishness, but even the language of the prologue supports his view that "honor" is a thing of outward show or performance, primarily dependent upon others' recognition. The prologue describes the "honorable" Callimaco, "Costui, fra tutti gli altri buon compagni, / A'segni ed a'vestigi / L'onor di gentilezza e regio porta" ("Above all other good men, he exhibits the signs and ways of a gentleman worthy of honor") (73). That which indicates his "gentilezza" (courtesy or gentility) as well as "regio" (princeliness or royalty) are "segni" (signs) or "vestigi" (marks or traces). The prologue does not imply that Callimaco's merits are intrinsic traits bestowed by birth. He *appears* to be worthy of honor, and, therefore, he is. In this framework, Nicia's obsession with these appearances or social niceties in the search for recognition is not so foolish as we might first presume. Later, when accompanied by Siro to collect his wife's urine sample for a fertility test, Nicia expresses his desire for status and laments the lack of respect that he receives from his peers: "Il fatto è che in questa terra chi dei nostri pari non è legato a quelli che hanno il potere, non trova un cane che gli abbaia" ("In this city, people without status, people like me, can't even find a dog willing to bark at them") (2.3, 84).[54] In order to have social standing, someone must talk about him, but Nicia's present situation is so dire that not even a dog will acknowledge him. One way for Nicia to expand his social influence and to create ties to those with power is to have a child of his own name, which, despite six years of marriage to a young, noble bride, has yet to be produced.

All of Nicia's emphasis on signs and status should throw into stark relief the song between the second and third acts, containing the basic premise "ignorance is bliss" that supposedly pertains to Nicia. The song remarks on the lucky Nicia, "L'ambizione non lo preme, / non lo muove il timore, / che sogliono esser seme / di noia e di dolore" ("Ambition does not press him, / Fear does not move him, / Which normally are the seeds / of suffering and pain") (89). Yet the claim that Nicia has no worries in life because he has no ambitions must be false given both the lengths to which he is willing to go for an heir and his hyperawareness of appearances and titles. Even if one grants that Nicia may be blissfully ignorant that the child will not be biologically his, there is certainly a case to be made that he is willing to take the risk that it is not.[55] Nicia is aware that a child's presence solidifies honor, though the particulars of that child's body might be inconsequential.

Moreover, though the opening frames the couple's desperation as something that Callimaco can exploit in order to bed Lucrezia, the play quickly brushes Callimaco's desire aside to emphasize the child's production. Lucrezia's future pregnancy is no longer a means to Callimaco's end, even though his desire is the impetus for the whole scheme. Instead, Callimaco's desire is the means to the procreative end. For even if Lucrezia's infertility may be a hoax, the justification made available by the biblical bed trick ceases to be a mere front. The difference is that in Machiavelli's transposition of this biblical bed trick, the trope ceases to prove the importance of bloodlines: that the child must be conceived as a vessel for some future divine purpose. Instead, bloodlines are abandoned in favor of pragmatic solutions to the concerns of wealth and social status; God's sanction merely provides a cover. As Ligurio explains to Callimaco, in order to overcome Lucrezia's virtue one must use "la voglia che lui e lei hanno di avere figliuoli, che, sendo stata sei anni a marito, e non n'avendo ancor fatti, ne hanno sendo ricchissimi, un desiderio ch muoiono" ("the longing that he and she share to have children. After being married six years without having any and being extremely rich, they're dying to have them") (1.1, 76). Since both Nicia and Lucrezia supposedly share this want, both may be ultimately convinced that the end justifies the means. They need "figliuoli" for succession. Since they are "ricchissimi" ("*extremely* rich"), they desperately need a child to inherit this wealth. Messer Nicia repeatedly affirms this desire and desperation, lamenting, "Pure io ho tanta voglio d'avere figliuoli, che io son per fare ogni cosa" ("Still, I want so much to have children that I'm ready to do anything") (1.2, 79).

He even tells Fra Timoteo, "Io non ho figliuoli, e vorre'ne, e per avere questa briga vengo a dare impaccio a voi" ("I do not have children and wish to have them, and in order to acquire this problem I have come to create problems for you") (2.2, 83). Children have become Nicia's sole preoccupation and it is a "impaccio" (trouble, fixation) that he will now pass on to Fra Timoteo.

In fact, one could say that the entire bed trick scheme functions on account of Nicia's single-minded fixation becoming contagious. One by one, each of the characters assents to accomplish this end. For instance, Lucrezia's mother, Sostrata, buys into the necessity of having children at all costs: "Se, ad avere figliuoli, voi non avete altro rimedio che questo, si vuole pigliarlo [take it], quando e'non si gravi la conscienzia" ("If in order to have children there's no other option, we must do this as long as we all agree that it doesn't weigh on anyone's conscience") (3.1, 89). She, like Nicia, has set aside moral qualms *so long as* it is assured to work. Fra Timoteo, too, agrees: "Noi condudereno questo parentado questa sera" ("We'll complete the coupling this evening") (3.8, 95). Here the term *parentado* denotes sexual relations, but the word connotes the creation of family or kin as its result. Fra Timoteo thus conveniently sets aside two small matters: first, that this coupling involves adultery and, second, that the "parentado" created that evening would not be biologically related to Nicia, which if God were to sanction this bed trick utterly defeats its purpose. Nicia in single-minded obsession sees past his own cuckolding,[56] staying up all night imagining "del bambino, che me lo pare tuttavia avere in braccio, el naccherino!" ("the baby boy, whom I can practically already feel in my arms, the little brat!") (5.2, 112).

Thus despite some talk of sons,[57] references to children throughout the play conspicuously lack any reference to consanguinity or concern for family lineage traditionally conceived. It becomes clear that the progeny merely fills a slot, fitting into a designated social role that does not require any biological specificity. The play thus completely disregards the moral justification of the bed trick in the biblical tradition based upon a particular meritorious bloodline. As Nicia yells at Lucrezia, "S'io credevo non avere figliuoli, io arei preso più tosto per moglie una contadina" ("If I'd known that we wouldn't have children, I would have rather married a peasant woman capable of bearing them") (2.5, 85). The translators here insert the relevant information to help modern audiences, but the original merely says "I would have taken a peasant woman as a wife." Nicia here underlines that the child is more important than a woman's honor and gentility. According to traditional blood discourses, Lucrezia's honor *should* matter a great deal to Nicia because it would ostensibly have a bearing on the character of his children. As we witnessed with Bertram, typically for men who are conscious of their honor and status, a peasant woman simply would not do as a mate since her inferior blood would contaminate the familial line.[58] But as Nicia, and Machiavelli's play at large, suggests, who really cares? Society's recognition of a woman's honorable station might be important from a legal and social standpoint, but a woman's nobility makes very little difference if she has no child, or put another way, if patriarchy cannot glean the benefits of that honor.

For as *La mandragola* proposes, signification—rather than nature or blood of divine design—determines identities. For example, Ligurio's scheme to test Fra Timoteo, though a seemingly unrelated digression, uncovers the way in which society's pragmatic signification disregards the physiological. Ligurio, in order to evaluate whether Fra Timoteo can be swayed to participate in their morally dubious plot, asks the friar to agree to an abortion in the name of "tutto l'onore di casa sua" ("the honor of his [Nicia's] house") (3.4. 92). First, the honor "of his house" highlights the fact that family matters are first social, a function of living in the same place, rather than consanguinity.[59] Second, Ligurio reasons that, in agreeing to this abortion, Fra Timoteo will save the honor of all involved and "voi non offendete altro che un pezzo di carne non nata, senza senso, che in mille modi si può sperdere" ("you'll be offending nothing but a chunk of unborn flesh, without feelings, which could be lost all the same in a thousand ways") (3.4, 93). Of course, Ligurio's description of a fetus would not be so shocking in the period; a fetus had no claim to personhood until it was "ensouled," formed, or "quick" (i.e., capable of movement).[60] There existed a kind of moral grey area in which "there was an indefinite but fairly certain time, a window of opportunity as it were, during which a woman could end what we call a pregnancy, and neither she nor her contemporaries regarded the act as an abortion. Taking a drug for delayed menstruation was just that and nothing more" (J. Riddle 27–28). After all, menstruation could end for all kinds of reasons, and Galenic medical theory emphasized the need to restore the natural flows of a blocked body (Terpstra 86). Yet this gray zone only further supports the point that signification matters first and foremost; *how* one refers to this fetus becomes all the more important. At this point, technically a month too late,[61] one would judge this woman pregnant, and terminating this pregnancy would be deemed immoral at this time. Referring to the fetus as "a chunk of unborn flesh," however, rhetorically permits the abortion. The performative utterance creates a loophole. Ligurio's description reveals that the hunk of flesh *only* matters on account of its signification.

Lucrezia then is the last to agree to the scheme due to her belief in the importance of God's blessing as well as her credence in a body's inherent worth or virtue—in this case, the worth of her chastity in God's eyes. The biblical bed trick emphasizes the import of virtuous bodies and bloodlines; and yet Lucrezia appears to recognize that even if those around her cite this tradition, she would compromise her virtuous body by participating in it. Other characters claim that her deep desire for children should promote a favorable attitude toward this plan. Nicia tells Siro, "E non è dire che la non abbi caro di fare figliuoli, che la ne ha più pensiero di me" (And one shouldn't say that it isn't dear to her to have children, and she has thought of them more than me) (2.5, 85).[62] Yet Nicia is hardly a trustworthy source, and he likely projects his own desire onto his wife, just as he has imposed it upon everyone else around him. Lucrezia may want a child, but she never explicitly says so. She does, on the other hand, plainly express reticence about the moral compromises involved in the *means* of begetting this child. When speaking to her mother and Fra Timoteo about the scheme, she exclaims, "Di avere a sottomettere el mio corpo a questo vituperio, ad essere cagione che un uomo muoia per vituperarmi; che io non crederrei, se io fussi sola rimasa nel mondo, e da me avessi resurgere l'umana natura, che mi fussi simile partito concesso" ("To give up my body to this dishonor and to cause the death of the man who dishonored me—even if I were the last woman and earth and it fell to me to start the human race all over again. I don't think I would be allowed to do it") (3.10.96). Here, Lucrezia explicitly questions the logical inconsistency that characterizes the biblical bed trick, at least as it would be applied in this case. Namely, a bloodline's import is deemed so important that it warrants a bed trick; yet it requires that the woman compromise her chastity, which God has likewise deemed morally valuable. Even in the case that God has divinely sanctioned this bedding as God did for Lot's daughters to repopulate the world—preserving maybe the most important of all bloodlines, that of the human race—the possibility of bodily dishonor renders its moral status suspect.[63] Lucrezia twice emphasizes her belief that this compromise would be a bodily one and, consequently, deeply personal: first to submit "mio corpo" (my body) and again "vituperarmi" (to vituperate/shame me). To submit her body is no small thing if, according to God's design, her bodily virtue has bearing on the status of her soul.

Fra Timoteo seeks to change Lucrezia's mind and solve this conundrum by revising the biblical bed trick so as to disregard bloodlines and divorce the soul from the body. He thus recasts the biblical bed trick as a way to procure not a specific child or bloodline deemed valuable by God, but rather a soul for God, lacking any specific traits. This revision is subversive for it evacuates the biblical bed trick of its central message: God's investment in specific bloodlines. Of course, Fra Timoteo ignores the bed trick's original purpose because it suits his own purposes to do so—having already taken money from Ligurio in return for his participation in the scheme. Instead, the friar will emphasize God's blessing for the sake of a child, though seemingly any child would do. As he explains to Lucrezia, the good of her child is a soul for God: "Che voi ingraviderete aquisterete una anima a Messer Domeneddio" ("You will become pregnant and gain a soul for God Our Father") (3.11, 97). Or, in the friar's mind, Lucrezia has two "ends" that should motivate her to acquiesce. He expounds, "Il fine si ha riguardare in tutte le cose. Il fine vostro è di riempire una sedia in paradiso, accontentare vostro marito" ("One must consider the end of every action: your goal is to fill a seat in paradise and to satisfy your husband") (3.11, 97). Although the soul of the child matters, the body that houses that soul might not. Both of these ends simply require a person to fill a seat, and Lucrezia can accomplish this end by bedding a person other than her husband. As Mera J. Flaumenthaft suggests, the conclusion functions as a kind of parody of the holy family, only with Callimaco instead of God acting as the father (54). And thus even the paramount bed trick, the miraculous conception of Christ, is not immune to society's pragmatism. The biblical bed trick is subjected to rhetorical manipulation, and further evacuating the import of blood's substance renders the entire ideology hollow.

Lucrezia's compromise of her bodily integrity, however, is not easily won. Although her thoughts and motives are notoriously difficult to unpack, she does not seem particularly happy about this pragmatic change in worldview. Those scholars who imagine that Lucrezia willingly embraces this new state of affairs based on the "signore, padrone, guida" ("lord, master, guide") speech (5.4, 113) should remember that this speech is recounted secondhand by Callimaco. And Callimaco, a less than reliable narrator, is incredibly interested in hearing precisely these words from Lucrezia to satisfy his own ego.[64] Richard Andrews goes so far as to suggest that "Machiavelli is not actually very interested in what Lucrezia thinks or feels about the transformation that she undergoes" ("Anti-feminism" 25–26). And he further claims, "I do not think that any real attention is paid to the possibility that a Lucrezia in real life, as opposed to a stage Lucrezia, might actually feel a personal repugnance and outrage at the rape which is effectively practiced on her. The principal significance of her moral collapse is the dishonour done to her husband and to patriarchal values" (26).

Yet it seems to me that patriarchy, though one devoid of a biological foundation, is precisely what is being upheld. And Lucrezia, by contrast, is now deprived of the only thing over which she had some measure of control, namely her body.[65] To obtain a better sense of Lucrezia's feelings about compromising her integrity, one should examine her only appearance onstage postcoitus (5.5). Nicia, after complaining that Lucrezia seems a little testy, marks his surprise since the previous night she seemed "mezza morta" ("half dead"). To this charge she responds, "È la vostra cortesia" ("That's thanks to you") (5.5, 114). One might say that part of Lucrezia did die, the chaste body she spent her entire life protecting. Or, alternatively, the part of her died that believed that her bodily honor meant something. Fra Timoteo upon greeting her, wishes, "Che Dio vi dia a fare un bello figliuolo maschio!" ("May God grant you a handsome son!") (5.6, 114). The friar's blessing appears to be the only thing that disagreeable Lucrezia agrees to that morning, marked by her reply, "Dio el voglia!" ("God willing!") (5.6, 114). The child, supposedly conceived by God's decree, is the only thing to which all parties agree. Lucrezia finally concedes, citing (though maybe ironically) God's blessing, but, like Boccaccio's Giletta, she has no real assurance that this is true and a strong suspicion that it is not.

Lucrezia's ultimate concession might be explained by yet another force besides Fra Timoteo: her mother. Sostrata's central lesson for Lucrezia is that it is always possible for a person to return to social a-signification, losing the status of personhood altogether. Sostrata urges, "Non vedi tu che una donna, che non ha figliuoli, non ha casa? Muorsi el marito, resta com'una bestia abandonata da ognuno" ("Don't you understand that a woman who doesn't have children doesn't have a secure home? If her husband dies, she is left like an animal, abandoned by everyone") (3.11, 97). Sostrata's reality check operates on two levels. On a superficial level, Sostrata highlights the harsh reality for a woman who does not produce children.[66] But beneath that, her description of a woman's transformation into an animal elucidates something more, namely that "ognuno" (everyone) surrounding a person is what determines their status as human, as something other than animal. Reminiscent of Nicia's earlier comment about the barking dog, Sostrata implies that without this social recognition, a person is nothing more than a beast, and the nature of said beast does not matter anymore than does Ligurio's imaginary chunk of unborn flesh. This reasoning, in conjunction with Fra Timoteo's promise of absolution, is the last scene prior to Lucrezia's acquiescence. Sostrata's plea appears to be the final key to Lucrezia's acceptance, prompting the realization that her bodily integrity will not matter if she is cast out and her society fails to recognize her chaste body.

La mandragola's scheme to impregnate Lucrezia effectively evacuates the biblical bed trick of its justification. The body of the heir and his blood should matter to sanction this deception, although here consanguinity clearly makes no difference and the child serves a functional role. The body is simply a locus for this signification in a world wherein appearances, not some nature of God's design, ultimately dictate one's place. Lucrezia's conservative belief that her bodily integrity/chastity dictates her honor, independent of rhetoric or social recognition, has no place in this world. *La mandragola* disturbs blood's ideology and its purported divine backing. It exposes that the sociopolitical status quo of "Firenze vostra" (your Florence) has a faulty premise, thereby undermining the social groups who depend upon a vocabulary of consanguinity to both establish and maintain social power (prologue 73).[67] The morally questionable characters of *La mandragola*'s Florence acknowledge the fluid and rhetorical value of this discourse in the maintenance of social boundaries. Yet the evidence of some reticence on Lucrezia's part highlights a complicated stance toward blood's ideology. As for Corneille's Chimène, blood's conservative ideology—in this case, the bodily foundation of chastity—holds for Lucrezia meaning and agency albeit in a rather limited respect. In short, it affords her some small power over her own body. In light of this new, pragmatic, and fundamentally rhetorical world, Lucrezia is seemingly forced to relinquish this autonomy. Although her motives (unlike Helena's) are clearly ambiguous, I remain unconvinced that Lucrezia's acceptance is a happy one, even if the young lover Callimaco may be some recompense. Regardless of Lucrezia's wishes, she ultimately submits to society's reality: the status of herself (her honor) and her children as socially signified, entirely divorced from bodies.

JUSTIFYING THE BED TRICK AND BLOOD EXCHANGED IN THE DARK

Both Shakespeare's *All's Well That Ends Well* and Machiavelli's *La mandragola* allude to this biblical bed trick to interrogate both God's sanction and its larger message, namely that it is blood's intrinsic worth that determines worthy persons. Tamar can deceive if she must in order to perpetuate the bloodline of the House of Israel; Giletta may scheme to conceive heirs for her reluctant husband and win his love by verifying their new consanguinity. Although just as the larger framework of Boccaccio's *Decameron* places pressure on this moral, these plays expose that blood has neither the worth nor the agency ascribed to it. Blood neither determines affinity or character (as with *All's Well*), nor the parameters/coherence of kinship structures (as in *La mandragola*). Instead, as Shakespeare implies and Machiavelli asserts, the biblical bed trick tradition is simply another tool in a rhetorical arsenal used to accomplish desired ends: an exit strategy for Bertram, a consummated marriage for Helena, sex for Callimaco, and a child and social recognition for Nicia. And if love and worth are not biologically determined and these features are rather socially signified, then this was clearly not of God's design as Tamar's story would suggest. In short, the biblical bed trick story, by upholding the value of particular bloodlines, served the early modern elite. And when these plays allude to this tradition, they tarnish blood's ideology and thereby undermine the elite's foundation for power. These plays leave God's endorsement of this system open to interrogation. This skepticism then might allow for a better, alternative system of social organization that does not naturalize, and render permanent, these differences and distinctions.

And yet, though rejecting blood discourses would appear to benefit both heroines, upon considering the plays, pathos for the heroines, the playwrights' stance toward this ideology becomes more complicated. Exposing blood as rhetorical would serve both heroines in the long run. In the case of *All's Well*, it would deprive Bertram of an excuse to object to Helena's love, for they would no longer be biologically incompatible. And for *La mandragola*'s Lucrezia, it would alleviate the generally oppressive strictures of female conduct. However, *All's Well's* uncomfortable, unromantic close implies that Helena has realized the sad, disappointing truth about her invocation of blood. It did not work to establish the love that this system promised. Helena may be misguided, but it is nonetheless unfortunate that she, in spite of all existing social barriers, found a way to satisfy her desires only to be disappointed. Ultimately, there was no higher power that could engender Bertram's love. Similarly, though Lucrezia might benefit from the bed trick scheme, affording her both a young lover and a child to ensure her social position, *La mandragola* encourages the audience to feel profoundly uneasy about a society that coerces Lucrezia without regard for her own feelings about the loss of her honor. After all, if she accepts that her bodily integrity does not matter, she renounces the little agency that she once had in this context, if nowhere else. Those in power have changed the rules, using her body as best befits the needs of patriarchal structures in that moment.

Women, it seems, can be collateral damage of both blood's ideology and, potentially, the new system that would replace it. Both Shakespeare and Machiavelli demonstrate not only the difficulty of relinquishing older vocabularies, as we see with the Countess and King of *All's Well*, but also the problems and losses for those disenfranchised—those who find a degree of agency in an otherwise oppressive social system. Blood's rhetorical status, although true, does not benefit Helena or Lucrezia in the world of these plays. Instead, those who already have power find ways to achieve their ends. The King troubles the existing discourse just enough to compel the marriage to recompense Helena; Callimaco beds the woman he desires; Fra Timoteo gets his payday. Recognizing the rhetoricity of these discourses as opposed to their naturalization has the potential to serve the largely powerless like Helena and Lucrezia (and even Lavatch), but it does not guarantee it. And both *All's Well* and *La mandragola* suggest that those in power can simply find new discourses to reassert their power. Interrogating the existing system here in hopes to dismantle it is only half the battle. In sum, thinking otherwise is not only difficult but might warrant attentive rethinking.

4

"Commend Me to My Kind Lord"

LOVE IN DIFFERENCE

This chapter's title derives from Desdemona's final words. After having been strangled, Desdemona requests this of Emilia (5.2.123), begging the question: Why is *kind* her last characterization of Othello? Our ordinary gloss—"having or showing a benevolent, friendly, or warm-hearted nature or disposition; ready to assist, or show consideration for, others; sympathetic, obliging, considerate" (*OED* 10a)—seems tragically misplaced. Of course, *kind* can also mark something or someone as of good quality, designating characteristics associated with nobility such as generosity, courtesy, graciousness, and the like (*OED* 6a, 6b). And it is possible that Desdemona seeks to reassert her husband's nobility or worthiness, saving him from defamation or slander, as a final act of love and devotion. Yet, for our purposes, I suggest that we import yet another meaning of *kind* one which I use throughout this book, as in of one's own kind as well as the now obsolete adjective denoting "related by kinship; of one's own kin or people" (*OED* 5). For our contemporary denotations of friendliness or affection appear to derive from the following definition: "Having or showing a normal, natural affection for one's family and close relatives; well-disposed towards one's kin" (*OED* 7a). In short, kindness is about treating others *as kin*. Thus, the central tenet of blood's ideology appears built into this very word: affection springs from kinship, liking based in likeness. So returning to Desdemona, the question remains, why call Othello kind? I submit that this is Desdemona's attempt to claim Othello as her kin, as with Helena, to render, at least rhetorically, the man of her liking her "like."

Having explored the ideology of blood thus far in the context of social class or narrower kinship groups, this chapter marks a departure from those before it. Both plays under consideration, Shakespeare's *Othello* and *The Merchant of Venice*,[1] address blood and difference in kind in more legible realms, conceived either as religious, ethnic, or racial (chromatic/somatic). And the more marked axis of difference heightens the anxiety around these romantic pairings, illustrating racist logics that may seem quite different from the discussion thus far. Arguably, the difficulty of perceiving differences in blood, as highlighted by the King of *All's Well*, grants blood's ideology a measure of flexibility. Namely, it permits the nobility to assert biological difference and distinction from the hoi polloi without any obvious visible marker to which to point.[2] Yet it likewise allows an individual to claim a bodily transformation upon becoming "one flesh"; one spouse becomes similar to the other as bloods mix despite the absence of any perceptible change—benefitting a character like Helena who claims just such a transformation. As is the case for *The Merchant of Venice*, Jewishness as an ethnic category bears some similarity to social class since it is likewise less visible. Janet Adelman and others have noted that although Jews were presumed to be visually distinct (darker skin, curly hair, big noses), the Fourth Lateran Council demanded markers in clothing precisely because bodily markers were insufficient and unstable. And the fact that Jessica, like Helena, presumes a conversion based upon her marriage highlights the similarity between these discourses. Of course, the comparison grows more complicated in the case of Othello and Desdemona.[3] Nonetheless, both plays stage a supposed impossibility according to blood's ideology, namely an attraction between *unlike* persons, an assumption that extends to other racial thinking in the period.

Thus, while examining blood's ideology at work in terms of class difference has been central to the contribution of *Thicker Than Water*, I hope to illustrate the translation or application of this historical discourse to the realm of race—that which lurks beneath every mention of "kind" throughout this book. This bloody discourse, which informs the assumptions built into Desdemona's last request, can help piece together a larger picture of racist ideologies, underlining a key element of these discourses: the policing of affinities. This chapter likewise marks a departure from the comparative methodology of those prior in order to convey this lens to otherwise well-trodden territory. Namely, it brings the framework of blood's ideology to the place where much of the debate about early modern race has taken place: Shakespeare's Venetian plays.

Conceptualizing race in the early modern period has been a source of contention, to put it mildly.[4] Fueled by scholarship placing the advent of contemporary race (understood as chromatic difference) in the eighteenth and nineteenth centuries,[5] some scholars have opted to trace other frameworks for understanding difference making in the period (e.g., Julia Reinhard Lupton on religion in *Citizen-Saints,* and Emily Bartels on national or ethnic diversity).[6] Others trace the origins of racial awareness in the construction of English, Christian, White identity to the early modern period, even if the fully "scientific" renditions of race as articulated by Kant's "ethnocontinental tetragon" had yet to emerge (e.g., Mary Floyd-Wilson's geohumoralism in *English Ethnicity*[7] or Ian Smith's "barbarism" in *Race and Rhetoric*).[8] Particularly important contributions have been made outlining the foundations of anti-Black racism by Kim Hall,[9] Peter Erickson, Ian Smith,[10] Lynda E. Boose,[11] Farah Karim-Cooper,[12] Ambereen Dadabhoy,[13] Arthur Little's "Re-Historicizing Race," and Elizabeth Spiller.[14] And for many scholars, blood has been central to the story of early modern race. They point to the usage of the word *race* in Shakespeare, which more often refers to ancestry than it does to either skin color or religion, designating "a notion of lineage that confers a specific social rank" (Gil Harris 201–2).

As underlined by Sujata Iyengar, race as chromatic difference can be more or less construed as a difference in blood, made evident in George Best's commentary on the book of Genesis (1578).[15] Iyengar summarizes this logic as follows: "Having identified black skin with Noah's curse, Best concludes that blackness must be an 'infection of blood' (7:264), transmitted through 'lineall discent' from father to child (7:263). Best's account links black skin to polluted and polluting 'sexual transgression,' implicitly associating the Ethiopian father in England and his 'faire' wife with the sexual disobedience of Cham and his spouse in the Ark" (9).[16] Thus, aligned with the work of Iyengar, Jean Feerick,[17] Ania Loomba,[18] Margo Hendricks,[19] Lara Bovilsky,[20] and Patricia Akhimie, racist discourses appear to be an extension of the logics central to blood's existing ideology and not some wholly different phenomenon.[21] And though this story of race's conceptual transformation persuades me personally, for the purposes of this project, I am less interested in litigating *what* race was in the period; instead, I want to outline *how* these racial discourses worked, and consequently, whether they could be disturbed or disrupted. To borrow the appeal of Ania Loomba and Jonathan Burton, "it makes less sense to try and settle upon a precise definition or indeed to locate a precise moment of origin for racial ideologies than to delineate the ways in which they order and delimit human possibilities through a wide range of conjoined discourses and practice" (2). Or, put another way, I see my work here as outlining the ideological construction of difference in the early modern period, one of the many practices of "race-making" (a term I take from Ayanna Thompson).[22]

Thus, in the context of this book, *Othello* and *Merchant* are of interest because difference in "kind" as a romantic obstacle acts as yet another limit case for blood's ideology. When examined through this lens, one explanation for the divergent ends for these couples, though certainly not the only one, is the degree to which they are susceptible to the ideology that rejects their love's possibility.[23] These plays ask, does love (both romantic and familial) only accommodate likeness as this system promotes? Can like only love its like, or could one love difference? And might this conception of love be preferable?

The answer to this question in *Othello* is fraught. Many of the characters—from Brabantio's claim that Othello must have used magic to sway Desdemona from her nature, to the linchpin of Iago's argument that Desdemona must be having an affair with Cassio—believe that Othello and Desdemona's love is impossible due to a fundamental biological incompatibility, conceived in terms of blood that denotes a difference in kind. And within this framework, the relationship is doomed for either (1) Desdemona will "return to her senses" and find her like match (such a one as Cassio) or (2) her nature is so errant that she could never be a faithful, loving wife (for if she could, she never would have married such a one as Othello). And like the readings of so many critics, Othello's downfall is the product of having internalized this ideology that makes similitude the prerequisite for love. But Othello imbibing Iago's poison in the "temptation scene" (3.3) is only part of the story, for the tragic end of the couple was initiated beforehand. Akin to so many other protagonists discussed thus far (Romeo and Juliet, Chimène, Helena, etc.), their desire to work within the existing ideological confines portends their tragedy. Othello and Desdemona both appear to conceptualize and rationalize their love by seeking points of similarity and aiming for a love of concord rather than embracing their differences, the latter of which might offer the possibility for a more favorable outcome.[24]

In contrast to the Othello-Desdemona relationship, *Merchant of Venice* proffers an alternative model for love—at least for one of its pairs. As Frank Whigham says in reference to the central romantic couple, Portia and Bassanio, "carefully restricted social and sexual intercourse played major roles in this ideology of harmony, which presented itself as a natural model of reciprocal interaction" (115).[25] Blood's ideology—a naturalized model for dictating affinity between persons and thereby encouraging endogamy—contributes to this picture of harmony and stability by perpetuating likeness over difference (and potentially discord). For instance, the Prince of Morocco appeals to this logic in an effort to woo Portia by demonstrating their similarity. He petitions, "And let us make incision for your love. / To prove whose blood is reddest, his [a competitor's] or mine" (2.1.6–7). He disregards his chromatic difference by proposing an alternative material that would authenticate his similitude and thus his worth, namely a nobility of blood.[26] His appeal fails, but Shakespeare's *Merchant* also presents an alternative to this vocabulary that has at least a chance of success.

Merchant critiques blood's ideology by first exposing its mechanism to the audience and then proffering an alternative, and potentially disruptive, mode of *thinking otherwise*. First, it invites the viewer to see this ideology's rhetorical nature with respect to Jessica's identity. Nearly all of the characters cite differences in her blood toward conflicting ends and none with much success. Second, the romantic relationship between Jessica and Lorenzo evidently does not function according to the logic of likeness, and in contrast to Othello and Desdemona, they love one another *as* different rather than despite difference. They likewise seem to embrace potential discord or friction that might occasionally surface when one is not in complete accord with one's partner. In this respect, *Merchant* offers one of the greatest critiques of blood's logic thus far; it entertains a love that falls outside of this ideology, one based in difference that can potentially disrupt the status quo.

TRIUMPHANT IDEOLOGIES, SEARCHING FOR SIMILITUDE IN *OTHELLO*

Analyses of Shakespeare's *Othello* in light of the protagonist's chromatic difference or blackness effectively ask one question: Is the play racist, or, more precisely, does it endorse the racist discourses that it presents? For some, the play emerges as an endorsement of racial stereotypes: either Iago's scheme has slowly unearthed Othello's "innate savagery,"[27] or Othello must assume Christian ethics and White manners to be deemed virtuous,[28] or Othello's exceptionalism further validates the contemporary stereotypes of Black Moors,[29] or the symbolism that equates Black with evil or sin persists despite Othello's virtue.[30] By contrast, others have maintained that the play critiques these very discourses, for either Othello's weakness is not racial but "human," what stereotypical "Moorish" traits apply to Iago not Othello, or Othello is clearly the victim of White racism and is treated as such.[31] Most readings fall somewhere in between, illustrating that the play asks the audience to remain skeptical about its judgements[32] or that *Othello* illuminates *how* racial discourses function, if not critiquing them outright.[33]

Yet other scholars highlight nonchromatic aspects of Othello's difference given the complicated denotation of the term *Moor* in the early modern period. Some emphasize Othello's religious difference as a Muslim,[34] or they gloss his ethnicity as explicitly Turkish in either the context of the contemporary Ottoman threat[35] or with respect to Venetian politics.[36] Chakravarty suggests that we revisit the play and Othello's "circumcisèd dog" by examining the relationship between race, natality (a corporeal and spiritual birth or rebirth), and theology. In this renewed frame, circumcision is about both religious and racial incorporation: it signifies a physical mark of difference from most but also belonging to a community. Chakravarty argues that these discourses play an important role in the transition from sanguineal theories of race to somatic (those based on some other physical morphology) to chromatic (those based in the color of skin pigmentation) ("Race, Natality, and Biopolitics"). According to most of these arguments, the play negotiates early modern anxieties about "the accommodation and alienation of mixed populations, mixed marriages, and mixed blood, and . . . otherwise negotiate[s] the extraordinary volatility of cultural change and exchange" (Bartels 17).

My own reading, like many others, ascribes Othello's tragedy to his internalization of a racist discourse, though it differs in the particulars.[37] This racist discourse stipulates a biological incompatibility between himself and Desdemona such that there must be something wrong with her, as her father, Brabantio, puts it, a "treason of the blood!" (1.1.167).[38] In this extension of blood's ideology, Desdemona's perceived fault that is responsible for her love is not a misogynist libel; rather, it is a problem with her individual person such that her nature would err from its normal course, not act in accordance with her kind, namely to be attracted to one *like* herself. Within this ideological framework, therefore, Desdemona is caught in a no-win situation for either she must return to her nature and find a suitable partner in one like Cassio, or she must have a "will most rank" for inclining to someone different in kind like Othello (3.3.236). This explication, of course, is well-trodden territory.[39] Yet this interpretation is only part of the story. Both Othello and Desdemona conceptualize their love by seeking points of similarity and striving for a love of concord, as opposed to embracing differences and the inevitable possibility of discord.[40] These attitudes—the pair's inability to escape the ideological framework that equates affinity and love with similarity—contribute to the tragic outcome.

Regardless of whether or not we deem the background assumptions of *Othello* racist in the modern sense, a ubiquitous belief that the two protagonists are different in kind (a sanguineal theory of race) and, therefore, are biologically incompatible permeates the play's Venice, reminiscent of the animal metaphors used throughout early modern treatises on marriage and nobility.[41] Iago's opening alarum to Brabantio that his daughter has eloped—"Even now, now, very now, an old black ram / Is tupping your white ewe!" (1.1.87–88)—emphasizes chromatic difference but does not have the appropriate effect. Thus, later in the scene, Iago doubles down, threatening instead an interspecies coupling with a Barbary horse that will engender monstrous, neighing progeny: "You'll have your daughter covered with a Barbary horse; you'll have your nephews neigh to you, you'll have coursers for cousins and jennets for germans!" (1.1.109–12). Brabantio then expresses distress at what he perceives as a bestial coupling that threatens his bloodline. In so doing, he clearly articulates the supposed rules of nature, or what I have called the premises of blood's ideology, on several occasions. He calls Desdemona's elopement with Othello a "treason of the blood!" (1.1.167). Blood is likewise the medium of Othello's supposed witchcraft, wherein "mixtures powerful o'er the blood" were "wrought upon" Desdemona (1.3.105–7). For within this framework, Desdemona's elopement is impossible without witchcraft: "For nature so preposterously to err / Being not deficient, blind, or lame of sense, / Sans witchcraft could not" (1.3.63–65). Brabantio repeats this logic only a few lines later, insisting that love or affinity accords with like natures, and thus Desdemona loving Othello is errant:

> she in spite of nature,
> Of years, of country, credit, everything,
> To fall in love with what she feared to look on?
> It is a judgement maimed and most imperfect
> That will confess perfection so could err
> Against all rules of nature. (1.3.97–102)

This belief in the "rules of nature" is so trenchant that, once it becomes clear that witchcraft was not used, the conditional clauses of Brabantio's statements take root. Namely, Desdemona's nature must have been "deficient, blind, or lame of sense," and Desdemona's "perfection" must have erred. Otherwise stated, this ideology is so pervasive that from this vantage point, one could never call the "rules of nature" into question, and instead, Desdemona's nature becomes subject to scrutiny.

Iago, of course, picks up this language and famously uses it against Othello by placing Desdemona in a double bind that was only intimated in Brabantio's proclamations. One possibility is that Desdemona's blood will compel her to return to her kind, which Iago promises Roderigo when first starting the rumor of Desdemona's affair. Iago assures him, "When the blood is made dull with the act of sport, there should be, again to inflame it, and to give satiety a fresh appetite, loveliness in favour, sympathy in years, manners and beauties, all of which the Moor is defective in" (2.1.224–28), and consequently, her "very nature will instruct her in it and compel her to some second choice" (2.1.231–33). Her blood will be "inflame[d]" by "sympathy" finding its *like* in years, manners, beauties, and, what goes unsaid here, kind. But then Iago articulates the second option that Desdemona has something distinctly wrong with her, for "If she [Desdemona] had been blest she never would have loved the Moor" (2.1.250–51). In the temptation scene, this logic comes to a head. Where Othello's initial thought is that "she had eyes and chose me" (3.3.192), his reasoning pivots from the central issue as one of some visible difference to a deeper biological incompatibility: "And yet how nature, erring from itself" (3.3.231). Iago springs on the moment to launch into the clearest recap of the principle of affinity from blood's ideology; and if the ideology is held fixed, then Desdemona remains with only the two options noted above, neither of them flattering to her nature nor favorable to her relationship with Othello:

Ay there's the point: as, to be bold with you,
Not to affect many proposed matches
Of her own clime, complexion and degree,
Whereto we see, in all things, nature tends—
Foh! one may smell in such a will most rank,
Foul disproportion, thoughts unnatural.
But pardon me, I do not in position
Distinctly speak of her, though I may fear
Her will, recoiling to her better judgement,
May fall to match you with her country forms,
And happily repent. (3.3.232–42)

The first half of this speech could easily be pulled from marriage treatises of the period that advocated for "matches" in "clime, complexion and degree," in short, the traits carried in blood.[42] And yet, despite his promise that he is not speaking of Desdemona "distinctly," the argument here is not simply that she will recoil to her better judgement but that she is "distinct" or different from what her nature ought to be, in this case, "foul" or "unnatural."

These two divergent characterizations of Desdemona stick with Othello even to the scene of her murder. Othello calls her "Thou cunning'st pattern of excelling nature" (5.2.11). On the one hand, for all of his enmity, Othello still believes in Desdemona's superiority and points to this as the reason for her supposed treachery. She might be a cunning pattern or *creation* of an excelling nature. But alternatively, one might gloss this description as the cunning pattern, in the sense of an *imitation,* of excelling nature that she is not. She is not blest as one would expect but has a "will most rank" as everyone, even her father, has said. Unfortunately for Desdemona, it does not much matter which of the two options Othello settles upon because she is found wanting and, thus, must die. For the beauty of the double bind is that there is no third option. If the ideology is held fixed and all assume that it is true, then there is no room for a love that does not fit the model. This is why the fact that neither Othello nor Desdemona reject the principle that love increases with similitude is so tragic. They aim to fit their love of difference into a promise of similarity, a square peg into a round hole.

Initially, diversity or difference seems to spark their love, though it is short-lived. Othello's opening account of their love story highlights Desdemona's love for that which is different, being taken in by his "passing strange" and "wondrous pitiful" story (1.3.161–63); and he concludes, "She loved me *for* the dangers I had passed / And I loved her that she did pity them" (1.3.168–69; italics added). Othello's difference or strangeness is highlighted by its repetition as is the emotion of "pity," an emotion that seems to recognize difference. Related emotions like sympathy or empathy presume a likeness between the subject and object,[43] denoting an ability for the emotional subject to occupy the object's position and emotional state, to feel *as they do*. But pity is explicitly other regarding, a tenderness or concern for another's misery, which does not presume any specific kind of relationship between the emotional subject and object (*OED* 2b).[44] While not obviously the most promising foundation for romantic love, Desdemona's pity at least acknowledges their differences and does not presume that she can understand his former life, the "dangers [he] had passed." And even if she cannot understand them, she can nevertheless love him *for* them. This love for difference has the potential to break the ideological assumption that love is only possible when the two parties are matched in "clime, complexion and degree."

Yet Desdemona's appreciation of difference quickly falls away; in the same scene, her rhetoric glosses over any difference. She rather employs a language of likeness and parity between herself and Othello in order to fit her love within the confines of the existing ideology.[45] When Desdemona speaks about why she wishes to accompany Othello to Cyprus, she highlights what she deems similar about their natures, especially when put in context with Brabantio's earlier adulatory description of Desdemona's character traits. She explains her love for Othello as follows:

> My heart's subdued
> Even to the very quality of my lord:
> I saw Othello's visage in his mind
> And to his honours and his valiant parts
> Did I my soul and fortunes consecrate. (1.3.251–55)

Desdemona's is a very different account of love from Othello's, based on similarity not strangeness. She loves his intrinsic worthiness, which is, she implies, *not* the same as his visage that is visibly different from her own. She glosses her love as one of similarity. As Edward Berry notes, "The implicit denial of physical attraction shows that Desdemona tries to separate

Othello's essential humanity from his appearance, but it also shows that she is sensitive to and disquieted by the insinuations that there must be something unnatural in such a love. She does not say that she found Othello's blackness beautiful but that she saw his visage in his mind" (321). This impulse is quite similar to the Duke's rhetoric when he assures Brabantio that his "son-in-law is far more fair than black" (1.3.291). This insistence by both Desdemona and the Duke, Othello's greatest allies, speaks to the limitations of society's ideology; namely, Othello cannot be embraced for his difference but must be *made similar* in order to find him worthy, to make sense of, and make acceptable, this pairing.

Both Desdemona and Othello then search for sympathy and concord in their relationship to fit it within the existing ideological framework. And their harping on agreement and harmony leaves very little room to accommodate difference or diversity in opinion. For instance, during the reunion of Othello and Desdemona in Cyprus, Othello declares,

> I cannot speak enough of this content,
> It stops me here, it is too much of joy.
> And this, and this the greatest discords be [*They kiss*]
> That e'er our hearts shall make. (2.1.194–97)

Herein lies a problem. Othello's "content" or happiness might, in fact, be too much joy. If a kiss is their greatest discord, as indicated by the stage direction, then this does not leave a great deal of room for difference in their "hearts" or opinions; and being occasionally out of tune or not perfectly in accord could be acceptable, even desirable (as we see with Jessica and Lorenzo). After all, Othello effects perfect concord simply by giving in to Desdemona—implied in the repetition of "I will deny thee nothing" (3.3.76, 83)—at the cost of his own will. Thus, when sympathy is not reached or worse yet Desdemona affirms feelings of sympathy for another, the relationship quickly falters. Here, Desdemona proclaims that she suffers with (sympathizes/empathizes with) Cassio rather than pities him—"he left part of his grief with me / To suffer with him. Good love, call him back" (3.3.53–54). Desdemona thus unwittingly makes a case for her affinity for Cassio based on likeness, thereby feeding Othello's jealousy. The language's ambiguity creates further issues. "Good love" is clearly a pet name for Othello as Desdemona addresses him in the following imperative. But the epithet could likewise be interpreted as the subject for "him" (Cassio) who is the "good love" in contrast to her "errant" love for Othello as cast by blood's ideology (and as Othello fears).

The couple's desire for similarity and accord then has its swan song as the two echo each other when discussing Desdemona's "moist" hand. First glossed as an indication of a "young and sweating devil" (3.4.42), full of lust so it would seem, Othello then applauds the hand for its honesty, purportedly in contrast to Desdemona's deception: "'Tis a good hand, / A frank one" (3.4.43–44). Desdemona subsequently completes his line with agreement—"You may indeed say so / For 'twas that hand that gave away my heart" (3.4.44–45)—which he also follows with a kind of assent: "A liberal hand / The hearts of old gave hands/ But our new heraldry is hands, not hearts" (3.4.46–47). This scene of consensus, fraught as it is, is the last of the couple's discourse without discord; for in the remainder of the scene, they proceed to talk past each other. Desdemona sues for Cassio, and Othello harps on the handkerchief (3.4.87–100). As the play proceeds, Desdemona repeatedly voices confusion about Othello's meaning, asking "My lord?" (4.1.226, 233, 237). In this repetition she both articulates her confusion and interrogates his identity, which she is distressed to no longer understand.[46]

But even as their relationship deteriorates, Desdemona makes excuses for Othello by using the vocabulary of the very ideology tearing them apart. She insists, "My love doth so approve him," which can be glossed not only as her love commending or speaking well of Othello but also as her assessment of his worth (4.3.17). Otherwise stated, Othello must be worthy because she loves him; since worthy persons love other worthy persons, so must her love be indicative of his worth and similarity to herself.[47] Even Desdemona's last words are ones of assent or consent, as if she is guilty of her own murder so as not to be at odds with her husband. To Emilia's question about who is responsible for her death, she responds, "Nobody. I myself. Farewell. / Commend me to my kind lord—O farewell!" (5.2.122–23). Is this a request to be entrusted to Othello in her death or to speak her praises to Othello? Though unclear, in either case she remains in agreement with him to the last. Finally, as noted in the opening of this chapter, her final characterization of Othello as "kind" speaks to casting him as alike, as her "kind."

The cost of this search for likeness—trying to accommodate a relationship like Othello and Desdemona's within the existing ideological framework—is that, though it may briefly succeed, the system remains in place. This system that deems Othello different in kind cannot embrace that difference. It can embrace him only insofar as he can assimilate (becoming similar) to Venetian society, an all but impossible task, especially given his chromatic difference marking a sanguineal one. Just as a kiss cannot be the only discord admitted in a relationship between partners, there must be some room left for difference and divergent opinions. For failing to do so means reproducing more of the same, precisely the aim of blood's ideology.

PROBLEMATIC JESSICA OR LOVE IN DIFFERENCE

Just as with Desdemona and Othello, it is quite difficult to square the love between Lorenzo and Jessica with blood's ideology. Lorenzo should, by all accounts, dislike the Jewish, alien Jessica.[48] And in general, Jessica's character has proven to be extremely problematic for critics. Jessica is a seemingly superfluous addition to Shakespeare's narrative source,[49] her status as a foil for Portia is a contentious one,[50] and, finally, she often strikes audiences as simply unlikeable.[51] Moreover, if Shylock's character and Jewishness remains problematic,[52] this especially applies to Jessica in adjudicating whether or not she can convert to Christianity. Much like the characters onstage, critics are split in their opinion of Jessica's Jewishness as persistent ethnic identity. One camp sees Jessica's conversion as successful, either insisting on her similarity to the Christians[53] or suggesting that because she is a woman she is capable of conversion unlike her father.[54] The other party maintains that her status as a Jew lingers in her blood,[55] activating all kinds of anxieties about the English, Christian subject.[56]

To my mind, both critical camps are correct in part. As we see throughout drama's interrogation of other limit cases, Shakespeare's *Merchant* exposes blood as fundamentally rhetorical, invoked by opposing sides around the issue of Jessica's ambiguous identity for competing ends. Lorenzo, Gratiano, Salarino, and Solanio talk of her remarkably similar body to their own, especially in complexion (both as temperament and as skin color) and blood or kind, to provide proof that she was already Christian-like. This similarity then explains Lorenzo's attraction to her. Others insist on Jessica's persistent bodily difference and consanguineous connection to her father; these include Shylock, Launcelot, and, in certain instances, Lorenzo. Lastly, Jessica herself appeals to the logic of the one-flesh model (much like Helena in *All's Well*) in order to sever her blood's connection with her Jewish father and thus become a fully Christian wife. Jessica hopes to determine her own affinity and identity via sexual consort.[57] The play seems to accentuate, however, that Jessica remains a problem. The vocabulary of blood's ideology fails on all fronts since she cannot be absorbed into the existing network.[58] This failure is made evident by the fact that, contrary to her desires, Jessica never becomes fully like her husband, Lorenzo; the two are never fully in accord, never truly alike. *Merchant* thus highlights that the principle of commixtio sanguinis fails to dissolve difference, but Jessica and Lorenzo, nevertheless, love one another. And if love exists between them, then it is a love not explained by the logic that love springs from likeness. Their love is rather grounded in difference.

Several characters insist on Jessica's physiological "Whiteness" or gentility such that she would be sufficiently *like* Lorenzo; and, in so doing, they shore up this premise of blood's ideology. The most infamous is Salarino's wine metaphor with which this book began: "There is more difference between thy flesh and hers than between jet and ivory, more between your bloods than there is between red wine and Rhenish" (3.1.35–38).[59] As Janet Adelman puts it, "[Jessica] must be rhetorically constructed as fair by Lorenzo and the other Venetians in order to enable her gentile-fication and thus Lorenzo's theft of her" (*Blood Relations* 83). These characters emphasize her white skin or "fairness" and gentility to insist that her body determined her virtuous (i.e., Christian) character. They thus declare that she is not only *not* a Jew but also deserving of Lorenzo's love. In other words, they insist that Jessica was always already Christian in order to explain Lorenzo's affinity for her. As Elizabeth Spiller describes, "Lorenzo believes that he can see inside Jessica for precisely the reasons and in precisely the terms that Morocco would see Portia inside the golden casket. Jessica's fairness becomes the romance confirmation of her innately Christian soul" ("From Imagination" 154). Upon receiving Jessica's letter, Lorenzo is the first to insist upon her Whiteness and fairness as evidence of virtue: "I know the hand. In faith, 'tis a fair hand, / and whiter than the paper it writ on / Is the fair hand that writ" (2.4.12–14). The first "fair" presumably refers to Jessica's penmanship meaning "attractive," and the second designates the pallor of her skin, so pale that it "out-whites" paper. Lorenzo's doubling of "fair" and the likely hyperbolic comparison of her skin to paper betray an insecurity or need to overcompensate, a need to prove (or rhetorically fashion) her worthy by making the supposed similarity of her blood visible.[60]

And the insistence on Jessica's similitude is not limited to fairness, for the characters speak more generally of her ethnic classification or kind, her biological membership to Lorenzo's group by dubbing her "gentle," alluding to the Jew/Gentile distinction. Lorenzo sends Lancelot with the directive, "Tell gentle Jessica / I will not fail her" (2.4.19–20). Gratiano, too, just prior to Jessica's elopement exclaims, "Now by my hood, a gentle and no Jew!" thus prompting Lorenzo's proclamation of love: "Beshrew me but I love her heartily!" (2.6.51–52). In this moment, Lorenzo's weak oath "Beshrew me but" hints at Jessica's difference, for the "but" indicates that the "I love her heartily" is in contrast to something else, presumably the aversion that he should feel for a Jew. This proclamation of love is, therefore, prompted by Gratiano's assent to Jessica's likeness. Or in other words, with Gratiano's affirmation of Jessica's acceptability—due to her membership in the group of "gentles" rather than Jews—Lorenzo feels licensed to express his love, for it is now appropriate.[61]

The declaration of Jessica's likeness, which upholds the logic of affinity based in similarity of kind, also takes the form of asserting the difference between her and Shylock. Here, advocates employ a vocabulary of distinct species. Solanio comments upon the fact that Jessica must have had a temperament to prompt her running away: "And Shylock for his own part knew the bird was fledge, and then it is the complexion of them all to leave the dam" (3.1.26–28). The first gloss of Solanio's metaphor depicts Jessica as a bird capable of flight, punning on her flight from Shylock's home by pointing to a "complexion" or temperament that dictated this action. Yet this metaphor simultaneously reveals the vocabulary of species that often characterizes blood's ideology. Jessica's body is so different, so unlike her father, that this bodily difference dictated her leaving him to find others of her kind. Or, to further Solanio's pun, one might say "birds of a feather flock together."

Although Lorenzo does not use this vocabulary of species, he, too, underlines Jessica's temperamental difference from Shylock. After the elopement, Lorenzo rejects Jessica's consanguinity with her father by refusing to refer to him as such; when later recounting their love story, Lorenzo simply says that she stole from "the wealthy Jew" (5.1.15).[62] And he suggests that a temperamental difference from Shylock dictated her flight and her love for himself, continuing: "And with unthrift love did run from Venice / As far as Belmont" (5.1.16–17). Jessica's "unthrift" love for Lorenzo is evidence that she did not inherit the trait of miserliness, as he sees it, stereotypical of Jewry like her father. Otherwise stated, Jessica might not only be prodigal in her love for Lorenzo, giving too much, but this trait of prodigality makes it possible both for her to love him *and* for him to requite this love. In sum, in an effort to make sense of this ideologically inappropriate romantic match, this group of characters contend that Jessica must be like them: fair, gentle, virtuous, and, consequently, Christian. For the alternative would be to acknowledge that this tenet of blood's ideology does not hold or at least to admit that there must be exceptions to this law of nature. And by calling attention to the likewise ineffectual citation of this ideology by the opposing camp, *Merchant* invites the audience to entertain the alternative explanation, namely that love is not determined by likeness after all.

Given Jessica and Shylock's consanguinity, others cite another tenet of blood's ideology—that shared blood should indicate shared character and affinity—to insist upon Jessica's Jewishness and persistent difference. Shylock especially continues to lay claim to Jessica in these terms after her flight, exclaiming, "My own flesh and blood to rebel!" and "I say my daughter is my flesh and blood" (3.1.31, 34). The first exclamation expresses his surprise, for how could one's own flesh rebel against itself? If consanguinity dictates affinity and loyalty, how is Jessica's rebellion even possible? And the fact that Shylock even has to question what should be an impossibility according to the laws of nature reveals its failure. Shylock's second exclamation then is a protest, reflecting his need to forcefully reassert their consanguinity with "I say"—a telling turn of phrase since it intimates a performative utterance. Shylock's contention comes immediately prior to Salarino's famous wine metaphor, and the phrase "I say," though meant to be an intensifier, betrays the fundamentally rhetorical nature of both Shylock's protest and Salarino's retort. Ultimately, this is a fight over Jessica's signification because no individual claim to bodily determinism on either side rings true. Jessica seemingly cannot be assimilated into either of these camps, Jewish or Christian.

But it is not only Shylock who insists upon Jessica's difference; despite all of the attempts to pronounce Jessica like the Christian Lorenzo, even after their marriage Jessica remains different, distinct, and apart. For one, she is called the "infidel" upon her entrance to Belmont (3.2.218). Gratiano asks that Nerissa welcome this "stranger": "Nerissa, cheer yond stranger, bid her welcome" (3.2.237). Notably, though also a newcomer to Belmont, Lorenzo does not merit the label of stranger, only the "her," Jessica. Additionally, there is no indication that Nerissa welcomes the stranger, either in a verbal response or a stage direction. Even in an attempt at integration, this gesture emphasizes Jessica's difference and her special need of welcome or inclusion. Lorenzo too has difficulty completely divorcing Jessica from her familial line and thus erasing her difference. Looking to highlight the contrast between Jessica and Shylock, Lorenzo simultaneously reasserts their consanguinity. He says of Shylock,

> If e'er the Jew her father come to heaven,
> It will be for his gentle daughter's sake;
> And never dare misfortune cross her foot,
> Unless she do it under this excuse,
> That she is issue to a faithless Jew. (2.4.33–37)

Shylock's salvation or Jessica's misfortune both hinge upon their consanguinity, a connection between them that purportedly cannot be undone. Jessica will always be "issue to a faithless Jew" regardless of either her intrinsic merit or her conversion. Finally, Lancelot's (potentially) humorous exchange with Jessica regarding her incapacity to be saved accentuates the anxiety surrounding her body's persistent, intrinsic Jewishness. As Lancelot asserts, "the sins of the father are to be laid upon the children," absenting a "bastard hope" that she is not Shylock's daughter (3.5.1–2). Yet, true to Galenic form that posits a more or less equal inheritance from both parents, Jessica notes that not even this would be enough: "So the sins of my mother should be visited upon me" (3.5.11–12). To this Lancelot responds, "Truly then, I fear you are damned both by father and mother. Thus when I shun Scylla your father, I fall into Charybdis your mother. Well you are gone both ways" (3.5.13–16).[63] By virtue of the principles of consanguinity, Jessica's Jewishness cannot be escaped. Her only hope is to extend this logic and invoke the principle of commixtio sanguinis, arguing that her body can become *like* Lorenzo's in marriage.

Like Helena of *All's Well*, Jessica hopes that marriage and its consummation could make her like Lorenzo and thus a Christian; and in making unlikes alike, marriage could subsequently divorce her blood from her father's. Jessica insists during her exchange with Lancelot, "I shall be saved by my husband. He hath made me a Christian" (3.5.17–18). And despite his joke about Lorenzo's culpability for the rising price of pork, Lancelot does not explicitly deny Jessica's claim, for it is based on the same logic as his Scylla and Charybdis metaphor.[64] Jessica's profession that she will assume Lorenzo's blood, and thus his likeness, in the act of marriage is fundamentally no different from asserting her similarity with her parents also based in consanguinity. Jessica's appeal to blood's ideology to serve her own end, seems no different (or less valid) from those who insist on her necessary similitude or her persistent difference. It just so happens that all of these appeals are rendered suspect. She nevertheless insists that her love will make her different from her father, and, throughout the play, her love for Lorenzo emphasizes this act of separation or escape.[65]

Jessica's love, in fact, foregrounds Lorenzo's difference in conjunction with her desire to sever herself from Shylock, seemingly both in space and in body.[66] She first envisions her elopement as an end to her imprisonment, an escape from Shylock's "house of hell" (2.3.2). Lorenzo also reports that as part of her love letter, "She hath directed / How I shall take her from her father's house" (2.4.29–30). Lancelot's rhyme too—coyly hinting at the elopement to come—succinctly glosses Jessica's attraction to Lorenzo: "There will come a Christian by / Will be worth a Jewès eye" (2.5.41–42).[67] Jessica's attraction and Lorenzo's worth are explained by this difference, namely his status as a Christian and not a Jew. As Adelman remarks, "Though [Jessica's] escape from her father's house to her lover fits conveniently into the conventions of a romance plot, her speech is not of love-longings of a typical romance heroine: Lorenzo is invoked not as the solution to the problem of Jessica's erotic desire but as the solution to the problem of being her father's daughter" (*Blood Relations* 71). For, after sending Lancelot away with the letter for Lorenzo, Jessica does not pine away for her love as we would expect in this moment. She instead speaks of her father, their problematic consanguinity, and her hope to foster a new connection:

> Alack what heinous sin is it in me
> To be ashamed to be my father's child.
> But though I am daughter to his blood,
> I am not to his manners. O Lorenzo,
> If thou keep promise, I shall end this strife,
> Become a Christian and thy loving wife. (2.3.16–21)

Jessica's monologue first acknowledges her own unnatural situation in light of blood's ideology: feeling ashamed of her own father. In order to make sense of this unnatural occurrence, she further aims to distinguish or divorce bodies/blood from actions/manners, namely greed and unsociability. While she maintains that her consanguinity does not determine her character, in her mind the mechanism for this detachment is also naturalized. The "strife" of her unnatural situation, her body and character out of sync, will be corrected by becoming a "loving wife." Becoming a "loving wife" of course hints at both copulation and its result, mixing bloods with her new husband. Jessica reiterates this hope for physiological separation from her father just prior to her elopement: "Farewell,—and if my fortune not be crost, / I have a father, you a daughter, lost" (2.5.55–56). Once she and Lorenzo are married, Jessica believes, she will be different enough from her father that these titles given to consanguineous relations need no longer apply. After her marriage then, in her report to Portia about Shylock's intent to take Antonio's pound of flesh, she distances herself from the Jewish people, calling them "his [Shylock's] countrymen" (3.2.285). Having supposedly transformed in the process of marriage's consummation, she is no longer of that tribe. She is no longer a Jew, in her mind, and no longer need be ashamed of a material connection with Shylock. She is now like her love as she believes it should be. Of course, her appeals to blood's ideology are equally as rhetorical as those of her opposition. Rejecting the label of "daughter" and disassociating herself from the Jewish "countrymen" are, after all, linguistic acts, not material ones.

Thus, the varied and contradictory appeals to blood's ideology throughout Shakespeare's *Merchant* expose its fundamentally rhetorical nature. In the end, the discourse is too pliable, and, as a result, it fails to accommodate Jessica's person into a network of *like* persons on either side of the divide. Yet the question still remains: if Jessica is not fully incorporated by Lorenzo's Christian blood and is not sufficiently like him, then what explains their love if love is limited to the attraction of like persons? The coupling of Jessica and Lorenzo in *Merchant* invites us to consider the possibility of a love based in difference rather than similitude, evidenced in part by Jessica's attraction. It is certainly plausible, as many scholars have suggested, that Lorenzo and Jessica do not love one another.[68] For instance, as Gratiano remarks regarding Lorenzo's tardiness for the elopement, "And it is marvel he outdwells his hour, / For lovers ever run before the clock" (2.6.3–4). It is also possible that Lorenzo is better suited to a woman more like himself. His exchanges with Portia, for example, seem to be more transparently romantic.[69] In general, the love between Lorenzo and Jessica remains ambiguous because there are very few outright declarations of love between them.

Nevertheless, there is evidence to suggest that Lorenzo embraces Jessica's difference, as she does his. His attraction to her is not an affinity for one like himself as the prevailing logic would dictate. Lorenzo explicates his love for Jessica as one of distinction: she is unlike anyone else, divorced from connections, having independent virtues revealed by her actions. He divulges to Gratiano,

> Beshrew me but I love her heartily!
> For she is wise, if I can judge her,
> And fair she is, if mine eyes be true,
> And true she is, as she hath proved herself;
> And therefore, like herself, wise, fair, and true,
> Shall she be placed in my constant soul. (2.6.52–57)

As Irene Middleton remarks, this scene in performance is "played lightheartedly, encouraging the audience to see Jessica as—following the text—an acceptable object of desire for a young Christian, and—in sympathy with modern views—still Jewish" (301). Although what might appeal to the modern audience about this passage is not Jessica's persistent Jewishness but the fact that Lorenzo loves her individuality. True, Lorenzo's pronouncements of Jessica's virtues are contingent, "if" he can judge her and "if" his eyes have not deceived him.[70] Though despite the hypotheticals, Lorenzo's rhetoric firmly founds his love upon his own judgments and not on anyone else's. Jessica's true nature is based upon her actions, she hath "proved herself," if not necessarily to anyone else at least to him. Finally, he concludes ("and therefore") that Jessica is not like anyone else but only "like herself," to then be joined with his constant soul. Lorenzo's love of Jessica's distinctiveness betrays a love of difference rather than similarity; and in accordance with that kind of love, it would be unfitting for that difference to dissolve or for her to be perfectly assimilated upon their marriage.

Lorenzo and Jessica's dissimilarity bears itself out in their interactions both before and after their marriage. Clearly Jessica and Lorenzo stand in stark contrast to Portia and Bassanio,[71] who, even in disputes, are poetically in accord. For example, their debate about whether or not to delay the casket test involves one person picking up the trope or poetic phrase put down by the other. After Portia's monologue asking Bassanio to "tarry," their exchange proceeds as follows:

> Bassanio: Let me choose
> For as I am, I *live* upon the *rack*.
> Portia: Upon the *rack*, Bassanio! then confess
> What *treason* there is mingled with your *love*.
> Bassanio: None but that ugly *treason* of mistrust,
> Which makes me *fear* the enjoying of my *love*:
> There may as well be amity and life
> 'Tween snow and fire, as *treason* and my *love*.
> Portia: Ay, but I fear you speak upon the *rack*,
> Where men enforced do speak anything.
> Bassanio: Promise me life, and I'll *confess* the truth.
> Portia: Well then, *confess* and *live*. (3.2.24–35, italics mine)

Portia and Bassanio pick up one another's expressions and even repeat key vocabulary, signaling that despite their quarrel they are of one mind. In yet another example, the heated disagreement over Bassanio's loss of the ring has both of them creating quartets that literally "ring" together ending all four lines with this word (5.1.193–97, 199–202). If in Portia and Bassanio's relationship the audience is invited to celebrate a love based in likeness, then Jessica and Lorenzo's subplot asks us to imagine an alternative or *otherwise*, illustrated by the very different manner in which these lovers relate to one another.

The lovers' mode of communication, a kind of tit-for-tat banter, arises not from affinity or agreement but conflict. Their love might, in fact, spring from this tension between opposites, made evident in their language of teasing questions, contradictions, and refusals to play the game according to the terms that the other has set. In their first exchange onstage, Jessica asks Lorenzo to confirm his identity, and when he identifies himself—"Lorenzo, and thy love"—she responds, "Lorenzo, certain, and my love indeed, / For who love I so much? And now who knows / But you, Lorenzo, whether I am yours?" (2.6.28–31). Lorenzo first confirms his identity and then their love, which Jessica does in kind; but she follows this verification with a rather awkward, teasing set of rhetorical questions. Her question "for who love I so much?" could be glossed in two different ways. Is she asking about whom she loves, which she has seemingly already affirmed with the pronouncement "indeed," or does she ask for whom she loves? Is she in love for Lorenzo's sake or for her own? Her second rhetorical question about her belonging to Lorenzo remains uncertain given this "whether," especially if we take the line break to indicate a significant interruption. Jessica might ponder, for a long awkward pause on "who knows" before recollecting herself to bring her attention back to "but you Lorenzo." Lastly, Lorenzo's response is not yes, I do know you are mine, but "Heaven and thy thoughts are witness that thou art" (2.6.32). Both Jessica and Lorenzo in this exchange tease by withholding direct confirmation of love. And rather than perfect concord, neither fully play along with the terms of exchange set by their partner. Neither Jessica nor Lorenzo in this dialogue take the other's bait as an opportunity to affirm their love but rather deflect, passing off the responsibility either to the other person—no one knows if I love you, but you—or to a higher power, "Heaven."

But this mode of discordant communication persists after their marriage, which poses a problem for Jessica's logic: that in becoming a "fully Christian wife," she would be like Lorenzo in the manner of *unanimitas* (those of one body also being of one soul or will in perfect harmony). And nowhere is this relationship's friction more evident than in the "In such a night" dialogue of act 5, scene 1. One could view this moment as two lovers collectively composing a lyric. Like Romeo and Juliet's sonnet construction at the masque, their "fittedness," albeit of opposites, finds poetic reflection in the fact that they share lines as these:

> Jessica: In such a night
> Did Thisby fearfully o'ertip the dew,
> And saw the lion's shadow ere himself,
> And ran dismayed away.
> Lorenzo: In such a night
> Stood Dido with a willow in her hand
> Upon the wild sea banks, and waft her love
> To come again to Carthage.
> Jessica: In such a night. (5.1.6–12)

Sharing poetic lines indicates a certain agreement, especially visually. But it could likewise indicate a brusque interruption on each lover's part in performance—one intrusion after another in a game of poetic exhibition. Moreover, in the selection of allusions, one assumes a kind of concord since both select tragic love stories. Yet Jessica's selections emphasize flight, while Lorenzo paints scenes of longing for return; thus, even the content is incongruous. This discord characterizes Jessica and Lorenzo's banter in this scene, becoming a one-up-manship of wit as Jessica's "I would outnight you" betrays (5.1.23).

The critical interpretation of this scene, which struggles with the ambiguous status of Jessica and Lorenzo's relationship, is quite polarizing. Many critics insist that the couple alludes to these love stories to mark their successful avoidance of the tragic outcomes.[72] Others believe that these allusions foreshadow a tragic ending for the mismatched, interfaith couple.[73] Of the latter opinion, Sigurd Burkhardt glosses the allusions from "In such a night" as follows, "what we are given is the genealogy of fly-by-night love: betrayal (Troilus and Cressida), disaster (Pyramus and Thisbe), desertion (Dido and Aeneas), sorcery (Medea) and theft (Jessica)" (255). As Catherine Belsey works out, the already suspicious allusions are made even more so because the scene does not "select from their tragic narratives moments of reciprocal happiness. On the contrary, Troilus is represented on the walls of Troy, sighing his soul towards the Greek camp and the absent Cressida. Thisbe is fearful and dismayed, Dido already deserted. Medea, gathering enchanted herbs, has not yet murdered her children in revenge for Jason's infidelity, but the text hints at her demonic powers and begins her characterization as a witch" ("Love in Venice" 42). But what seems most evident about "In such a night" is that all of the stories present beautiful visuals in the snapshots, if not reflecting particularly happy moments or particularly happy endings. Additionally, all of these stories capture couples that should not have loved for one reason or another, but loved nonetheless, thereby highlighting numerous counterexamples to the "like likes like" logic of blood's ideology. These stories may not have ended well, but the insistence that love is based in similitude cannot accommodate them either.

Jessica and Lorenzo's perpetual difference further extends to different renditions of their own love story. What begins as light teasing turns into a more substantial disagreement since the pair has conflicting narratives of their elopement. Lorenzo recounts the night as one of theft from a wealthy Jew to which Jessica responds by accusing him of false oaths of love[74]:

> In such a night
> Did young Lorenzo swear he loved her well,
> Stealing her soul with many vows of faith
> And ne'er a true one. (5.1.17–20)

In recalling the previous scene, the audience never witnessed these "vows of faith" so it is questionable whether or not they occurred; but it is telling of their dynamic that Jessica's knee-jerk reaction is an accusatory one. Lorenzo, of course, replies in kind. He promises forgiveness in the present night for the sin of slander, of which he accuses her in the same breath: "In such a night / Did pretty Jessica, like a little shrew, / Slander her love, and he forgave it her" (5.1.20–22).[75] If *Merchant* offers a depiction of Lorenzo and Jessica's love, these lines are undoubtedly it: an image of two strong-willed people bickering while gazing at the shining moon only to be interrupted by "the footing of a man" (5.1.24). This alternative love is not contented concord in which kissing is the heart's only disagreement but born of discord. Lorenzo even expresses a certain willingness that they might not be incorporated into the larger community, for he suggests that they remain outside upon the group's return to Belmont: "Let's in, and there expect their coming. / And yet no matter: why should we go in?" (5.1.49–50). Or to rephrase Lorenzo's question, why interrupt this moment?

However, even in this teasing, and possibly playful moment, there is no peace or perfect integration for the two lovers. At the conclusion of the "in such a night exchange," Lorenzo recasts this moment as one of concord and peace. He launches into revery,

> How sweet the moonlight sleeps upon this bank!
> Here we will sit and let the sounds of music
> Become the touches of sweet harmony.
> Sit, Jessica. Look how the floor of heaven . . . (5.1.54–58)

Lorenzo depicts this moment as one of togetherness and harmony, both looking at the moon and sky and listening to music while seated together. But the imperative, "Sit, Jessica," indicates that she did not initially accept his invitation to sit together in "we will sit," which, if she had done so, would reflect a shared will and mind. Instead she resists, and one could easily stage her standing. Moreover, her response, "I am never merry when I hear sweet music," indicates, as many scholars suggest, a genuine incompatibility (5.1.69).[76] This incompatibility absolutely reflects the lovers' dissimilarity, but it does not necessarily indicate that they cannot love. In fact, maybe Jessica dislikes "sweet music" because she dislikes harmony; she might prefer the discordant flavor of her love with Lorenzo over one of perfect, though potentially more boring, concord. In short, it is not simply that they began unlike, she a Jew and he a Christian, and marriage must make them alike in kind. Their dissimilarity endures and so might an alternative, *other* form of love than that which blood's ideology advocates.

As Jessica remains apart and unassimilable by blood's discourse cited throughout the play, *Merchant* critiques blood's ideology by asking the audience to see its function as rhetorical, not factual, exposed as such by this unique case. Consanguinity does not determine agreement or harmony. Just as blood failed to dictate agreement or affinity with her father, Shylock, so Jessica's relationship with Lorenzo is no different when the marriage bond supposedly mixes their bloods. Furthermore, by staging a relationship wherein the lovers are continually at odds, *Merchant* presents a love that defies blood's ideology, raising the potential for disruptive romantic relations and, consequently, disruptive social relations. Blood's ideology fails to determine affinity and thus fails to bestow Jessica with a proper place of belonging. Yet, a love unconnected to blood might also permit greater personal freedom and choice, potentially generating disruption and discord as well as the dissolution of clear, "biological" boundaries.

Of course, one caveat for this reading is that Jessica and Lorenzo's relationship occupies the place of a secondary (or even tertiary) plot. Portia and Bassanio are the play's golden children; seemingly perfectly suited, their love involves no such disruption. The presence of Jessica and Lorenzo, however, draws attention to the fact that there is some difference between Portia and Bassanio, too, for the "perfect" pairing from a socioeconomic standpoint would be Antonio and Portia. Or, as queer criticism has noted, there could be more affinity (similarity of character) between Antonio and Bassanio. Consequently, the impact of religious/ethnic/racial difference for Jessica and Lorenzo calls us to see how the dominant logic of Venice magnifies some differences while glossing over others in the interest of reproducing the status quo. Embracing a love found in difference, this alternative proffered by Jessica and Lorenzo is obviously not an easy task. After all, this attempt fails for Othello and Desdemona. The degree to which this alternative is tenable in the face of ideological pressures remains an open question, for it might only be permitted at the margins. Jessica might be able to determine the object of her love and affection, her body might not dictate it for her, but even when proven false, the ideology may persevere.

Epilogue

THE LOVING BONDS WE CHOOSE

This book begins with the premise that within the parameters of blood's ideology—both appealing to the aphoristic quality of "like likes like" and presuming that blood has agency—privileged persons were able to divest themselves of responsibility for existing structures of power. Yet all of the drama examined here offers some critique of this bloody ideology, heavily veiled or no, by showcasing its limits. The plots of blood feuds, sibling incest, biblical bed tricks, and incompatible love have staged how blood *in fact* functions, not biologically but rhetorically. Consanguinity does not itself determine love or affinity, but the *belief* that it does perpetuates hegemonic class structures by supporting the literal reproduction of the elite's kinship networks. Examining "kin" plays with similar or related sources reveal this shared critical project across early modern European drama, despite cultural or national boundaries. These plays, when examined together along the lines of particular limit cases, suggest a shared testing of an ideology's bounds in the early modern period—a period that amid massive sociological change had the nobility more forcefully asserting these bloody logics that supported their power.

The blood feud of *Romeo and Juliet* and *Le Cid* demonstrates that to assume the inheritance of affinity and to fetishize consanguinity can only perpetuate violence. Assuming this system, even when a character aims to extend existing kinship networks and to establish new affinal bonds, cannot help but lead to destruction or heartbreak. Namely, Juliet must pay for her family's bad blood for blood fails to determine affinity as promised. And Chimène suffers, for wed to her consanguineous duty, she seems loathe to abandon it even if romantically inclined to Rodrigue; after all, would one truly wish to marry the man who murdered their father? And monarchical power does not offer an escape from this system, even if King Fernand wields it toward a romantic end. For, though he insists on an end to the aristocracy's blood feud, he advocates blood shed on behalf of his nation, solidifying his own claim to power.

The sibling incest plays of chapter 2 illustrate the conceptual limits of this ideology, for it would appear to support incest. For if similarity in blood dictates both familial affinity and romantic love, then little differentiates the love between brother and sister from that between husband and wife (achieved via commixtio sanguinis). Sibling incest-averted comedies like *La Sorella* and *No Wit* trade on this confusion, staging long-lost "siblings" who purportedly confuse stirrings of erotic longing with the cri du sang that attracts them to their kin. The comedies shed light on the rhetorical nature of this system by magically solving the problem of incest via crib-switching ex machina. Yet Middleton's play especially points to the underlying motivations of the existing bloody system. Here, the initial crib-switching illustrates just how important it is to be a specific person's child (or at least perceived as such), impacting one's life and opportunities since wealth typically correlates with bloodlines. However, wealth and privilege could easily be divorced from this system. Ford's tragic rendition of the topos foregrounds that blood's ideology is incestuous by nature. Giovanni's passion is not the perversion of this logic but its perfect instantiation. Thus, any disgust or moral shock that the audience directs toward Giovanni—made plain in the spectacle of Giovanni claiming ownership of his sister's heart on a platter—should not be directed at him as an individual but rather at the system that produced him and sustains the elite.

The divinely sanctioned, biblical bed trick is the limit case meant to illustrate blood's import, not only as a matter of fact but as a vital part of God's creation. Namely, certain bloodlines (e.g., the House of Israel) are so important that God will sanction an otherwise immoral, deceptive act. And this logic seemingly extends to the naturalized hierarchies established by blood's ideology, or at least the elites would have it so. Yet the dramatic reimaginings of this trope, by way of Boccaccio's novella, undermine this premise, albeit in very different ways. In *All's Well*, Helena's use of Giletta's bed trick, meant to produce a child of Bertram's blood, remains secondary to her hope that a different force will activate. Namely, upon the consummation of their marriage and the attendant commixtio sanguinis, their shared blood would compel Bertram's love, overcoming their supposed biological incompatibility. Helena, however, remains sorely disappointed. Though the Countess and the King try to rhetorically manipulate this ideology in order to incorporate Helena, they ultimately rely on blood's vocabulary due to their privileged positions that the ideology affords. Lavatch's comedic foil, however, points toward the ridiculous implications of the logic that blood materially determines love and worth. In *La mandragola*, by contrast, *All's Well's* critique is flipped on its head. In Machiavelli's Florence, the *child* matters but not the bloodline. As articulated by the desperate Nicia, kinship networks are extremely important for class status, though these networks are not fundamentally biological; the social signification of family does the work and not the blood at all. While this could be freeing, it does not work for those unprivileged. Just as Helena found a loophole in the existing system in an attempt to make it work for her, Lucrezia locates her status and self-worth in her chaste, ennobled body. The play's message, I would argue somewhat cruelly, denies Lucrezia this solace; as her mother, Sostrata, tells her, her body does not have value intrinsically but only instrumentally for the production of children.

Lastly, in a detour from the book's comparative project, Shakespeare's Venetian plays further highlight the function of blood's ideology to maintain hierarchies and to perpetuate systems of power when its mechanisms are applied to the realms of ethnicity or race, other "differences in kind." Othello and Desdemona seek acceptance for their love within the existing framework, trying to circumvent the issue of perceived difference by maintaining that their love is founded in similitude, understanding, and concord. Yet their acceptance of this existing logic for love apparently dooms their relationship to failure. Venice's logic that like must love its like, articulated most plainly by the scheming Iago, means that their counternarrative will falter. Othello then becomes undone by his fear that Desdemona will return to her kind. Jessica and Lorenzo's love, by contrast, suggests that love can exist not only despite difference but can appreciate difference. Yet their example may be only mildly satisfying. After all, much of the play's action is dedicated to negotiating Jessica's fitness to become similar and thus suitable, both as Lorenzo's partner and as a member of the larger Christian kinship network. While this debate around Jessica's blood underlines its rhetoricity, it might likewise temper the audience's optimism for this project.

In fact, to the extent that any of the plays discussed thus far offer alternatives to this ubiquitous ideology, they do so in a limited, tentative fashion. Most of the critiques offered come to us courtesy of minor or marginalized characters on account of their gender or station (Juliet, Elvire, Trasimaco, Pickadille, Putana, Lavatch, Sostrata, and Jessica). Their critiques are almost questions: maybe calling for more violence is pointless; maybe the "whore" that views love as indiscriminate is onto something; maybe the virtuous bloodline of the House of Israel is the exception, and the rule is that we all are equally sinful; and maybe love is not rooted in finding similarity in "kind" after all. But these voices and their questions are unsurprisingly pushed to the margins themselves. One might ask, then, if early modern drama is equipped to provide its audience with a genuine alternative to this bloody discourse.

The answer that I offer here is a rather qualified yes, at least in the realm of romantic or erotic love, if not the affine bonds of kin. For instance, the figure of Cupid's arrow casts erotic love as random, fickle, wayward, not the result of one's body dictating a perfectly compatible mate. By way of example, in John Lyly's pastoral *Galatea* (c. 1584), Cupid, wishing to assert his power and superior hunting skills, inflicts wounds of love on the followers of Diana (1.2). Though the nymphs experience love as if for a very specific object, evidenced by their squabble among themselves to claim particular beloveds (3.1), Diana insists that this love lacks discrimination and is ultimately superficial. Upon berating her nymphs for risking their virtue, she asks, "How is your love placed? Upon pelting boys, perhaps base of birth, without doubt weak of discretion. Ay, but they are fair!" (3.4.56–58). To be "weak of discretion" is to lack judgment. Diana underlines that there are no means to adjudicate between love objects; their attraction seems based solely on appearance with nothing more to it. According to her, if bodies matter at all, it is only on the surface.[1]

Moreover, any audience member of *A Midsummer Night's Dream* remembers the love potion that projects love on any object—made evident by every high schooler's concern that Demetrius is still under its spell, never to be cured. This vision of love likewise arises in *The Two Gentlemen of Verona,* wherein the homoerotic relationship between two friends overtakes the concern for their respective love interests. At the play's close, Valentine offers to bestow his beloved Sylvia to his friend Proteus, even after Proteus has just threatened to rape her in the woods (5.4.82–83). Though Valentine gets to keep his love, audiences find the resolution (understandably) unsatisfying. The notion that one could simply transfer love, or trade objects, is rather disturbing in a Western tradition that traditionally grants the utmost importance to a specific beloved. Yet this suggestion that love need not be preoccupied with the specifics of its object might offer our best option for *thinking otherwise* or thinking beyond this bloody ideology. For in this case, a lover's similarity to, or difference from, their beloved matters not. To illustrate this alternative picture more forcefully, I turn to Maddalena Campiglia's pastoral play *Flori* (1588). For, as a female author, she places voices that were previously in the margins of drama at the center of her play.

"WHOEVER IT IS, I SHALL TELL HER THAT I LOVE HER ALONE"

As highlighted by feminist scholars over the last few decades, Campiglia's *Flori* is iconoclastic in a number of ways, of which the arbitrariness of love objects might be the least interesting. The unconventional life of its author has undoubtedly added to the text's appeal. In 1576, Maddalena Campiglia married at age twenty-three the noble Dionisio Colzè. The marriage was a failure, but we do not know why. We only know that they did not have children, and in the early 1580s, they separated and she returned to her father's home (Cox and Sampson 3). As Lori J. Ultsch and others have underlined, Campiglia would have been deemed a *dimessa,* a woman who fell into what is often called the "third estate," living a respectable life of secular celibacy between the estate of marriage and that of the cloister.[2] After her separation from her husband, Campiglia threw herself into the literary society of Vicenza. An example of early modern literary self-promotion, she had connections with both the local literary academy in Vicenza (the Accademia Olimpica) and corresponded with other poets throughout Italy, including Curzio Gonzaga and, most notably, Torquato Tasso. She acquired a patroness, the marchioness Isabella Pallavicino Lupi, and eventually had her works published by the Vicentine publisher and bookseller Perin Libraro (Cox and Sampson 4).[3] Campiglia even chose an emblem for herself. It depicted a phoenix rising from the ashes, modestly interpreted by scholars as a proclamation of her uniqueness and aspirations of literary immortality—and even more boldly deemed "a feminine regeneration from male subjugation" via poetry (Perrone 29, my translation).

Flori is just as unusual as its author, but I admit that I cannot recommend it as a particularly enjoyable read. The play is tragicomic bordering on melodrama with lengthy soliloquies, grammatically complex verse, moralizing sententiae, and a generally pious tone. Hearing the plot, however, a rather different work emerges, especially when placed alongside contemporary Italian comedy. Due in part to formal pressures of the Aristotelian unities, *commedia erudita* set its dramatic action in a piazza. Consequently, this genre placed female characters at the literal margins of the action, in windows or on balconies, and often had their speech narrated by male characters. High-status female characters like Machiavelli's Lucrezia, albeit at the center of the plot as the target or prize of the *ingengo* (trick), could not themselves be present in these spaces.[4] By contrast, the pastoral genre generally offered female characters more agency. As Lisa Sampson notes, pastoral drama that foregrounded the private emotions of love was appealing to women writers since they could draw on personal experience. Moreover, the idealized setting "allowed the representation of subjects removed from political affairs and characters less bound by (patriarchal) social and familial restraints" (104). It thus offered a space for veiled social complaints, exploring the alternative, female-oriented society of Diana's band of nymphs.

Yet, even in this context, Campiglia's pastoral is unorthodox. The play features several romantic plotlines, but the simplest synopsis is as follows. The eponymous heroine, "mad" with love for her now dead beloved nymph, Amaranta, violently rejects all suitors including the other mad lover Androgeo, who is beloved by her best friend, Licori. At the play's climax, the shepherds and nymphs gather for a sacrifice that will "heal" these mad lovers, which it does, but Flori has been fated to fall in love again with the first person she sees at the sacrifice's conclusion. In this case, it is the foreign shepherd Alessi who, on passing through Arcadia, watches the sacrifice and likewise falls madly in love with Flori. The play concludes with a stark revision of the recalcitrant nymph topos (the chaste member of Diana's band who is eventually persuaded by love to marry).[5] Despite agreeing to marry, Flori advocates for her own version of the "third estate"; she and Alessi will have a chaste marriage in which they will forgo physical passion in order to devote their lives to Diana, not dissimilar to the marriage of Mary and Joseph lauded by Campiglia's first devotional text *Discorso sopra l'Annonciatione della Beata Vergine . . .* (Discourse on the Annunciation of the Blessed Virgin Mary) (1585). This conclusion thus conjoins Flori's previous life as a virginal nymph, a lifestyle more appropriate for individual, intellectual pursuits, with the conventional demands of the pastoral genre, necessitating the protagonist's conversion to embrace love and marriage. As Lori Ultsch underlines, the marriage is a "decidedly nontraditional one that celebrates not the future progeny of two flesh become one but Flori's creation . . . of a state where the secular virgin can dedicate herself to learning and devotion" ("*Epithalamium Interruptum*" 85–86). Thus, while the play's depiction of love's arbitrary quality is possibly the least interesting quality in a play that includes lesbianism, gendering the Neoplatonic gaze female, and a celibate marriage, it is in this context that love's arbitrariness has most purchase as a critique of blood's ideology. Namely, once we need no longer be concerned with future progeny and the mingling of bloods, then the import of compatible bloodlines falls away. The language of love, of course, preserves the assumption that love has a singular, perfect object. Yet *Flori* also self-consciously undercuts this language.

The fickle nature of love is treated most explicitly in an embedded narrative that Licori, best friend of Flori and lover of Androgeo, recounts to another nymph beset by unrequited love. Licori's story features Flori telling off a Casanova shepherd by the name of Titoro; in it, Flori playfully asks him,

> Se'l cor è un sol, se una sol cosa data
> Già non si deve più pigliar, com'io
> Mai crederò che'l tuo languir sia vero?
> Se a mille una sol cosa doni e togli,
> E tolta la ridoni, e poi donata
> La ripigli di novo e a mille a un tempo
> Involi doni, e ancor donato furi? (2.2.92–98)

> (If the heart is a single thing, and that thing, once given, should not be taken back, how, then, can I believe that what you say of your languishing is sincere, when you give your one heart to a thousand nymphs, and then take it back, give it again, and take it back once more, only to give it anew?)[6]

This question could be posed to all the lovers of *Flori*, including the speaker, and it seems true that giving one's heart away is an extremely misleading way to understand the nature of love; instead, the emotion finds its object arbitrarily, for she is simply one nymph among so many other possible love objects. This is why Titoro's response illustrates just how ridiculous typical love language is; in true Petrarchan fashion, he claims, "io te sola amo" ("I love you alone") (2.2.99), and promises

> mirar potesser entro
> A questo seno mio,
> So che l'imago tua vedrebbon sola
> Star, par mano d'Amor nel mezo incisa. (2.2.102–5)
>
> (if those eyes could look into my inner self, it would see nothing but your own image, etched within by the hand of Love.)

Petrarchan love lyric might encourage a conception of love that has a singular, special object that has been impressed on one's heart, metaphorically if not literally. But the context of this language renders it empty of substance. One could simply deem Titoro a cad, a foil to the true lovers of the central plot, but if this were the case, the play need not have a double to reiterate this claim.

Another minor shepherd, Serrano, likewise acknowledges the ability to transfer affections. Despite his shenanigans attempting to obtain Flori's affections, his love appears to have little to do with Flori as such. For one, Serrano does not insist upon the eternal constancy of his love, for he blatantly says, "divenir gia pazzo / Non voglio, ch'altre ancor ninfe saranno / Che non mi sprezzeran (forse)" ("I certainly do not wish to remain mad with love for her, for there are many other nymphs, perhaps, who will not despise my suit") (2.3.33–35). Later in that same act, having given up on Flori in order to honor his brother Androgeo's suit, he holds that he is still in love, though it awaits a direction:

> E poscia Amor renderà forse in fine
> A lo mio merto il guiderdone uguale.
> Licori alfin non mancherammi, ed altre
> Ancor. Ma che? Non mi rammento ch'oggi
> Quella a cui diedi aita,
> Mi mirò dolcemente?
> Amerò quella, e se non quella un'altra;
> E dirò a questa, a quella: "io t'amo sola." (2.5.76–83)
>
> (then perhaps Love will in the end give me some reward for this forbearance. After all, there is always Licori, and there may be others. . . . But how could I have forgotten? Only today, that nymph

> I saved looked so sweetly on me. I shall love her—or if not her, another, and whoever it is, I shall tell her that I love her alone.)

Amor or Love as a proper noun, as a god or greater force, exists independently of a love object. Though in the end he promises to love only one woman, explicitly echoing the language of Casanova Titoro, it matters not which one. His tone is nonchalant—"Amerò quella, e se non quella un'altra" ("I will love that one, if not that one, another"). The specifics of the love object do not matter.

And the heroine's love likewise proves rather fickle, even though the prologue promises the workings of destiny behind her transfer of affection. Flori articulates her love in the same hyperbolic, Petrarchan fashion as did Titoro. She opens the play lamenting, "e'l core / Che tutt'altro aborrisce, lassa alquanto, / Che si consoli in dolce vista, e amara." ("Let my heart, which abhors all other objects, seek solace where it may—in a sight, for me, as sweet as it is bitter") (1.1.2–4). From the start, the audience knows that her heart will not abhor all other objects besides her now perished beloved Amaranta and will rather be moved to her destined beloved Alessi. Her inconstancy is made clear after the sacrifice at the play's climax. In her extensive monologue, she renounces her former love, only to embrace another, acknowledging full well that she loved Amaranta but could not hold to her previous oaths to love her till death: "Né già con brame più d'affetto calde / Alcun amante il suo pregiato oggetto" ("Nor has any lover ever followed the object of his longing with keener desire") (3.5.142–43). Furthermore, Flori's opening of act 5 sounds remarkably like the opening of act 1, making her hyperbolic statement of everlasting love for Alessi ring disingenuous, or, at least, more flat:

> Ma pur, lassa si parta,
> Che la memoria sua
> Giamai fin ch'avrò vita
> Non partirà dal cor, dove il bel foco
> Primo giunse d'Amor, ch'a questo petto
> L'ultimo fia che poco o assai lo scalde.
> Il mio fatale e volontario affetto
> Non avrà fin giamai: sì vuole Amore,
> Tal di questo mio cor fido è 'l costume. (5.1.4–12)

(as long as I live the memory of him will always be etched on this heart of mine where first love's fires were kindled; for no other man will henceforth stir the least breath of desire in this breast. My fatal chosen passion shall see no end, for Love wills it—well do I know the habits of this faithful heart of mine.)

Flori's affection, somehow simultaneously chosen or voluntary ("volontario") and fated ("fatale") will supposedly never end, though there appears a caveat, namely "sì vuole Amore" ("If Love wants it").

Finally, the pastoral's conclusion with all of the shepherds and nymphs happily mated depends on love being arbitrary. Flori must successfully reorient the love of her former suitors toward new objects: Androgeo to Licori and Serrano to Urania, which are (apparently) successful. Despite having emerged from his maddened state, Androgeo still proclaims his love for Flori, "Ardo ancor per disio di poter fare / Cosa che ti sia grata" ("I do still burn with desire to do anything that may please you") (5.3.17–18). Flori immediately takes advantage of his promise by arranging the match for her friend, effectively telling him to project his love onto Licori: "E ben dir si potrebbe / Tre cori uniti una sol volglia regge" ("Thus, through your marriage, we may become three hearts united in a single desire") (5.3.55–56). And this seems easily accomplished, as Androgeo replies,

Ch'io le sia amante e sposo, che un sol cenno
De'tuoi m'è legge espressa, oltra che sempre
Dopo te non vidi io ninfa veruna
Che più di lei piacesse a gli occhi miei. (5.3.61–64)

(I shall readily declare myself her lover and her husband, for your every wish is my command; besides which, I never saw a nymph besides yourself so pleasing to my eyes as her.)

And, even if it is for Flori's sake, Androgeo promises utmost fidelity: "So che di fede almeno passerò inanzi / A ogni marito e amante" ("I can swear that in fidelity at least I shall outstrip all other husbands and lovers") (5.3.127–28). These lofty, romantic proclamations seem all well and good, but the play quickly brings us down to Earth with a potentially more honest depiction of how or why one can redirect love in this way.

At the end of this scene, the two central love plots are finished, the *epithalamium* celebrating the impending nuptials has been sung, and this could have been the end of the play. But Campiglia's text returns to the minor characters' plotlines, and we end with the sentiments of Serrano. For his own part regarding love, he resolves, "Poiché sogliamo noi dire in ogni modo / Ch'al fine, al fine, elle son tutte donne" ("for, as we men like to say, when it comes down to it, one woman is very like another!") (5.4.47–48). Then finally, but with very little reflection, Serrano selects his object: "Purché Urania contenti, ecco di fede / Il pegno, la mano porgi. Altra mai ninfa / Non mi sia donna o sposa" ("If it please Urania, here is my hand on it as a pledge of faith: she alone shall be my lady and my wife") (5.5.101–3). Ending the play with Serrano seemingly undermines the prophetic, fatalistic frame of the prologue wherein Cupid foretells Flori and Alessi's famous love. Ending with Serrano's sentiments forces the audience to project this frame back upon the central protagonists, to view their love as equally arbitrary.

Campiglia's pastoral might, therefore, give us more hope for an escape from blood's ideological limits for, albeit in the confines of a fictional pastoral, matings may be a matter of convenience. Moreover, the nymphs of Arcadia, otherwise marginal characters who are ordinarily the objects of male pursuit, are here brought to the center of the dramatic action. They are the primary drivers of their narratives in contrast to the marginal voices whose critiques consequently remain in the margins. The limit cases featured here have shown the existing power structures steeped in blood's ideology to be oppressive and destructive, a false promise, or simply a sham. Yet the question remains to what extent an early modern audience might embrace difference (as is the case for Shakespeare's Jessica and Lorenzo) or grant love's arbitrariness (in the fashion of Campiglia's *Flori*) as a viable alternative to the existing system.

But finding a conceptual way out of this ideology might be as simple as returning to the beginning. In this case, a return to the origin of the proverb "blood is thicker than water" with which this book begins. The conventional interpretation that underscores the importance of consanguinity is based upon a misunderstanding, a clear misreading of the original biblical passage due to its abbreviation. The original passage from Matthew reads, "The blood of the covenant is thicker than the water of the womb" (26:27–28). The source does not advocate for honoring kinship relations, denoted as "the water of the womb." In this passage, blood does not signify kin but rather ties made by swearing oaths during the performance of a blood sacrifice (as we might call them, "blood brothers"). Unlike the contemporary proverb, the biblical version bestows agency *not* upon the blood of one's body that is innately attracted to those of its "kind" but on the agent's will. In short, the relations that matter are not those of biological similarity, but those that we enact, those that we *choose*.

NOTES

Introduction

1. Reference from proverb's entry in the *Oxford Dictionary of Proverbs*, 6th ed. (2015).

2. All references to Shakespeare refer to the *The Complete Pelican Shakespeare* (2002) unless otherwise noted.

3. It is possible to stage this difference as a chromatic one in casting, but this is probably not a choice made very often—and, in my mind, with good reason—since Salarino's statement seems more rhetorical than anything else.

4. Sujata Iyengar's study of skin color and race in early modern England advocates using Raymond William's "structures of feeling" over the term *ideology*. Iyengar maintains that this alternative framework highlights "the places where seemingly explicit, overtly political institutions and seemingly hidden, private beliefs overlap" (3). I will favor the more quotidian *ideology*, as have many other scholars, though I acknowledge the benefits of William's framework.

5. As proof, Mullaney cites Thomas Heywood on the ideological power of the English history play as that which "hath power to new mold the harts of the spectators and fashion them to the shape of any noble and notable attempt" (quoted in Mullaney 30).

6. This definition is based upon that of Tamás Demeter, who uses very similar language to describe the ideologies of "knowledge-making practices" in early modern Europe: "More or less explicit ways of making sense of these practices, justifying their relevance, and circumscribing the sphere of legitimate inquiry and its applications" (1). This definition of *ideology* seems to capture what many other early modern scholars mean when using the term to pick out a particular belief system relating to power structures, be it scientific, political, religious, or cultural. See for instance, the collection of essays *Politics, Ideology and the Law in Early Modern Europe*, ed. Adriana E. Bakas (1994); Peter Dear's "What Is the History of Science the History Of?"; another edited collection *Conflicting Values of Inquiry: Ideologies of Epistemology in Early Modern Europe*, ed. Tamás Demeter, Kathryn Murphy, and Claus Zittel (Leiden, Netherlands: Brill, 2014); and Patricia Canning's *Style in the Renaissance.*

7. As I explain in more detail in the historical section, blood was mutable and often very volatile with respect to an individual's temperament or complexion. Quality of blood that marked bloodlines, however, appeared to be less changeable except through sexual intercourse. Comparisons to animals and diverse species in discussions of cross-class relationships were not purely metaphorical in this sense.

8. I use the term *affinity* here in the more ordinary sense of liking or sympathy based on sameness, but obviously, this term has a special meaning in the early modern natural philosophical context. This more particularized meaning of *affinity* relies on the belief that there were hidden resemblances between all matter due to the assumed relationship between microcosms and the macrocosm. Individual bodies of matter could supposedly act on one another at a distance by virtue of these affinities through the forces of "sympathy"

or attraction. Clearly, the assumption that I have outlined fits into this more general, natural philosophical framework.

9. It is not surprising that this ideology would play out in the realm of marriage and mating, for, as Ania Loomba reminds us, "sexuality is central to the idea of 'race' understood as lineage, or a bloodline, because the idea of racial purity depends upon the strict control of lineage. In the early modern period, the term *race* could indicate a family or household, a nation, a religion, a class, or even an imaginary group, but in every case, the boundaries of this group could only be guarded or expanded by carefully regulating sexual behavior, especially that of women" (*Shakespeare, Race, and Colonialism* 32).

10. See Jean Howard and Margo Hendricks especially.

11. As Shannon outlines, perfect consent in friendship required an equity of persons and their agency, requiring both likeness and agreement in the grammatical sense (7–8).

12. Howard justifies this view, which runs counter to the work of Francis Barker, Michel Foucault, Jonathan Goldberg, Stephen Greenblatt, and Christopher Pye, to name only a few, by pointing to two very salient points. First, the crown's censorship did not always work. Periodic "clampdowns" suggest that the theater often tried to promote interests that were antithetical to those of the state (14). Second, the public theater surely represented popular viewpoints as well since, as a group-defined commodity, it required the economic support of a primarily popular audience. One distinction between my view and Howard's is that Howard keeps her discussion of stage plays limited to the realm of politics, asking what or whom a particular representation would benefit (8). I, on the other hand, and maybe naively so, still wish to see a possible ethical injunction or judgment on the part of the playwright, and the tone of my project reflects that attitude.

13. In this respect, my work resembles the methodology advocated by Paolo Valesio in "literary semiotics" (11–12). Although I advocate for socially conscious (and even critical) playwrights, I may not, like Chris Fitter, want to fully commit to a "radical Shakespeare" since I believe that the bard's stance toward blood's ideology is somewhat more complicated and tempered. And unlike Fitter's discovery of a radical Shakespeare, my study examines not specific political issues but rather a larger structure of hegemony, resembling in this respect Jonathan Dollimore's *Radical Tragedy* (Durham, NC: Duke University Press, 1984). Nevertheless, my own inquiry undoubtedly gives Shakespeare a forward-thinking, critical stance.

14. Greenblatt attributes this similarity to an awareness in the period of what he calls "self-fashioning," but one could likewise use the vocabulary of performance or even participation to describe the practice of citing particular codes or discourses.

15. See Mary Thomas Crane's reading of *As You Like It*, wherein she argues that the play makes the maneuvers of the dominant ideology visible even if the play contributes to it.

16. These usually minor characters function similarly to what Paolo Valesio calls the *interprétant* in his literary semiotics. Specifically, that there is a kind of roman à clef for the text, a moment of suggestion that gets close to the "unsaid": that which cannot be said given a particular ideological framework (17).

17. One should not confuse my method, however, with the equally important body of scholarship aimed at the historical and literary study of the institution of service in an effort to, as Iman Sheeha recently put it, "recover . . . the stories of the marginalized, voiceless and the vanished" (3). My analyses of these characters point to their largely formal purpose in early modern theater, acting as a mouthpiece for the critical, controversial, or iconoclastic views of the playwright. My analyses attend to what these characters can say

about hegemonic structures and logics due to their exclusion from them in a way that the plays' privileged protagonists often cannot. For highlights from the scholarship on service, see Mark Thornton Burnett, *Masters and Servants in English Renaissance Drama and Culture: Authority and Obedience* (London: Palgrave Macmillan, 1997); David Evett, *Discourses of Service in Shakespeare's England* (London: Palgrave Macmillan, 2005); Linda Anderson, *A Place in the Story: Servants and Service in Shakespeare's Plays* (Newark, NJ: University of Delaware Press, 2005); and Judith Weil *Servants and Dependency in Shakespeare's Plays* (Cambridge: Cambridge University Press, 2005).

18. Although I focus primarily on blood as a discursive ideology, this dramatic critique often involves material blood, and, consequently, the use of stage blood helps to illustrate the ideology's violent consequences. I agree with Lucy Munro, who suggests that the use of realistic stage blood is neither for the sake of spectacle alone nor the product of Elizabethan protonaturalism but rather has a "particular narrative end" (85). Theater historians have disagreed about the materials used for stage blood with diverse arguments for animal blood, vermillion or cosmetics, vinegar, red wine, and paint. See Leo Kirschbaum, Lucy Munro, Andrea Stevens, and especially the published proceedings from the Stage Blood Roundtable at the Shakespeare's Globe Theater History Seminar in 2006.

19. Greenblatt's *Renaissance Self-Fashioning* is the most well-known, though one should also include the important work of Joel Fineman's *Shakespeare's Perjured Eye.* Related to this line of research are those studies that examine how authors cite a variety of cultural codes to create a character with a sense of individuality and psychological depth. See J. Leeds Barroll on the early modern character in conversation with theological discourses, Elizabeth Fowler's study of layered "social persons," and Julia Lupton's study of Shakespeare's character types ("Character Criticism"). Then there is also a camp that rejects any sense of a "clearly bounded self" in the Renaissance, be it Burkhardtian or postmodern in formulation, as for instance John Jeffries Martin's work *Myths of Renaissance Individualism.* For an extensive review of the scholarship on identity, see A. J. Piesse's entry in *A New Companion to English Renaissance Literature and Culture.*

20. Katherine Eisaman Maus's seminal work, *Inwardness and Theater in the English Renaissance,* explores early modern society's anxiety about an inability to perceive the "true" person, a hidden interior in contrast to a theatricalized, artificial exterior. She cites both the early modern desire to know invisible religious difference in the aftermath of the Reformation and the new urban space populated by strangers (the threat of "coneys" or con-artists, etc.) in which it is difficult to know another.

21. Jonathan Sawday cites the emergence of "the very sense of selfhood" in early modern dissection and a history of the body as a field of inquiry. He successfully demonstrates that interiority appears to be fashioned paradoxically by looking into another's body. He points to the definition of *autopsy,* which meant both the inspection of a dead body and more generally "seeing with one's own eyes, personal observation" (6–7). Katherine Park's work on "holy autopsies" shows one way in which body parts became signs for saintly character in a new culture of dissection.

22. Michael Schoenfeldt tracks the desire to "express psychological inwardness materially" by highlighting how early modern poets pointed to the body, and especially humoral psychology, to articulate psychological states (1–2). He places himself opposite Foucauldian-heavy New Historicist models in order to emphasize that this discourse gave individuals agency or self-control in a literal sense. The humoral body was not merely Paster's "pathological leaky

body" but one that can be controlled by the individual with temperance (11). I happen to agree with *both* Paster *and* Schoenfeldt on this matter. Like most discourses, Galen's humors were invoked as an instrument of societal repression as well as individual agency. My own study embraces this tension.

23. In attending to early modern models of contagion and the porous humoral body, Langley's *Shakespeare's Contagious Sympathies* posits the modern subject as the knee-jerk, anxiety-induced reaction to this model in a time of plague and disease. As he puts it, "an awareness of immanent surface solubility necessitates compensatory fictions of profundity, coherence, and depth: anxieties concerning fluency must, it seems, be answered by counteractive philosophies which promote notions of intrinsic identity and ipseic integrity" (28–29).

24. David Hillman and Carla Mazzio's edited collection, titled *The Body in Parts*, includes work from the editors as well as from Nancy Vickers, Marjorie Garber, Stephen Greenblatt, Katherine Park, and Katherine Rowe, to name only a few. In line with this project, Amy Kenny's *Humoral Wombs* adds to this scholarship.

25. Regarding the use of primary historical sources like household manuals, domestic guides, treatises, sermons, conduct books, and advice literature, Iman Sheeha importantly draws attention to the limitations of such materials because "they offer an image of the ideal, rather than the real; the desired, rather than the actual and lived" (3). But, for my purposes, these materials nevertheless provide a framework for early modern behavior and illustrate what the culture believed *should* be the case, even if it was not.

26. Solomon's "You've Got to Have Soul" complicates the early modern view of the soul by explaining its existence along a spectrum between a Galenist viewpoint and an Aristotelian-Thomist one. The Aristotelian-Thomist soul seemingly has more agency or control than the materially determined Galenic soul, as the work of Paster or Schoenfeldt assumes.

27. Langley explains, "*Pathopoeia* communicates pathos, extending and transmitting emotion via enforcing sympathetic accord: therefore to study em*pathy* or sym*pathy* is, unavoidably, to describe both a disease *path*ology and semiosis, as Renaissance medical thought—enduringly grounded in long-standing Galenist presumptions even as his absolute authority is challenged—is predicated upon a prevalent conception of what could be described as *sympathology* (i.e. disease transmission via pathetic exchange, abetted by communication)" (*Shakespeare's Contagious Sympathies* 10).

28. This assumption led twelfth-century scholastic philosopher William of Conches to conclude that a sanguine or balanced temperament was practically impossible to achieve, and, consequently, all animals were melancholic, choleric, or phlegmatic (Conticelli 61–62).

29. Even in an Aristotelian interpretation of blood, Andreas Cesalpino distinguishes between blood the humor and blood the material in the veins, the second of which "was considered to be a sanguineous mass consisting of a mixture of pure blood with a lesser proportion of the other three humours. . . . This was a concoction that combined blood and spirit, the latter drawn by the lungs from the air and brought into the heart. He sometimes called this 'perfected' or 'matured' blood" (Santing 432).

30. Galen outlines in book 1, chapter 10, of *On Natural Faculties* that food transforms to blood, which then becomes flesh, bone, and so on, in order to be assimilated. The long process of alteration necessary for this assimilation, he explains, is why there are so many different organs involved in digestion (35). In an interesting side note, the assumption that like can only be assimilated with its like led him to advocate a diet of meat over vegetables because the

former is already of flesh and blood (36). As far as I know, advocates of the Atkins or Paleo diets have not taken up this line of reasoning.

31. I purposely avoid the discussion of blood in the context of Christian theology, especially the eucharist and Catholic-Protestant debates of this period. Not that this context for blood is unimportant to the early modern imagination, but it stands sufficiently apart from the blood that dictates social identity and affinity, which is my central interest. For a wonderful starting point for this research, see Caroline Bynum's *Wonderful Blood*.

32. The belief in blood's intrinsic power as a material life force is evident, for example, in the discussion of witchcraft trials outlined in Albrecht Dürer's personal diary. It was commonly held that "suspected witches, once captured, could be rendered powerless... by extracting their blood directly... [since] blood was the presumed medium of magical potential" (Linke 45).

33. One can gloss Galen's view on the body/soul relationship according to two theses, namely that "the soul (or its faculties) *follows* the mixture" or that it *is* the mixture (Singer 33). The most conservative version to attribute to Galen is that of the "reciprocal effect of bodily state, usually conceived in terms of mixture, and of mental capacity or activity" (Singer 28). Of course, some scholars do attribute to Galen the second view, a relatively determinist picture, even though the individual could take action to manage their bodily mixtures (Donini 196, 201–2). And, in my mind, this second version seems to be the one that the early moderns favored.

34. Other scholars, of course, have countered this basic premise. For instance, Richard Strier problematizes conflating the literal and metaphorical, arguing that they may not have been understood as one in the same. Moreover, he reminds us that the Galenic discourse was likely not a "master discourse" but actively questioned in the period (Strier and Mazzio "Two Responses" 16). Carla Mazzio also asks us to consider traditions of mind/body split that already existed in the early modern period (Platonic, Christian, and ascetic traditions) when considering the body (18).

35. For more on the effect of Harvey's discovery on cultural and philosophical understandings of the body politic, see Robert Appelbaum's "Flowing or Pumping?" Yet, to my mind, even though Harvey is often depicted as a sole iconoclastic genius, using empiricism to fly in the face of medical and philosophical authorities, Harvey frames his own work as an explanation of what Galen had *already* found. Moreover, his contribution to the discourse about blood is, in many ways, confined to its motion as opposed to addressing the nature of blood itself. For instance, he believed, as Galen did, that blood was the source of vitality (Miller 106). Whether his contribution could accommodate the belief that blood is particular to an individual and their nature is unclear. Harvey only draws distinctions between bloods on the level of species.

36. Certainly, many thinkers contested this notion. Boethius provides the classical critique of hereditary nobility in his *The Consolation of Philosophy*, which claims that humankind must all have the same origin (Crouch 148). Later, Dante Alighieri adamantly argues in the *Convivio* (*The Banquet*) that nobility could not be inherited by blood "for the divine seed does not descend into a stock or family-line; it descends rather, into individual people, and... it is not a family-line that makes individuals noble, but individuals who ennoble a family-line" (quoted in Pearce 39). And Baldesar Castiglione debates this point in *Il Cortegiano*. For a more extended discussion of the individual, unique character of these lineages, or "dynastic identities," see Geevers and Marini's edited collection *Dynastic Identity in Early Modern Europe*.

37. Even Aristotle's view should not be simplified to male form and female pure matter; instead, "as with semen, the menstrual blood contained a potential version of the organs abstracted from blood, but a potential of a lower, more material order than that of the seminal *dynamis*. If the female made no contribution in terms of form then all offspring would always resemble the male parent, but Aristotle was only too well aware that this was not the case" (Russell 27–28).

38. Historians David Sabean and Simon Teuscher, however, cite several problems with using this model as the sole means of understanding kinship relations in the period. For example, they point out that the Fourth Lateran Council's (1215) refusal to distinguish between prohibitions about marrying kin based on consanguinity or affinity means that this distinction is not as clear or clean as one might expect (5).

39. David Sabean maintains that come the seventeenth century any sense of an equal blood-mixing disappeared in favor of protecting patrilineal descent (163). Valeria Finucci likewise cites the notion that blood-mixing was unequal across gendered lines with women's blood more affected than their male counterparts. She explains that women's "flesh was thought more humid, wet, porous, and penetrable than men's, and they had too many orifices; thus they were more able to pollute and be polluted. But then male semen, being refined blood, was theoretically always putrifiable (female semen, of course, was already putrefied as menstrual blood); and blood too could become corrupted through mixture with, say, bile" (26).

40. Of course, this notion that mothers would breastfeed their children in light of these considerations was an ideal rather than a reality since wet-nursing was common practice in the period. Evidence from *ricordanze* (diaries) of Florentine merchants (1300–1530) suggests that there is little concern over principles of contamination or the inheritance of unwanted character traits from wet nurses' breast milk: "The first criteria in the choice of a nurse were not her moral qualities. Girls who had been seduced, 'bestial' Tartar slavewomen, or mothers who had abandoned their children all made good nurses if their milk was 'young' and abundant. In fact, the social identity of the nurse often seems of little importance to parents" (Klapsich-Zuber 141). One reason why wet-nursing was so popular, in spite of the advice literature, might have to do with the downplaying of the mother's role and material contribution to the child in the first place (Klapsich-Zuber 162).

41. For instance, theologian William Gouge, author of the treatise *Of Domestical Duties*, argues that man and wife were one flesh only when all of the rituals in the Book of Common Prayer were followed to the letter (Cleland 13–14).

42. As many historians note, the precise beginning of marriage has a long history of intense theological, political, and social negotiation: Did it begin at the ceremony or upon its consummation? In the early Middle Ages, Papal Law said that marriage only involved mutual mental consent and rejected the relevance of physical consummation whatsoever, but this would change over the following centuries (Brundage 12). The Decretals of Pope Alexander III (c. 1159)—the marriage doctrine that remained practically unchanged until 1563—viewed sexual intercourse as essential to binding a marriage. As historian James Brundage notes, "A consummated marriage was indissoluble, but an unconsummated exchange in consent, even in the present tense, might be voided by subsequent actions of the parties. Sexual intercourse between spouses changed the nature of their union from consensual but still potentially dissoluble marriage into a more perfect union that was in principle indissoluble" (72).

In the social sphere of late medieval and early modern England, the completion of marriage seems to be understood as consummation in practically every case; what was in dispute was whether or not sex becomes licit after the spousal, after the public church ceremony, or at a date specified much later when the bride was escorted to the groom's house (and sometimes not even then) (Peters 84–86; Adair 146). The Protestant Reformation, on the one hand, looked to make the marriage contract a purely verbal affair, one that must be spoken in the present tense. As Balizet remarks, "the 'reformation' of marriage had the effect of replacing, in the eyes of the Church, the private act of sexual consummation with the public performance of solemnization as evidence of the marital union" (*Blood and Home* 28). Nonetheless, sex remained a central component of marriage even in the Protestant context, leading "Luther to advocate divorce in the case of impotence, adultery, desertion, absolute incompatibility, or the refusal of a spouse to have sex" (Wiesner-Hanks 63). And certainly consummation seems to be at marriage's center as it was not uncommon to sue for divorce on the basis of nonconsummation, even after very lengthy legal marriages (Adair 177). The implication seems to be that regardless of the Catholic/Protestant divide, a marriage was not really a marriage without the material act of consummation.

43. This romantic or idealized vocabulary of unity in marriage is fraught, as feminist critics encourage us to remember. For instance, Fran Dolan highlights that Genesis provided the idealized vision of union in the one-flesh model and "two become one, we are told, but only by means of compromise, friction, and loss. Since equality is understood as encouraging battles of the will, only a hierarchy, according to which one submits to the other, can resolve conflicts so that the occasional or apparent achievement of union becomes possible" (*Marriage and Violence* 26). Catherine Belsey likewise notes a dark hue in the Edenic imagery cited in discussions of marriage: "The comparison offers an ideal and idealizing account of the conjugal relation. At the same time, however, it brings with it an excess of meaning [namely, the threat of woman's betrayal and the Fall] which calls into question the very ideal it proclaims" (*Shakespeare and the Loss of Eden* 59).

44. Dolan notes that even Puritan writers viewed "sexual intercourse as the wellspring of marital plenitude, the primary means by which two become one, and yet each is more rather than less" (*Marriage and Violence* 47).

45. J. Bertaut likewise refers to the strict tie of blood that unites man and wife in his *Les oeuvres poetiques* (Paris, 1611).

46. The full title of Jean Pontas's text is as follows: *Dictionnaire de cas de conscience ou decisions des plus considerable difficultez touchant la morale et la discipline ecclesiastique, tirées de l'ecriture, des conciles, des decretales des papes, des peres, et des plus célebres théologiens et canonistes*, 2 vols. with supp. (Paris, 1715; with supp., 1718).

47. Historians debate the extent to which this advice about ideal marriage pairings actually bore out. Ann Jennalie Cook highlights "the Elizabethans' increasing difficulty in adhering to either the ideal of intrinsic worth or the ideal of absolute equity in the choice of a mate. . . . The increasing costs of their rank led peers to seek heirs or heiresses as spouses. As for the ambitious, wedding the right mate offered a ready shortcut to status and power" (46). Alan Macfarlane advocates the predominance of individual choice in coupling especially given the absence of codified rules (329). However, the absence of explicit rules does not mean that these persons did not face strong ideological pressures that were naturalized in blood. As historian Lawrence Stone mentions, marriage partners deemed acceptable (i.e., adequately similar) by kinship networks likely led to relatively successful marriages, "provided it is not undertaken purely for mercenary considerations and that there is not too great a

discrepancy in age, physical attractiveness or temperament, may well work out not too badly" (104).

48. Floyd-Wilson's *Occult Knowledge* seemingly does not place affinity in blood, insisting that the humoral framework is capacious, though it will not explain all emotions, "one's humoral complexion cannot function as an indicator of hidden sympathies and antipathies" (7). Yet, as I outline, blood seems not only to embody humoral complexion but also an extra quality including, as she describes, the hidden "sympathies found in nature [that] determine the enigmatically close connection one may feel with another person" (7). As Lemnius glosses in *The Secret Miracles of Nature*, knowledge of these affinities must be gained by experience since appearances are insufficient for their discovery; but there is nevertheless an assumption that these like natures will come together (Floyd-Wilson *Occult Knowledge* 7).

By contrast, for discussion on a contemporary *anxiety* around likeness and sympathy in this period due to Paracelsan notions of contagion and disease, see Langley's *Shakespeare's Contagious Sympathies* and Floyd-Wilson's "'Angry Mab with Blisters Plague.'"

49. Of course, as Patricia Crawford notes, the connection between shared blood and natural affection was complicated due to concerns for legitimacy: "Although shared blood made for 'natural' affection between father and child, the metaphor disguised complex paternal attitudes. Unless a man were married to the mother, he usually found it easier to deny any responsibility. Even though the father had a blood connection to an illegitimate child, neither he nor his contemporaries would expect the same natural affection as for his legitimate child" (131).

50. Katie Barclay points to additional evidence for this claim in correspondence by Sir Robert Montgomerie of Skelmorlie, who "explicitly tied the physical activity of breastfeeding to an emotional bond when he asked his gardener [John Todd] to send his wife 'to foster the chyld' in 1665" (87). Her work on the early modern Scottish family similarly concludes that "the centrality of affection to the construction of family [its embodiment in blood] placed emotion at the heart of national heritage" (93–94).

51. David Cressy provides many examples of colonists appealing to very remote or affinal kinship relations in their attempts to establish roots and curry favors in New England (47).

52. Zygmunt Bauman's theoretical work *Liquid Love* discusses the etymology of *affinity* in German and its distinction from kinship, but the word nevertheless gestures toward a desire to collapse the two concepts: "'Affinity' is kinship qualified—kinship but. . . . Choice is the qualifying factor: it transforms kinship into affinity. It also, however, betrays affinity's ambition: its intention to be like kinship, as unconditional, irrevocable and unbreakable as kinship is (eventually, affinity will weave into lineage and become indistinguishable from the rest of the kinship network; the affinity of one generation will turn into the kinship of the next)" (28).

53. Similarly, André Du Laurens's essay "Of Another Kinde of Melancholie" points to eye contact as the source or cause of love, which then circulates in the body from the head to the heart, liver, and genitals through blood and heat (1597).

54. Love madness was likewise cured by physiological means: "Bewitchment could only be cured by draining off the poisoned blood through phlebotomy, by purging of the black biles, or by coitus to draw off the offending surfeit of sperm. . . . Finally there was therapeutic marriage, or at least the physical joining of the lovers that automatically cures the disease" (Beecher "Quatrocento Views" 56–57). Langley likewise notes that this logic of falling in love via sight bore similarities to theories of vision in general. The extramissive vision theory

deriving from Platonic and Galenic philosophies "emphasized the likeness and sympathy between the subject's visual ray and the object's effluvia, describing how for vision to occur the twin subject and object beams must *correspond*, must interweave and lovingly mingle, forging a very material thread of connection or contagion between the viewer and viewed" (*Shakespeare's Contagious Sympathies* 67; italics mine). The marked similarity between the process of falling in love and that involved in the cri du sang helps to explain some later incest plots wherein characters claim that they unfortunately misunderstood kin recognition with falling in love at first sight.

55. Anita Guerreau-Jalabert posits that a renewed attention to blood in the mid-fourteenth century might mean that "an evolving medieval society tackled the problems of defining groups and of establishing criteria for delimiting their contours (with the illusion of being able to 'close' groups) with new vigor" (75). David Crouch places the trend of a self-conscious assertion of nobility even earlier: "The change in levels of literacy and symbolism meant that the noble man and noble woman of 1300 were much more self-conscious in their search for individual distinction" (123).

56. These diagrams first emerged in the manuscripts of the *Etymologie* by Isidor of Seville in the late twelfth century; the diagrams accompanied canon law in order to illustrate the nature of kinship ties as well as incest prohibitions.

57. Interestingly enough, in addition to proclaiming that their blood was royal and, therefore, meritorious, the Hapsburgs understood their blood to be melancholic, which they believed to be a suitable temperament for a ruler, military commander, artist, and scholar (Kugler 132).

58. A strange and interesting consequence of this genre is the privileging of nobility over religion since this audience valued antiquity and noble lineage. As D'Elia notes, "Better to trace one's roots back to Attila the Hun than to a Christian nobody" (D'Elia 72).

59. Padgett explains that "multiple, inconsistent stratification logics [wealth, political office, and age of family] made it very hard demographically to find someone to marry who was similar to you on all three dimensions at once. . . . Florentine elites, like elites everywhere, tried to close social mobility doors behind them. But their inability to converge on a single dominant criterion of status twisted their inconsistent closed-elite intentions into open-elite mobility for others. This created political turmoil, but it also renewed Renaissance Florence elites" (359).

60. Ellery Schalk's *From Valor to Pedigree* emphasizes the role of demonstrated military prowess and other virtuous activities in the Middle Ages as more indicative of nobility than biological lineage or pedigree, which she believes only became explicit in the 1590s. The nobility continued to be martially important during both the Hundred Years' War and the Wars of Religion. However, as many scholars have since noted, biological membership may have been simply taken for granted in the Middle Ages, though this became complicated in the early modern period with the increasing ranks of the *noblesse de robe*. Constance Bouchard, for instance, tracks the rise of this concern as early as the twelfth century, when nobility begin to avoid unequal marriage pairings because "knights and burghers began to pose a challenge to the integrity of the noble class which viscounts and castellans had not posed a century earlier" (56). Yet she holds that each order, generally speaking, would look to one order below them (royalty to counts, counts to viscounts, etc.) in order to avoid incest prohibitions but usually only for daughters (57). For a history of one particularly notable upstart Italian family in France, the Gondi, see the recent monograph by Joanna Milstein.

61. Art historian Rebecca Zorach, on the other hand, notices a roughly simultaneous push against this ideology in France's iconography of blood in the period. She maintains, "The association of blood with nobility—a heretofore facile equation—would change with the growth of *noblesse de robe* (noble status gained through administrative posts, necessary to maintain a growing governmental apparatus) and the sale of offices. Voracious popular devotion to the materiality of the ever-abundant blood of Christ's wounds caused consternation for reform-minded theologians" (33).

Devyver's work traces this discourse of Frankish blood that in the nineteenth and twentieth centuries was translated as anti-Semitism. According to this theory, the purified blood of nobility becomes attached to a racial myth of Aryan superiority (10). Uli Linke also marks blood's continued import as a sign of power and privilege into the modern era's biological/racist system of heredity, especially looking toward Nazi Germany.

62. See historians Robert Descimon and Elie Hadddad on the debate over from whence nobility came and the question of imprescriptibility; namely, did nobility arise from the office, or was this nobility imprinted on the person (i.e., prescribed) as a result of holding the office, which could then become hereditary? In the first camp, Gilles-André de la Rocque claims that those nobles of the office can live ennobled, but they die as non-nobles since these are not permanent elevations in status and does not change the man's nature (Descimon 100). Moreover, scholar Jacques de la Guesle (1598) insisted that one should be of hereditary nobility (*ex genere*) rather than noble by mere office (ex officio) in order to be a judge, for otherwise it would be unseemly for them to judge the actions of those above their original (nonelevated) station (Descimon 99–100). This issue is later complicated further by the notion of gradual nobility; offices given by the chancery did not bestow true nobility on the holders but those from sovereign decrees did, and bestowing "dignity" could then be hereditary (Descimon 106–7). For a recent study on the role of literary production and its relation to families' construction of nobility in Ancien Régime France, see Neil Kenny's *Born to Write.*

63. Many thanks to Deborah Shuger who pointed me toward this treatise.

64. Ferne does, however, mention a few complicating factors for the nobility of England particularly in the second portion of this text, "Lacies Nobility." This section outlines the genealogy of a particular family, and here, he reminds readers that there were effectively two types of nobles in England: the old nobility of English stock and the nobility created by the Norman conquerors. As a result of the displacement of the English, there might have been some people lurking in the commonwealth that had this older, noble blood (28–29).

65. Galen endorses this view of Hippocrates: "I too praise Hippocrates. For who does not observe that the bodies and the souls of people in the northern regions have a quite opposite disposition to those of people near to the fiery zone? Or who does not know that those between the two—those who inhabit the well-mixed country—are better in their bodies, in the character traits of their souls, their intelligence and discernment, than those people?" (*Capacities* 400).

66. Ruth Hill notes the connection between blood and early modern racial or ethnic ideologies in her study of the *blanqueamiento* or "whitening" equations like that of the Jesuit José Gumilla—namely a plan for reestablishing pure European bloodlines in colonial Spanish America. She makes the case that even if these charts only discuss color, they are clearly speaking about blood that is the causal agent of color, for "whiteness is the perceptible effect of an assumed cause or essence (European blood)" (54).

67. Greene maintains that older meanings of blood, such as nobility, sacrifice, heroism, and emotions, like love and passion, do persist in the early modern period, but their authority would begin to be questioned with the downfall of the particular discourses that supported them: the decline of feudalism, the Reformation, and the increasing popularity of experimental science (in this case the "debunking" of Galen through the discovery of the circulation of blood). Greene glosses the transformation as a move from ipseity, concern of how the self fits into a universal order/natural system, to identity, or how one is the same as or similar to others, which inspires a material attention to blood in new symbolic regimes of family, class, and race (110).

68. Neely highlights that, in discourses of lovesickness, Mediterranean bodies become more prone because of their naturally hotter humoral complexions, thus resulting in a trope of transgressive love. See also Lara Bovilsky who examines blood as one of her terms of "boundary marking language" (33), signifying a shift in the signification of race in the early modern period.

69. Dennis Britton, in his monograph *Becoming Christian*, makes a similar argument about non-Christian categories becoming more explicitly racialized in early modern England, for English Protestant theologians needed to assert that race or an inheritable faith could be enough for salvation in order to quell anxieties about young children who were too young to adequately protest faith in God (9).

70. It announces its project as follows, "In this volume theatrical practice and medical practice are found to converge in their approach to the regulation of blood as a source of identity and truth; medieval civic life intersects with seventeenth-century science and philosophy; the categories of class, race, gender and sexuality find in the language of blood as many mechanisms for differentiation as for homogeneity; and fields as disparate as pedagogical theory, alchemical cultures, phlebotomy, wet-nursing, and wine production emerge as historically and intellectually analogous" (Johnson and Decamp 2–3).

71. Feerick's "'Rude Uncivill Blood'" reads the pastorals of John Fletcher (*The Faithful Shepherdess*) and John Milton's masque (*Comus*) as critiques of the system of hereditary class structures. Craik's contribution to the edited collection *Blood Matters* examines the history plays *Henry IV* 2 and *Henry V* to show how blood's discourse functions in war settings. Namely, when aristocrats die and spill noble blood, they emerge as individual heroes (48), and by contrast "common soldiers' personhoods are instead yoked together in ways that make them appear plural, composite—and vulnerable. . . . These soldiers risk bleeding into one another, their identities valuable only insofar as they are adjuncts" (49–50). The work of Feerick and Craik directly counters the older, conservative criticism treating blood difference in Shakespeare by David S. Berkeley. He argues that Shakespeare's characters obey Galenic geohumoralism with regard to degrees of blood and class rank to the letter and, moreover, Shakespeare's plays align with and endorse this worldview. This conclusion is off-putting for those of us who want to see in Shakespeare a social critique of bodily determinist discourses.

72. The editors of the collection *Blood Matters* describe the phenomenon in this way: "Blood both does and does not do what is asked of it. While blood's circulation enables us to construct ideas about the organization of communities and the movements of their resources, it also denies the collective impulse by instead closing off the relationship and wealth to those whose racial, class, or familial blood sets them apart" (Johnson and Decamp 12).

73. See the authors included in the edited collections of Britton and Walker (*Rethinking Shakespeare Source Study*) and Marrapodi (*Shakespeare and Intertextuality*), many of whom

partake in the Theater without Borders research group. A small collection is listed here alphabetically, though it is by no means exhaustive: Catharine Belsey, Silvia Bigliazzi, Dennis Britton, Pamela Brown, Kent Cartwright, Brett Greatley-Hirsch, Laurie Johnson, Sarah Klann, Laurie Maguire, Michele Marrapodi, Robert Miola, Lori Humphrey Newcomb, Charlotte Pressler, Emma Smith, Melissa Walter, and Susanne Wofford.

74. Of course, other comparative scholars have done work in this way for years. For example, see Lorna Hutson's article, "On Not Being Deceived: Rhetoric and the Body in *Twelfth Night*," *Texas Studies in Literature and Language* 38, no.2 (1996): 140–174, which compares the Shakespeare play with its Italian forebear, the Academy of Siena's *Gl'Ingannati*. Another example is Hutson's extensive study of Ariosto's *La Cassaria* and *I Suppositi*—examined at length in their own right—as important precursors/points of influence for Jonson's *Volpone* and Shakespeare's *Taming of the Shrew* ("'Che indizio, che prova'").

75. Walter extends Clubb's theatergrams to include those plot devices or topoi that originate from novellas, what she dubs "novellagrams," which "transport narrative and symbolic material across national and generic boundaries (among Italy, France, Spain, and England, and from narrative to drama)" ("Dramatic Bodies" 63). Walter's *Italian Novella* examines Shakespeare's rendition of Italian(ate) sources, especially Boccaccio's *Decameron* and its imitators (Matteo Bandello, Marguerite de Navarre, William Painter) toward explicitly feminist ends, "draw[ing] on the explicitly dialogic form of novella collections, in which women's voices participate, and on a characteristically novellesque poetics of space, to imagine female comic character and authority" (3–4). Others, too, have appealed to the formal unit as a basis of comparison since it opens up a more flexible means for comparing national traditions (e.g., Rosalie Colie's "genre-theory" in *Resources of Kind* and Claudio Guillén's "literature as system"). Walter Cohen likewise notes the advent of similar genres and motifs at times of comparable cultural change across Eurasia—for example, the move to vernacular literatures and the novella tradition in conjunction with periods of economic recovery and urbanization.

76. Alluding to Miola's intertextuality and Clubb's theatergrams, Kent Cartwright maintains that these two models "make good on the methods of *imitatio* and *contaminatio*—imitating classical models and combining them in innovative ways—that were central to Renaissance humanism. Even better, they take us out of the elephant's boneyard of specific literary allusions and into the rich, expansive vistas of trans-European culture and literary tradition, now freshly available for comparative exploration" (193). Yet these models are not all that dissimilar from that used in Rosalie Colie's final book, *Shakespeare's Living Art*.

77. See Douglas Bruster's definition of New Historicism in *Quoting Shakespeare* (1–30).

78. Bruster defines *bricolage* as a collection of quotations that can include "such discrete components as words, events, and identities" (4). The study of these quotations and their patterns in a given text provides a fuller sense of an author's "position" or cultural world.

79. I do not presume to be the first to use this metaphor of familial relation, though I might be the first to explicitly connect it to blood. In "Looking Back," Clubb, reflecting on her work on "Shakespeare's kinship with Italy's theater," says that reading the Italian plays of the Renaissance "reveals a plurality of generations and combinations of features that suggests more complicated genetic connections and illuminates a method of composition different from imitation of discrete sources, one based instead on the existence of a repertory of theatergrams and of a default structure for each genre in which to combine them" (16–7). She hints at a kinship metaphor several other times throughout the article as well.

80. Elizabeth Fowler articulates the older, and somewhat looser, concept of family

resemblance as applied to genre theory as follows: "Representations of a genre may then be regarded as making up a family whose septs and individual members are related in various ways, without necessarily having any single feature shared in common by all" (Fowler quoted in Fishelov 123).

81. Carla Freccero's work on Rabelais, *Father Figures*, is one interesting application of this father-son metaphor to textual relations, using Lacanian psychoanalysis to demonstrate this relationship.

82. Cartwright describes the two models of relationality as follows: "One model, the diachronic, focuses on predecessor texts that leave distinctive linguistic traces in subsequent works as markers of allusion, borrowing, or adaptation; this is the field of traditional source study. The other model, the synchronic, attends more broadly to the loose matrix of plot devices, character types, and themes that . . . adumbrate Renaissance comedy as a culturally recombinant and expansive contemporaneous genre; this is the field that we shall treat as 'intertextuality'" (184). My method, in one sense, has it both ways: triangulating sources or looking for diachronic forebears to then compare synchronically.

83. The bed trick is a topos wherein a person is tricked into sleeping with someone by thinking that it is someone else.

84. This particular subset of the bed trick seemingly has divine sanction due to the importance of the progeny produced by a particular union. The most well-known example is Tamar tricking Judah into conceiving a child and thus founding the House of Israel (Gen. 38).

Chapter 1

1. As Catherine Belling tells us, "Metaphors of social or political bleeding were underpinned by the sense in which blood was a literally continuous link between individual human bodies and public states or hereditary bloodlines. The humoral system as it was understood in the Renaissance was a coherent part of the Neoplatonic paradigm in which the matter of each human person—each microcosm—was continuous with the matter of both social and wider, natural macrocosms. Individual and social bodies could be analogous because of a quite literal interpermeability of bodies across different scales in a humoral cosmology" ("Infectious Rape" 123).

2. The closest I could find to such a comparison was an offhand comment made by French scholar Georges Couton in his discussion of *Le Cid*, remarking, "Mais nul doute que Rodrigue épousera Chimène: Roméo et Juliette, pour une fois, connaîtront le bonheur, un jour, pas trop lointain" (But no doubt that Rodrigue will wed Chimène: Romeo and Juliet, for one instance, will know happiness, one day, not too far off) (20, my translation).

3. See Isabelle Schwartz-Gastine's "*Romeo and Juliet*." It was not until a new generation of artists, the Romantics, who looked to push against this aesthetic tradition, that this attitude changed—for instance, Eugène Delacroix and historian/politician François Guizot, who published Shakespeare's complete works in 1821 (85).

4. See Schwartz-Gastine (78).

5. See John Doran's "Shakespeare in France."

6. Daria Perocco points to this version as the first and ultimate source of the basic plot, though she maintains that Shakespeare likely was not familiar with this version. She furthers that Shakespeare bears more similarity to da Porto than Bandello because Boaistuau's translation of Bandello removes the moralist or didactic frame, warning of the excesses of love and desire (49–52). Silvia Bigliazzi likewise makes the case that Shakespeare's play most closely

resembles da Porto's version rather than the intermediaries of Bandello and Boaistuau, pointing to their treatment of gender and masculinity as a case study.

7. For more information on the development of this plot, see Jill Levenson's "Romeo and Juliet before Shakespeare." For references to other possible novella sources prior to the da Porto version, see Luigi Monga's "Romeo and Juliet Revisited."

8. By putting translation in scare quotes I mean to indicate translation both in its most literal sense (in this case translated into French by Belleforest and Boaistuau), and in the larger sense of a reinterpretation or rendition but that can involve, especially in this instance, quite substantial changes in content.

9. Alternatively, Schwartz-Gastine argues that Sevin translated Boccaccio's *Il Filocolo* instead, a Romeo and Juliet plot with a comic ending that allows the protagonists to marry (79).

10. Olin Moore's information on Sevin as Corneille's precursor comes from Henri Hauvette. Hauvette notes that this Juliet, named Burglipha, is more torn between love and duty than the Shakespearean iteration (334). She, unlike Shakespeare's Juliet, is distinctly angry: she calls him a traitor and says that he merits strangling. It is in this context that Hauvette points to Sevin's Burglipha as a likely predecessor for Corneille's Chimène (334).

11. Gustave L. van Roosbroeck argues in *The Cid Theme in France in 1600* that the love versus honor trope, as well as the psychological emphasis, is already present in a novel by A. du Périer *La Haine et l'amour d'Arnoul et de Clayremonde* and that this is the true origin of both Castro and Corneille. Roosbroeck's findings, however, do not disturb my larger point, which is that a Romeo and Juliet myth influenced *Le Cid*'s plot. Périer's novel, in one sense, is closer to Romeo and Juliet's premise because the violence is set in the context of a long-standing blood feud rather than begun by an act of political competition.

12. Translations provided here are for the ease of those who speak neither Italian nor French. All translations for this section are my own unless otherwise noted. Many will sound a bit awkward in English, but I have done so in order to give the reader the closest sense of the original.

13. Bandello sets the scene,"Furono giá al tempo dei signori de la Scala due famiglie in Verona tra l'altre di nobilitá e ricchezze molto famose, cioè i Montecchi e i Capelletti, le quali tra loro, che che se ne fosse cagione ebbero fiera e sanguinolente nemicizia, di modo che in diverse mischie, essendo ciascuna potente, molti ci morirono cosí di Montecchi e Capelletti come di seguaci che a quelli s'accostarono; il che di piú in piú i lor odii accrebbe" (There were already of the time of the Scala lords, two families in Verona, among the others of nobility and wealth very famous, namely the Montecchi and the Capelletti, who among themselves, that when there was not cause for there to be a proud and bloody enmity of the manner, in various quarrels, each one being powerful, many Montecchi and Capelletti died in this way, as with their followers as with them, they accosted each other; that more and more their hatred increased) (370–71).

14. Boaistuau describes the tragic turn as Fate betraying the two lovers, "La fortune (envieuse de leur prosperité) tourna sa roue pour les faire trebucher en tel abisme qu'il luy payerent l'usure de leurs plaisirs passez par une trescruelle et trespitoyable mort" (Fortune [envious of their prosperity] turned her wheel in order to make them stumble into such an abyss that would pay for the use of their past pleasures with a very cruel and pitiable death) (82).

15. Boaistuau narrates, "Mais ainsi que le plus souvent il y a envie entre ceux qui sont en pareil degré d'honneur, aussi survint-il quelque inimitié entr'eux, et combien que l'origine en fust leger et assez mal fondé, si est-ce que par intervalle de temps s'enflamma si bien

qu'en diverses menées qui se dresserent d'une part et d'autre, plusieurs y laisserent la vie" (But just as most often there is envy among those who are of similar rank so too there occurs some enmity between them, and regardless of how light and ill-founded its origin, if over an interval of time it is aggravated in many intrigues that one faction and the other rise up against each other, many will lose their lives to it) (64). While this might be implied in Shakespeare's prologue since the two families are "alike in dignity," it is certainly not spelled out to this degree.

16. For instance in Bandello's novella, Giulietta muses, "Io ho pure piu volte udito dire che per gli sposalizi fatti, non solamente tra privati cittadini e gentiluomini si sono de le paci fatte" (I have many times heard it said that with marriages made not only between private citizens and gentlemen was there peace established) (377). Here it is a simple social fact that marriages sometimes work to effect peace between rival factions. The tale's ending likewise illuminates the subordinate character of making peace between the two families. The resolution of the blood feud, though cited as motivation, receives little emphasis. It says only, "Il che fu cagione che tra i Montecchi e Capelletti si fece la pace, ben che non molto da poi durasse" (This was the cause/occasion that between the Montecchi and Capelletti peace was made, well that not much after then survived) (408). In Boaistuau, too, the peace effected between the two families is cited as a secondary reason for the lover's union: "Car peut estre que ceste nouvelle alliance engendrera une perpetuelle paix et amitié entre sa famille et la mienne" (Because maybe this new alliance will engender a perpetual peace and friendship between his family and mine) (73).

17. In Boaistuau's version, in fact, the first time that blood is mentioned it is as "bloodthirsty." The narrator recounts that everything will change for these lovers at the Easter holiday since "les hommes sanguinaires sont volontiers coustumiers, apres les bonnes festes, commettre les meschantes oeuvres" (bloodthirsty men are willingly accustomed, after nice feasts, to commit evil deeds) (82). The next mention of blood appears in a long list of body parts strewn about after the great fight: "La terre estoit tout couverte de bras, de jambes, de cuisses et de sang" (The earth was covered with arms, legs, thighs, and blood) (83). Moreover, when Bandello's Romeo tries to stop "la zuffa" (the quarrel, brawl) between the two families, he pleads with Tebaldo that they be good citizens and not spill any more blood "ché pur troppo sangue s'è sparso" (because too much blood is already spilt) (383). Boaistuau's Rhomeo, too, in trying to break up the fight, does not mention any love that he holds for the opposing side. He rather mentions the great shame and scandal that they should feel for putting the republic in disorder and asks Thibault to be content with the amount of blood already spilt: "Sois content de tant de sang respandu et de tant de meurtres commis le passé" (Be content with so much blood spilt and so many murders committed past) (83–84). Neither Romeo character proclaims a newfound kinship with Thibault nor hints that his marriage with a Capulet should have changed the status quo. Instead, Romeo could have just as easily cited this justification for peace without loving Giulietta/Julliette.

18. See the introduction for more information about blood mixing upon falling in love through the eyes, outlined for example by Marsilio Ficino. This moment in Bandello might also point toward Rodrigue's offer to die by Chimène's hands in order to salvage her honor, which he has tainted by killing her father.

19. This lament provides evidence for Boaistuau's Julliette as a predecessor for Corneille's Chimène, thus providing more textual evidence for a possible borrowing.

20. Lucy Munro continues that more than 150 plays contain directions for wounding of

some kind, which suggests the use of stage blood, *especially* when dialogue explicitly calls attention to it ("'They Eat Each Others Arms'" 78–79).

21. For example, in Baz Luhrmann's popular *Romeo + Juliet*, the presiding image is Juliet dressed in a pristine white shroud surrounded by lit candles (even after she has shot herself in the head). R. A. Foakes, writing on the violence in this play, likewise pays little attention to blood, which is unsurprising given that he analyzes the play alongside Luhrmann's film.

22. I am, in fact, reviving a very old intervention and reading made by Leo Kirschbaum on Shakespeare's use of stage blood, though he examines *Julius Caesar* and *Coriolanus*; he argues that when blood does appear onstage we should take it seriously, as important as poetry, character, or plot, and not write it off as spectacular, popular excess.

23. To elaborate, Paul Kottman's "Defying the Stars" aims to revise this common interpretive paradigm that the tragedy is "a conflict between lovers' individual desires and the reigning demands of family, civic, and social norms in relation to which those desires are formed" (1). He instead maintains that "Romeo and Juliet is the drama of a struggle for individual freedom and self-realization, and this drama has a tragic structure" (5–6). For a similar reading, see Harry Levin's "Form and Formality" and Lawrence Stone's *Family, Sex, and Marriage.* Finally, Ruth Nevo argues that the lovers are subjected to the violent chaos of the political setting (see "Tragic Form in *Romeo and Juliet*"). All of the interpretations above set aside the foundation of Romeo and Juliet's conflict of interest, namely the blood feud.

24. See Coppélia Kahn's "Coming of Age in Verona" and *Man's Estate*; Marianne Novy on "Violence, Love and Gender"; and Robert Appelbaum's "'Standing to the Wall.'"

25. On political riots and class conflicts, see Chris Fitter's "'The Quarrel"; on definitions of *civility*, see Glen Clark's article "Civil Mutinies."

26. Snyder details that ideology creates meaning by fabricating differences in *name* to solidify certain social imperatives. See Peter Holbrook (217) for the most recent version of Snyder's argument. Dympna C. Callaghan likewise attends to ideology in her reading, though hers emphasizes the play's role in ideological production rather than its critique. It bears repeating that I do not disagree with Snyder's claim about how ideology *actually* works; I contend that her argument minimizes the physical, literal stakes that underpin the feud's logic for the play's characters. We agree on the basic point that "the norms [of society] themselves bring about the tragedy" (Snyder "Ideology and the Feud" 95).

27. Matthew Spellberg operates with a premise entirely antithetical to my own. For him, "in Shakespeare's Verona, the means of understanding the world has been largely divorced from its material referentiality. Language itself has been lifted away from bodies and sensations; instead it hinges upon itself for definition" (63). In this account, the lovers Romeo and Juliet, by using a vocabulary of dreams, reinvigorate language with sensation and feeling; but though this project escapes the arbitrariness of the blood feud, it is also a naturally violent proposition. Like Snyder, I agree with Spellberg that "language twists and manipulates nature. The predominance of speech over touch is a central concern in a play that is tragically about the determinism of naming" (76). However, as I hope to show, the characters fail to realize this problem as such; if the problem is one of language, then, like Juliet, they can acknowledge that names can be changed. However, the characters appear to see their bodies acting out a conflict beyond their control.

28. Bonnie Johnson makes a similar appeal to examine discourses of blood in a more literal fashion, examining the role of corrupted blood in the lovers' green-sickness, thereby articulating cultural anxieties about "youth, desire, and dangers attending sexual maturation"

(135). She argues that the problem for Romeo and Juliet that explains their unsuitable desires is one of "bad weaning" (136). This application of Galenic theory to this play, however, does not accommodate our sympathy for the protagonists or their attempted rebellion; their bodies on this view are to blame for their tragic end and, intuitively, it seems that Veronese society is to blame.

29. Rebecca Totaro, by contrast, argues that the play disperses agency among the "actants" in "Fair Verona," which she labels an "epidemic assemblage." She maintains that the play troubles the neat and tidy readings that locate causal responsibility for the tragedy in any particular character, Fate, or the "ancient grudge" (which I locate in blood). However, locating agency in blood (or at least presumed agency in blood) helps to explain both the spectacle of Juliet's body and many of the characters' motivations (Tybalt's hatred, and why on God's green earth Friar Laurence believes in the power of the protagonists' marriage). Yet it is true that both the ubiquity of this discourse and the inability to locate blood's purported agency with any one individual or to any singular end highlights what I take to be the play's criticism of this bloody framework.

Other scholars have explored this bloody enmity in the context of early modern discourses of plague or contagion. Floyd-Wilson reads the "pernicious rage" as plague that is passed to the protagonists who are "death-marked" ("Angry Mab" 405–10). Floyd-Wilson's reading also accommodates anxieties around the principle of "natural affinity" put forward by natural philosophers Fracastoro and Paracelsus. But in this case, a person's sinful nature attracts its like in the contagion of plague ("Angry Mab" 411). Kathryn Schwarz in "Held in Common" reads plague as both literal and metaphorical to investigate the queer relationship between desire and plague or love and morbidity. Lastly, see Evelyn Tribble on a discussion of contagion as a larger metaphor for the malleability of memory.

30. For, as David Crouch notes, it was through the medium of blood that one expected the qualities of ancestors to be reproduced in their offspring, which included being the "*mortels ennemis*" of a rival house (127, 141). This belief gives further justification for Tybalt's anger, which does not view Romeo as an individual agent but merely as a vessel for Montague blood.

31. I would be willing to add Lord Capulet to this list based on Robert Watson's work "Lord Capulet's Lost Compromise." According to Watson, Capulet likely comes to Romeo's defense even after his murder of Tybalt (3.1.184–88 in Q2), though this line is ignored in most editions. But, of course, Lord Capulet's compromise like the friar's might be well-intentioned, but it does not escape society's existing, flawed discourse as his daughter Juliet's speech attempts. Watson agrees, however, that "Shakespeare was inviting mistrust of the formative consciousness underlying the feud and its analogues by provoking mistrust of, and impatience with, the rhetorical antitheses used to express it" (80).

32. Lynette Hunter likewise reads the play as Friar Laurence's attempt to cure the "canker" of Verona (the feuding families) with the salve of the protagonists' marriage, thereby illustrating the cultural tension between Galenic and Paracelsan methods in the period. Ariane Balizet interestingly glosses this marriage similarly, although her concern is with an ideology of domesticity as it pertains to kinship and larger social structures. Balizet reads the marriage of Romeo and Juliet as an effort to join (pun joint-ure) and heal dismembered physical, domestic, political bodies (i.e., the blood feud), and this marriage comes into effect after the lovers' deaths when the two patriarchs join hands in a second marriage ritual. Yet, as she notes, "Romeo and Juliet's dead bodies . . . underscore the fundamental fallibility and very real dangers presented by fantasies of early modern domestic order" ("Enamoured" par. 31).

33. Wendy Wall points to other aspects that might trouble the audience's satisfaction with the families' reconciliation. She remarks that in both Q1 and Q2 this "passage concerns competitive expenditure, each ratifies the protagonists' marriage after the fact, and each asserts fatherly control in a moment of grieving chaos" (162). While I agree with some of these observations, I disagree with the nature of the lovers' failure. Namely, I believe that the problem is their investment in a faulty ideology.

34. For an example of the argument that attributes the tragic plot to the workings of Fate as a cosmic force, emphasizing dreams, portents, and so on, see D. Douglas Waters, "Fate and Fortune in *Romeo and Juliet*." For formalist scholarship that views the lovers pulled by their tragic genre, see William C. Carroll, Martha Tuck Rozett, and Harry Levin.

35. Belling's article importantly highlights that women are typically the focus of tragic catharsis, for "when individual phlebotomy overlaps with curative measures applied to the body politic, only some bodies continue to function as microcosms of the social macrocosm; others come to be identified as expendable or excremental" ("Infectious Rape" 127).

36. This play was available to an early modern audience via Erasmus's Latin translation where it was published alongside *Hecuba* with the title *Hecuba et Iphigenia in Aulis* (1509). The story was likewise available in book 12 of Ovid's *Metamorphoses*.

37. Douglas Trevor's work aims to expose the problems with a strictly humoral analysis of the play, and Wolter Seuntjens, like myself, favors interpreting references to humoralism *literally*. Although a historical overview of this scientific framework and its connection to theories of passion constitutes the bulk of Seuntjens's article, he uses Romeo's description of love as smoke as his inroads to this discussion. For a completely different take on the body's symbolism as a "grotesque body," see Roland Knowles, "Carnival and Death," or, for an examination of the competing forces of medicine in the play, see William Kerwin's reading in *Beyond the Body*.

38. As Tiffany Stern notes, opening choruses were, more often than not, designed primarily for the purpose of early runs of the performance. They told the audience the play's plot just in case the actors faltered. Most of them, therefore, were cut upon publication. Hence, the survival of this chorus merits attention. Thank you to Elizabeth Tavares for pointing me to this reference.

39. An alternative way to gloss this line is not as a similarity in class or station but as a similarity by virtue of hating. As Dympna Callaghan argues, "In this passage the symmetry between the houses suggests an ominous familial resemblance. The ancient blood they share is bloodshed of enmity" (75).

40. The generational aspect of the blood feud is clearer in this text, signified in the third line, "From ancient grudge break to new mutiny," in contrast to Q1's version, "From civil broils broke into enmity," which does not emphasize a tradition of infighting. A few notes on Q1: I will occasionally record differences in Q1 in the notes. I have modernized the spellings from Q1 for the sake of ease. These quotations come from the Huntington's copy printed by John Danter [and Edward Allde] in 1597. On the whole, Q2 has a greater emphasis on blood than does Q1.

41. The Oxford English Dictionary's entry for *civil* in fact cites this line from *Romeo and Juliet*. Yet, interestingly, its citation substitutes *civil warre* for *civil blood*. It is unclear from which version this substitution originates, since even the recent Folger digital edition does not gloss *blood* as a word that varies between editions. But if, in fact, *warre* was at some point changed to *blood*, then the word should carry greater weight, encouraging the audience to

envision blood literally covering the families' hands, rather than treating the blood feud abstractly. For a longer consideration of early modern civility, see Glen Clark.

42. I might likewise proffer that "loins," taking the adjective "fatal," assume responsibility for this fatality rather than the "stars" of the following line. This complicates the picture for those who would argue for Fate's agency in this tragedy.

43. The prologue in Q1 is not only shorter (two lines short of a sonnet, in fact), but it has a very different order to the lines and does not use a vocabulary of removal. Instead, in reference to the familial anger it states: "And death-marked passage of their Parent's rage, / Is now the two hours traffic of our Stage" (9–10).

44. I would like to thank the undergraduates of Rebecca Lemon's senior Shakespeare seminar (Fall 2014) for this metrical analysis.

45. Edward Muir's study of the early modern Venetian vendetta, *Mad Blood Stirring*, similarly cites a biological basis for the revenge narrative: "When sufficiently endangered or provoked, a man's mad blood stirred, producing an irresistible flare of choler or anger, that emotion biologically induced by what we would call the fight response" (xxvi).

46. François Laroque places Benvolio's appeal in the context of early modern astrology. He underlines that one of Shakespeare's important alterations to the sources is setting the play's action in midsummer rather than Christmas time. By placing the action during the "dog days" of summer, under the influence of Cainus, would account for geohumoral "correspondences between the heat of the sun and the brightness of the stars, on the one hand, and the concomitant fury, violence, sexual excitement and combination of hate and love, on the other" (35).

47. Prince Escalus's mention of a "price" might also reference the law of the *wergild* or "man-price" as a means to control and limit revenge killings. This law required that the offending party pay a price for the man killed in order to satisfy the victim's family without blood for blood. However, the prince does not expressly establish this practice in the play's context; and certainly the call of blood for blood by the Capulet family means that such a custom is not already in place. This moment is absent in Q1. As an aside, the prince's worldview resembles that of *Le Cid*'s King Fernand, which I demonstrate later in this chapter.

48. Todd Pettigrew's recent work on medical practice in Shakespeare demonstrates that Friar Laurence would have been read as a particularly irresponsible and incapable medical practitioner (109–10). I agree with Pettigrew's reading in that Friar Laurence certainly fails to fully understand the consequences of the discourse he cites, still reducing persons' identities to their bodies in a way that is ultimately destructive.

49. Arguably, the background belief that blood marks the change in Juliet's identity—from girl to woman in Lord Capulet's mention of her "green sickness" (3.5.156)—contributes to the larger ideological framework (Johnson).

50. Paul Siegel maintains that the sexual love between Romeo and Juliet is one instance of love's larger cosmic force that will ultimately expel the hatred and disorder in Verona's society. For him, Romeo's attempt to make peace with Tybalt can be explained by this Christian, cosmic love emanating from his relationship with Juliet (386).

51. Q1's speech differs somewhat: "Tis but thy name that is mine enemy, / What's Montague? It is nor hand nor foot, / Nor arm nor face nor any other part. / What's in a name? . . ." (2.1). It then immediately launches into the rose sequence and omits any kind of circular definition for Romeo.

52. As Capulet cries, "How, how, how, how, chopt-logic? What is this? / 'Proud' and 'I

thank you' and 'I thank you not' / And yet 'not proud'? Mistress minion you, / Thank me no thankings nor proud me no prouds" (3.5.149–52). Heather James has done a more extensive reading of this scene and Capulet's hatred of language in her examination of Q2 Juliet.

53. As I show in the next section, Juliet is guilty of precisely what Chimène seeks to avoid. Due to her shared blood, Juliet is guilty of her own cousin's murder. In some sense, Juliet has her own blood on her hands.

54. As we have seen previously, Q1 is generally less bloody in its language than Q2. This likewise applies to Friar Laurence's monologue here: "What blood is this that stains the entrance / Of this marble stony monument? / What means these masterless and gory weapons? / Ah me I doubt, whose here? what Romeo dead? / Who and Paris too? what unlucky hour / Is accessory to so foul a sin?"

55. Again, Q1 lacks quite the same emphasis on blood. Lord Capulet's reaction attends to the dagger in his daughter's body, rather than the bloody result: "See Wife, this dagger hath mistook: / For (lo) the back is empty of young Montague / And it is sheathed in our Daughter's breast." Interesting about this moment, too, is that Lord Capulet does not appear to be moving toward peace as he intimates that this dagger *should* be in Romeo's breast. Without Q2's exclamation then, the Watchman's observation of Juliet's bleeding body is the final mention of blood in the play.

56. Boaistuau's translation might prefigure this humoral purging, though it remains a suggestion. Here the *sadness* of the two families washes away the family feud: "Les Montesches et les Capellets rendirent tant de larmes qu'avec leurs pleurs il evacuerent leurs coleres, de sorte que deslors il furent reconciliez, et ceux qui n'avoient peu estre moderez par aucune prudence ou conseil humain furent en fin vaincuz et reduicts par pitié" (The Montesches and the Capellets render so many tears that with their crying they expelled their anger, so that from that moment they were reconciled, and those who had not been curbed by any caution or human counsel were finally defeated and reduced by pity) (119).

57. This closure could be manufactured, of course, by stagings that render Juliet's death more palatable, if not aesthetic.

58. I will set the genre debate aside for the purposes of this project, but its ambiguous character is a matter of contention. The fact that Corneille changed the genre label, from a tragicomedy to a tragedy after the *Querelle du Cid* only heightens this debate. For more information on this play's formal relationship to the mixed genre tragicomedy, see Peter Bürger's article as well as Serge Doubrovsky's Lacanian reading. Robert J. Nelson characterizes *Le Cid* as a romance, and Cordell Black contends that the ambiguity of genre mirrors the play's central theme, "the fragility and impermanence of the state of happiness seemingly achieved by the protagonists" (73). Moreover, aside from generic considerations, much hinges upon one's reading of this ending: Whether the marriage ever takes place, and, if it does, is it a happy occasion? If one interprets Chimène as actively desiring the marriage with Rodrigue and only voicing dissent due to social pressure, then one can read into *Le Cid* the desired, albeit morally scandalous, comedic conclusion (see William Kibler, R. C. Knight, Georges May, and Mohammed Kowsar). Alternatively, one can highlight the inconclusive nature of the marriage as well as Chimène's concern and silence at the play's close, likely signifying a lack of consent (see Octave Nadal, H. C. Ault, David Clarke, Mitchell Greenberg's *Corneille, Classicism, and the Ruses of Symmetry*, and John Trethewey). Some critics also hold that the reading may vary depending upon the edition (1637 or the revised, final edition of 1660). While these critics emphasize the differences between the two versions due to the *Querelle*,

I am inclined to agree with C. J. Gossip, who argues that the changes made to the dénouement in the second version do not amount to very much. This is especially the case for my purposes since the attention to blood and conflict remains the same in both texts. I therefore use the 1660 edition as Corneille's final word on the subject.

59. This English translation and all others, unless otherwise noted, come from Roscoe Mongan's translation, originally published by Hinds and Noble Publishers in 1896 and now available on Project Gutenburg. I have chosen Mongan's prose translation since it is the most literal, but, on the whole, I recommend Richard Wilbur's translation, *"Le Cid" and "The Liar"* (Boston: Mariner Books, 2009) for a more enjoyable read that best approximates Corneille's lyricism.

60. Certainly, other scholars like Ralph Albanese have made this point—viewing in her silence "l'impossibilité radicale de réconcilier à ses yeux le parricide et le mariage" (the radical impossibility to reconcile in her eyes the patricide and the marriage) —though he does not understand it in the visceral way that Chimène seems to (369).

61. See Kristen Gager on the growing importance of blood to the definition of family, at least nominally, in early modern France. This emphasis coincided with legal restrictions on adoption, including denying adopted children the same inheritance rights afforded to biological ones (1–15). Moreover, Malbranche explains of the cri du sang phenomenon in his *De la recherche de la vérité*: "On voit dans une compagnie une personne dont l'air et les manières on de secrètes alliances avec la disposition présente de notre corps, sa vue nous touche et nous pénètre, nous sommes portés sans réflexion à l'aimer et à lui vouloir du bien" (quoted in Cherpack 8).

62. On this small point, I agree with Peter Bornedal. The quotation is his, but the comparison to *Romeo and Juliet* is mine. Generally speaking, Bornedal views King Fernand's command to Chimène as one of ultimate benevolence, the deus ex machina that allows the characters to escape the tragic consequences of their own code. But if we take the role of blood literally, and the mandate to maintain it, as Chimène surely does, then the King's proclamation does not look so generous.

63. See Georges Couton as well as Hélène Merlin-Kajman's monograph on absolutism for a rendition of the king as the new patriarch (211–18). For a study of the performance of fidelity in *Le Cid*'s politics, see Christy Pichichero ("Words"). By contrast, historian Kristen Neuschel's understanding of nobility and honor is likewise material or embodied, though not in blood or physiology so much as in embodied experience of interdependence. But, while her study tracks the transformation of honor from based in action to based in words, I would say that honor is not without a material foundation in blood that persists even after her period of study, the sixteenth century.

64. See Bénichou's *Morales du grand siècle* (68–73). See also John D. Lyons's *Tragedy of Origins* in which he maintains, "*Le Cid* exemplifies a tragic structure in which attention is drawn to the modification of values" (9). David Clarke expresses a similar view: "*Le Cid* offers neither an unqualified apology for the pretensions of a centralizing monarchical order nor an exaltation of the values of those who oppose that process," although the lovers represent an idealism that is admirable, though untenable (164–66). Greenberg's "Stage Whispers" likewise casts the play in this light, though this resistance is isolated to Chimène. Paul Scott views Don Fernand as an enfeebled monarch according to traditional measures (political, military, and judicial) but still reads his role as sympathetic since the contemporary audience would have identified him with Louis XIII, who successfully ruled nonetheless.

65. Other critics who consider blood in *Le Cid* generally hint toward my reading but do not explore its fullest implications; for instance, see Octave Nadal (163). Additionally, David Sabean says, "A central aspect of the rhetoric in Corneille's representation of the dynamics of kinship has to do with blood as a vector of alliance. Although sometimes this can be understood metaphorically, the widespread assumptions about the exchange of fluids in intercourse always point to a substantial, carnal, physical link that carries moral weight" (156). Sabean's observation does not lead, however, to a new reading of the play so much as provides evidence for a larger historical claim. Or, he does not attend to why it matters that this moral weight is carnal or physical rather than metaphorical. Scott also speaks about blood when discussing Rodrigue's fighting against the Moors, but he takes it in a spiritual, sacrificial direction (297).

66. This story cites various reasons for this decline but most include pressure from the rise of the bourgeoisie, the nobility's impoverishment, and the emergence of a strong centralized monarchy in Versailles court culture.

67. See Romaniello and Lipp's edited collection, *Contested Spaces of Nobility*. Additionally Christy Pichichero's "Pierre Corneille and Military Drama" argues that King Fernand successfully wields the system of *mérite*—indicating both personal worthiness and a kind of debt or mutual obligation in a patron-client relationship—to maintain his position in the "power negotiations between the principal agents of sovereignty, the monarch and the *grand capitaine*" (1093).

68. Keller's point is the foundation for his larger argument that "Rodrigue experiences the unconditional subjection to the king and the defense of Seville against the Moorish invasion as a liberation from the constraints of the traditional code of honor. He transfers his allegiance from his father and aristocratic family to the king and to Castile, which becomes his new symbolic family. Rodrigue's aristocratic identity and his strong sense of filial allegiance predestine him to become the hero of Castile and to personify a form of patriotism that becomes the condition of an increasingly centralized monarchical state and a growing national consciousness among Corneille's imaginary Castilians" (131). In general, Keller sees Corneille's *Le Cid* as ideologically participating in the creation of an absolutist nation-state. This is another difference between our readings, for I see Corneille's portrayal of blood's discourse as quite critical.

69. See the work of Pierre Giuliani for a history of blood onstage. He outlines its dramaturgic evolution from truly bloody violence to primarily verbal representations of suffering and the force of passions. Blood onstage—marked by the stage direction "ensanglanter la scene"—was standard practice in drama during Henry IV's era but faded during the time of Louis XIII and was officially prohibited by the time of Racine's theater. Giuliani notes that the *Querelle* may have effectively put an end to blood onstage (308). Hence, though Giuliani never says so explicitly, blood at least *could* have been present in early stagings of *Le Cid*. For work on this earlier "bloody" theater, see the monograph by Christian Biet.

70. This comparison is not a perfect one. Rodrigue's murder of the comte occurs before a prospective marriage to Chimène, whereas Romeo does not kill Tybalt until after he and Juliet have married. However, it is still true that Romeo would be guilty via blood inheritance of any and all Capulet blood spilt even prior to his personal act of violence. The predicament for Juliet should be roughly the same.

71. Katherine Ibbett's *Style of the State* underscores that early modern French audiences had a practice of identifying with passive female characters informed by the tradition of martyr plays from the first half of the seventeenth century (55).

72. Contributing to the blood discourse's atmospheric quality, and thus its oppressive force, is the fact that the characters cite other discourses that attach persons' characters to their bodies, for instance, metoposcopy, the science of reading a person's character in the lines on his/her forehead. Le Comte references Don Diègue's lines as evidence of his valor when considering whether Diègue's son Rodrigue is an appropriate suitor for Chimène, "Ses rides sur son front ont gravé ses exploits, / Et nous disent encore ce qu'il fut autrefois"("The furrows on his brow bear witness to [lit. have engraved his] exploits / and tell us still what he formerly was") (1.1.21–22). This is only one example, though I will refrain from listing all of them here.

73. It is difficult to identify the precise historical moment of the popularization of *vous* as formal address distinct from *tu*. As linguistic scholar Johannes Helmbrecht notes, "It is plausible to assume that this convention was developed during the Middle Ages around the courts of the French crown and the aristocracy" (190). Others locate its increased usage in the seventeenth-century French courts. In either case, it is likely that this distinction should hold in the cultural imaginary of Corneille's audience.

74. Rodrigue seemingly supports Chimène's reaction to the blood feud, agreeing that he should die, and further maintains that he should die by her hands. He offers himself for sacrifice until the very last moment, right before Chimène asks King Fernand if she must marry him.

75. William Goode casts this phenomenon as Chimène's inability to fulfill the "ethic of gloire," Corneille's conception of heroism that combines hand (action/deed), heart (feeling/desire), and mind (rational/ethical deliberation/duty) in a unified whole. See also Milorad Margitic's argument for a similar point and Mitchell Greenberg on the absolutist repression of femininity. Claire Carlin claims that Chimène attempts to express this female voice or an "ethic of care," although "her partial espousal of the male model based on the aristocratic code of honor places her in an awkward position" (24). However, it seems that the characters' values differ more along lines of class than of gender. For it seems that this "female" ethic does not properly belong to Chimène but more so to Elvire. Notably, Elvire is not a concern for most feminist readings that focus their attention on both Chimène and the Infante. Lastly, for a literature review of other feminist readings of *Le Cid*, see Claire Carlin (8–12).

76. Lesel Dawson recounts several different natural philosophical or magical explanations for this purported phenomenon, including that (1) the corpse retains some consciousness and bleeds in fury/desire for revenge, (2) the murderer's guilt causes rays to emanate from their eyes, which causes the remaining blood to warm and thus flow fresh, and (3) in the moment of death/murder, the perpetrator and victim exchange *spiritus*, much like lovers do with eye contact, and consequently blood flows due to the blood trying to make its way back to the murderer. Dawson importantly reminds us that "texts frequently depict the blood itself as crying for vengeance, probably alluding to God's words to Cain that 'The voice of thy brothers bloud crieth unto me out of the ground' (Genesis 4. 10)" ("In Every Wound" 156).

77. See especially Octave Nadal's reading of Rodrigue's love in *Le sentiment de l'amour* and Mary Jo Muratore's *Evolution of the Cornelian Heroine*.

78. My translation.

79. In *Contested Spaces*, Romaniello and Lipp make a very similar point in the introduction regarding the struggles of the nobility in early modern Europe more generally: "If a monarch could claim charismatic authority with signs and symbols of power, so too could the nobles" (10).

80. Marina Perkins's article "Message from the Margins" argues for a closer examination

of the Infante who has largely been treated as superfluous or a mere foil to the central protagonists, ultimately maintaining that her role "forges a connection with, and invites reflection upon, the world of the audience. Her misery, derived primarily from isolation and passivity, resonates with court culture in the nascent absolutist state of Louis XIII and his chief minister, Cardinal Richelieu" (520). I agree that though the Infante plays an important role in the play's social critique, at least when examining the play through the lens of blood's discourse, much of her role seems to illuminate how one might try to think creatively within the confines of blood's ideology rather than offer a fully alternative viewpoint as does another equally passive female character, Elvire.

81. Don Diègue likewise emphasizes this value when he tells Rodrigue that he has a greater duty to his country, to spill Moorish blood, than to protect the honor of his beloved, his "vaillantes mains / Se tremperont bien mieux au sang des Africains" ("their valiant hands will be more nobly steeped in the blood of Africans") (3.6.1093–94).

82. Marcus Keller similarly notes, "Not only does he [Fernand] consider himself the head of the body politic, he also perceives his kingdom as a collective household whose most precious good is the blood of his subjects. The king protects this blood which flows in each member of Castile and through which the Castilian community is bound together" (135). Yet one should add that King Fernand does not value blood *as such*, but as blood which can *fight* and *be shed* for him and his body politic. In other words, this is not a purely benevolent act, but a rather self-interested one. After all, King Fernand is willing to risk a great deal of violence to his body politic, these noble bodies, for his sake.

83. My translation.

84. Rodrigue seemingly invests in the king's economy of blood in this moment, too, crediting King Fernand and his empire for his own blood: "Je sais trop que je dois au bien de vostre Empire / Et le sang qui m'anime et l'air que je respire, / Et quand je les perdrai pour si digne objet, / Je ferai seulement le devoir d'un sujet" ("I know well that I must for the good of your Empire, and the blood that animates me and the air that I breathe and when I will lose them for so dignified an end, I will only do the duty of a subject") (4.3.1243–46).

85. This is John Lyon's argument in *Tragedy of Origins*.

86. For more information on these historical narratives, see Kristen Neuschel and Arlette Jouanna's *Saint Bartholomew's Day Massacre*. See also Penny Roberts on the king as physician of the body politic, a metaphor that became particularly relevant during the Wars of Religion because Catholics characterized Protestants as bad humors that required purgation.

87. In this moment, Agamemnon staunchly opposes Pyrrhe, who is intent on sacrificing Polyxène to fulfill his father Achilles's will. Agamemnon exclaims, "Quelle execrable horreur? Qui veit jamais cela / Qu'un homme trespassé dans sa tombe eust envie / D'un autre homme vivant, de son sang, de sa vie? / Vous rendriez vostre pere à chacun odieux, / Le voulant honorer d'actes injurieux" (What an appalling horror? Who could ever want that, that a man trespasses in his tomb would want another man living, of his blood, of his life? You would give back your father to someone odious in wanting to honor injurious acts) (3.1451–56).

88. Elvire and Juliet's Nurse, of course, would seem the more natural comparison to make. Both act as affectionate confidants to their respective mistresses, and both are largely pragmatic characters. However, their attitudes toward the dominant ideology of blood diverge due partly to the divergent attitudes of the heroines. Elvire acts as the counterpoint to Chimène's insistence that she honor her father's blood with bloodshed. Juliet's nurse,

by contrast, suggests that it might be alright to conform with her family's expectations and marry her father's chosen ally Paris after all (3.5.214–27).

Chapter 2

1. Neoplatonism founded by Plotinus in the classical period fused Plato's philosophy with religious, Pythagorean, and other doctrines. With regards to love, Neoplatonism suggests that an individual could become one with their beloved, integrating their souls together in a manner that allows one to have some access to the One (the omnipresent, transcendental Good). They, furthermore, achieve this union by virtue of their likeness. As Marsilio Ficino outlines in *Commentary on Plato's Symposium*, "Likeness is a certain nature which is the same in several things. For if I am like you, you also are necessarily like me. Therefore the same likeness which compels me to love you also forces you to love me. Moreover, the lover removes himself from himself and gives himself to the beloved. Therefore the beloved takes care of him as his own possession" (speech 2, ch. 8, 57).

2. Katherine Cleland's book asks a similar question though she places it in the context of contemporary debates around clandestine marriage. She notes interestingly that "the discourse of wedded love that many scholars now view as one of the period's defining legacies, a discourse that contributed to the rise of companionate marriage, thus fuels the first portrayal of sibling incest on the early modern stage" (137).

3. It is easy to see how comments about love based in likeness of persons, like Ludovico Dolce's *Della institution delle donne* (1545), could only be improved by perfect likeness. See introduction for more details.

4. Undoubtedly, blood could function similarly in other iterations of incest. The Oedipus myth, for instance, could easily be retold as the hero mistaking a cri du sang (his blood's natural affinity for his kin, in this case, his mother) for an erotic attraction. However, I am specifically interested in sibling incest because it captures the limit of this logic: blood's attraction to its exact likeness. This project will thus set aside parent-child forms of incest that are of interest to Freudian or psychoanalytic readings.

5. For a similar reading that takes this in the direction of promoting sameness for a homonationalist project (preventing foreign, alien bodies), see Chakravarty's "More Than Kin, Less Than Kind."

6. The connection between Middleton and Della Porta went unnoticed until D. J. Gordon's comparative article in 1941 (Clubb *Giambattista Della Porta* 289). Douglas Radcliff-Ulmstead cites strolling Italian companies in London to explain Della Porta's influence on English drama (238). Clubb has likewise attempted to trace *La Sorella*'s presence in England. First, she points to the appearance of these *commedia dell'arte* troupes as well as a scenario version of this plot available in the Correr Collection (*Giambattista Della Porta* 274). Clubb nonetheless admits that it is unlikely that this scenario version accounts for the actual transmission of this play, since the most obvious adaptations contain far more detail and proximity to the original source than the scenarios would have provided (275). John Jowett, in his recent introduction to the play in the collected works of Middleton, comments that Middleton likely read Della Porta in the original Italian given the many linguistic echoes (780). This dramatic link between Italy and England is especially important when looking toward Ford, since most scholars believe that his primary influences were English. If Ford were unfamiliar with Della Porta, then Middleton's play would be the natural vehicle for the sibling incest premise. Lisa Hopkins in her introduction to *'Tis Pity She's a Whore: A Critical Guide*

and Simon Barker, by contrast, cite Middleton's *Women Beware Women* and the relationship between the character Isabella and her uncle Hippolito as the inspiration for the incest plot (Barker 7). Hopkins mentions *No Wit* briefly, though she insists on the difference between the diffused threat in the comedy and the explicit and voluntary incest in the tragedy, making it uniquely threatening (*Critical Guide* 6). Yet, rather than discount this comparison due to the unrealized versus realized distinction, I hope to show that by focusing on the premise of sibling incest we might draw some interesting conclusions.

7. This story is not a perfect example of the sibling incest topos tracked throughout this chapter because the story involves incest twice over; they are both siblings (they share the same mother), but the woman is also the product of mother-son incest and thus she is his daughter. Additionally, unlike the dramatic cases examined here, the incest is never revealed to the protagonists, but instead, their mother is prompted by religious counsel to hide her shame and the sin from her children since they are already married—the deed is done and a sin unknown is not a sin committed. I would like to thank Susanne Wofford who pointed me in the direction of this tale in conversation.

8. My translation here.

9. Though this is one of the story's closing comments, it is not the central moral of this tale and acts more as an aside. The mother figure is at the tale's center, and consequently, the lesson pertains to the fruitlessness of trying to control human desire and the lengths that it will go to exert itself upon the human flesh, at least without God's help.

10. In fairness, I should note that Pullini's comments are pulled from an article in which he claims that Della Porta's innovation is the "stile di transizione" (style of transition) between the sixteenth and seventeenth centuries, characterized by an "espressione più viva" (more lively expression) (299).

11. My translation.

12. This scholarship begins with Clubb's translation, *Gli Duoi Fratelli Rivalli: The Two Rival Brothers* (Berkeley: University of California Press, 1980) and includes work by Robert S. Miola, *Shakespeare and Classical Comedy: The Influence of Plautus and Terence* (Oxford: Clarendon Press, 1994); Michele Marrapodi, *Shakespeare and Renaissance Literary Theories: Anglo-Italian Transactions* (Farnham: Ashgate, 2011); and Salvatore Di Maria, *The Poetics of Imitation in the Italian Theatre of the Renaissance* (Toronto: University of Toronto Press, 2013). A dissertation by Nicole Prunster treats Della Porta's theater in more detail in a structuralist examination of his oeuvre. However, she only alludes to *La Sorella* in passing.

13. Clubb, in addition to her primary goal of examining influence, glosses Della Porta's use of incest in this way: "Although Della Porta has not yet moved from marvelous situations and caricatures to marvels of virtue and feeling, he packs the various reconciliations in *Sorella* with pathos and serious emotions, and develops the threat of incest hanging over the lover melodramatically, if not tragically" (*Giambattista Della Porta* 201). Donald Beecher's introduction to the play's translation likewise emphasizes its formal elements including Della Porta's use of discovery (*anagnorisis*) and reversal (*peripetia*) as his means to incite emotion and wonder in the audience. He further argues that incest would have served as one of the greatest obstacles appropriate for inciting emotion, and, moreover, that Della Porta would have expected his audience to feel horrified at its discovery ("Introduction to *The Sister*" 379). The fact that Della Porta's use of incest was likely one of the first comedies to do so supports Beecher's reading. Rachel Poulsen ultimately follows this tradition, arguing that Della Porta's mannerist, baroque work is a formal experiment in which he tests the limits of mimesis (190).

14. While I agree with Gurney's more general thesis about the role of war-time drama, I do not see evidence for this reasoning in the play. Though involvement with the Ottomans is the occasion for the mistaken incest threat, it is unclear the extent to which it can be blamed. For one, Attilio and Cleria/Sofia are never actually in Constantinople during the play's action; second, Trinca seems to be clearly reaching for this as an excuse for the siblings' behavior when speaking to their father, Pardo, and thus, I am not sure that the audience is meant to take it seriously.

15. All quotations of *La Sorella* in the original Italian are from the Gutenburg Project and, consequently, are without page or line numbers. The translations provided are my own in order to provide a more literal rendition of the Italian, though they are based on Bruno Ferraro and Donald Beecher's translation in *Renaissance Comedy.*

16. As Gordon underscores, one of the key differences between Della Porta's text and Middleton's rendition in *No Wit* is that Sunset does not know about this trick whereas Orgio does, and Sunset, therefore, would only have allowed the incest to go forward unknowingly (413).

17. According to the play's lovers, the body bears testimony of one's emotional state, character, and love for another, though these appeals generally flop. Erotico assures Balia of his love for Sulpizia by insisting upon her astonishing beauty that corresponds to her soul: "Nella maestà del suo volto vi riluce una spezie d'imperio reale, che mi risveglia l'animo a gran desidèri di gloria e m'innalza con gli occhi dell'intelletto a considerar quella dell'animo suo senza pari" (In the majesty of her face shines in a manner of imperial and regal that awakens my soul to great hopes of glory and raises me with the eyes of my intellect to contemplate this from her soul without equal) (1.2). Erotico's invocation of the physiognomic trope *kalokagathia,* that is, that beauty necessarily signifies inner virtue, is meant in all seriousness. Della Porta's references to physiognomy are unsurprising given his scientific endeavors including one of his most famous works, *De Humana Physiognomonia* (1586). However, while Della Porta likely invested in this scientific theory, his drama is often in tension with it, suggesting that these scientific logics might be rhetorical rather than purely descriptive. And as a common trope of love poetry, Erotico's citation betrays its status as rhetorical rather than scientific fact.

18. It does seem strange that the lovers would be so concerned with having the appropriate legal status of marriage given the fact that the couplings of both sets of lovers have been consummated (as Erotico explains, "I nostri amori non son stati sterili," literally, our loves did not stay sterile [1. 3]). This might imply that the legal/social recognition of marriage is ultimately more important than the bodily foundations, the consummation and the newfound kinship relation, that it is meant to reflect or signify.

19. The language here, of course, likewise highlights the easy confusion of familial affinity with erotic desire: "desiderio" (desire or longing) then emphasizes carnality "carne" (flesh) and "pelle" (skin). Many thanks to the anonymous reviewer for this connection.

20. On the whole, Middleton's rendition makes only a few changes with respect to the original incest-averted plot, but it shifts the setting/context significantly. Firstly, the change from Italy to England means that the Dutch have replaced the Ottomans. The role of the nurse is removed as well as the secondary clowning plot between the parasite and the braggart soldier; the latter is replaced by another comic character, the fool Weatherwise who overly invests in the almanac. Lastly, and most significantly, Middleton adds the primary plot involving the wealthy widow of Sir Avarice Goldenfleece and her many suitors. Kate

Low-water, arguably the play's protagonist, decides to cross-dress and suit for the lady's hand in order to obtain the inheritance that has been denied to her husband, the widow's relative.

21. Following Muir's observation, there have been some very interesting New Historicist studies (e.g., Mario DiGangi's *Homoerotics*; David Bergeron's "Middleton's *No Wit*"; and Marjorie Rubright's *Doppelgänger Dilemmas*), but most are tangential to this project since they do not address the Twilight plot in great detail. For the play's literary, editorial, and performance history, see John Jowett's article "Middleton's *No Wit* at the Fortune" and introduction to the play in Taylor et al., *Thomas Middleton*. For readings that focus on the play's Wooing Widow plot, see Caroline Lockett Cherry's work examining the play's female characters, as well as DiGangi's *Homoerotics*, which examines the relationship between the Widow and Mistress Low-water. For an interesting reading of the influence of civic pageantry on this play, see Bergeron, "Middleton's *No Wit*." And lastly, for an explication of the Dutch merchant, see Marianne Montgomery's "Listening to the Emissary" and chapter 3 of Marjorie Rubright's *Döppelganger Dilemmas*.

22. Most previous studies grapple, in one way or another, with the uncomfortable tone of Middleton's play, which I gloss with respect to incest, though scholars have explored other alternatives, especially the play's genre. See Michelle O'Callaghan, George Rowe, and Swapan Chakravorty.

23. In fact, one might suggest that Middleton's play embraces a certain love between siblings that flirts with incest since it outweighs conjugal love. In the play's conclusion, Beveril, instead of exclaiming how fortunate he is to be bound to his love the Lady Goldenfleece, expresses how much he loves his sister Kate: "Whoever knew more happiness in less compass? / Ne'er was poor gentleman so bound to a sister / As I am" (5.1.471–73).

24. One can argue that this critique of bodily determinism does not end here but likewise applies to the parody of astrological determinism embodied in the character Weatherwise. See Katherine Walker's "Daring to Pry," where she analyzes mock almanacs.

25. See the historical section of the introduction for information on kin or affine relations made by virtue of sharing a nursemaid.

26. It is true that Lady Goldenfleece is ignorant of the impending marriage plans that would effect incestuous relations, which somewhat absolves her in comparison to Della Porta's Orgio.

27. Kate Low-water intuits this motivation from the very beginning of the play in her attempt to untie the mysterious riddle of the two girls to which Lady Goldenfleece hints. She says to her "cousin" Jane, "And now I'm put i'th'mind on't, I believe / It was some piece of land or money given / By some departing friend upon their deathbed, / Perhaps to yourself, and Sir Oliver's daughter / May wrongfully enjoy it" (1.2.49–52). Kate's suggestion is not far from the truth, though what has been stolen is even more serious than she imagined, for it is a stolen heritage, birthright, or bloodline. Later, in an aside to herself, Kate murmurs, "Sure in that oath of hers there sleeps some wrong / Done to my kinswoman" (1.2.63–64). And this is true; however, her real kinswoman benefited from this swap, and she refers to the wrong girl unknowingly.

28. Kate Low-water, albeit falsely, attributes a similar motive to Lady Goldenfleece with regards to the secret about the girls' swap. She says to Jane, "The course of her whole life, and her dead husband's, / Was ever full of such dishonest riddles / To keep right heirs from knowledge of their own" (1.2.45–47). Citing Lady Goldenfleece's greed as motivation is reminiscent of Orgio's in the source text. And though Goldenfleece is innocent with regards

to this plot, Mistress Low-water is right to complain of the widow's greed that overrides her duty to her kin when distributing her late husband's wealth.

29. Sir Gilbert, of course, is not only nefarious by dint of this avaricious ambition but also because he intends to be unfaithful to Lady Goldenfleece after marrying her, taking Kate Low-water as his mistress.

30. See Lawrence Stone and Alan Macfarlane on this change in the marriage market in early modern England.

31. Even wit is defined in terms of wealth as if this is the common denominator to which all things, persons, even virtues can be reduced. Wit here is defined as thriftiness. Mistress Low-water, when looking for a wit, asks Pickadille for advice about where to find one; he says he will go to an ordinary: "Because those that go to an ordinary dine better for twelve pence than he that goes to a tavern for his five shillings; and I think those have the best wits that can save four shillings and fare better too" (2.3.232–35).

32. Admittedly, I mention this to highlight a further point of comparison, though my analysis will not take up Annabella's unborn baby in detail. See Claudine Defaye on how Annabella's pregnancy creates a problem for Giovanni's incestuous love (37).

33. Literary scholarship has demonstrated many other possible influences. For instance, Ford's play might include elements of city comedy (see Verna Foster and the influence of Middleton's *Women Beware Women*). For comments on Ford's choice of setting, see Simon Barker on Italy as a place of excess and appetite, Janet Clare on private revenge, and Keir Elam on courtesans. Hopkins's introduction to *A Critical Guide* connects *'Tis Pity* to many historical sources such as a rumor of incest in Ford's family history, the incestuous ill-doings of the Borgia and Hapsburg families, Sannazaro's encomium on Venice, and William Harvey's *De Motu Cordis* (1628). This list of potential sources also includes numerous Shakespeare plays: *Romeo and Juliet* (Kenneth Muir "Case of John Ford," Simon Barker, Robyn Bolam, Raymond Powell, and Hopkins "Speaking Sweat"), *Othello* (Raymond Powell), *Richard III* (Hopkins, "Wooing Scenes"), and even *A Midsummer Night's Dream* (Hopkins *A Critical Guide*).

34. My gloss remains in tension with other more romantic (and sympathetic) characterizations of the protagonists' love. See Arthur L. Kistner and M. K. Kistner, Alan Brissenden, and Dorothy Farr. I do not necessarily look to morally condemn Giovanni any more than the plays' other characters, although, make no mistake, I dislike him; but I maintain that the assumptions that support his love are the more problematic.

35. Christoph Ehland analyzes Giovanni's character in terms of discourses around early modern happiness, struggling between happiness defined as salvation versus earthly individual happiness. Ehland argues, "Giovanni repeatedly tries to prove that he acts virtuously when he follows his inner passions, passions he identifies as his fate, the expression of his inner self, though repudiated by society and the church" (102). Thus, though maybe perverse, on this reading Giovanni becomes a "harbinger of modern individualism" (98). Kibrina Davey chooses to diagnose the lovers with an immoderate passion deemed pathological as in Richard Burton's description of heroical love melancholy. Lastly, for a reading of the play as a political critique of Catholicism in the Carolinian court via its presentation of clandestine marriage, see Cleland's conclusion in *Irregular Unions*.

36. For a more extensive demonstration of Giovanni's citation of the "cult of love" and the (Neo)Platonic discourse, see G. F. Sensabaugh to whom my reading is indebted. Sensabaugh argues, however, that Ford himself firmly believed in this discourse of love above all, which excuses all manner of sins (211). This reading, however, would imply that the

audience should sympathize and fully excuse the characters' actions, which seems unlikely. On a related note, narcissism undoubtedly underlies Neoplatonic love since to love someone most like yourself is not all that different from loving yourself in another. Many scholars take up this comparison; see Richard McCabe, Charles Forker, and Lesel Dawson's *Lovesickness*.

37. Like many critics, I focus my attention on Giovanni's character more so than on Annabella's, but there is some wonderful feminist criticism of this play. See especially Susannah Mintz, Nathaniel Strout, and Susan Wiseman.

38. Clearly Giovanni's rhetoric has a religious bent, and many scholars have explored Ford's play in relation to historical religious commentary. See Bruce Boehrer's "'Nice Philosophy,'" Laurel Amtower, Gilles Monsarrat, and Richard McCabe. In consort with these readings, one could argue that the lovers search for religion where there otherwise is none; see Robert Ornstein, Larry Champion, and William Hamlin.

39. As Rachel Poulsen points out in her discussion of *La Sorella*, early modern persons regarded incest as "natural law," though it was likewise known that its actual boundaries were determined by Church and State depending on the societal need for either exogamy or kinship ties. She reiterates that "it was not unheard-of for dynastic families to grant special permission from the pope to solemnize marriages that were technically against the law" (185). Moreover, the Church already has a vexed relationship to incestuous relationships due to their prevalence in Genesis: Abraham, Lot, Noah, Judah, Amnon, Cain, Abel, and so forth (see Bauman 15–16).

40. Richard McCabe cites this corruption as the reason for our sympathy with the protagonists, who are merely scapegoats for Parma's corrupt world. See also Verna Foster and Bruce Boehrer's *Monarchy and Incest*.

41. Obviously, blood's relation to class distinctions plays a large role in the competition for Annabella's hand. The Cardinal even pardons Grimaldi and his mistaken murder of Bergetto on account of his blood, "He is no common man, but nobly born; / Of princes' blood," whereas those of lesser blood like Vasques have no such considerations (3.9.57–58). Terri Clerico's reading takes up this aspect of the blood discourse at length, so I will leave that to her.

42. My argument is thus in conversation with existing criticism on the problem of epistemology in *'Tis Pity* and searching for truth in the body (see Hopkins's "Knowing Their Loves" and Zenón Luis-Martínez). For the play's relationship to dissection and the anatomy theater, see Hillary Nunn and Christian Billing.

43. Clerico views Friar Bonaventura's suggestion of bloodletting as metaphorically equated with exogamous marriage rather than repentance (430). Nonetheless, our arguments are potentially complementary since exogamous marriage would change the nature of one's blood and thus potentially cleanse its sinful character in the process. I would note, too, the rather uncanny resemblance of this friar to that of Friar Lawrence in *Romeo and Juliet*, not only in their character functions but also in their philosophical positions.

44. Hillary Nunn and Pompa Banerjee both gloss this moment as the extension of Giovanni's Neoplatonic fantasy, incorporating Annabella's substance, her heart, into himself by eating it. For more extensive readings on the heart's iconography and emblematic status, see Dawson's *Lovesickness*, Amtower, Hopkins's "Speaking Sweat," Donald K. Anderson Jr., Denis Gauer, and William Slights.

45. Many scholars also cite this moment of stabbing as a violent, symbolic representation of their sexual relationship (see Nunn 146).

46. For a more extended consideration of the Anabella-Putana relationship contextualized in historical codes of servitude, see Iman Sheeha's "'I Have Paid the Duty.'"

Chapter 3

1. One should note that this is not Lee Edelman's Child with a capital "C." Though it may bear some similarities—after all, the biblical tradition has this child ensuring the *future* of this bloodline's manifest destiny to produce a line of kings—this child is not, in and of itself, the ultimate good (embodying the virtues of hope or futurity). Instead, this child carries a *particular* familial lineage into the future, conceptualized as a continuity in blood.

2. Seemingly limited historical work has been done to trace either *La mandragola*'s reception in Tudor England or an explanation for the lack thereof. Ronald Martinez, for instance, cites the play as "a remote source of Shakespeare's *Cymbeline*," but he does not give any indication of how it got there ("Comedian, Tragedian" 206). Victoria Kahn has recently done a great deal of work on Machiavelli's reception in Europe, though she makes no mention of the play, focusing instead on *The Prince* and *The Discourses*. Though as historian Michael Wyatt notes, the publisher John Wolfe printed the *Mandragola* in England in a collection including Machiavelli's *L'asino d'oro* in 1588 (191). One complicating factor is that Wolfe's Italian works, while published in England, may have been designed primarily for export because Machiavelli was on the Prohibited Index of 1559 (Moulton 305). Of course, if most copies were exported, then this would explain why *La mandragola* might not have influenced English drama as one might expect. Michael Wyatt further claims that the comedy influenced John Florio's English-Italian dictionary, *Worlde of Wordes*, though it is unclear how much of this is conjecture (231). Robert Williams makes a case for the play's influence on Middleton's Touchwood Senior plot in *The Chaste Maid of Cheapside* (c. 1610), but this case is based on the plays' similarities (385). Lastly, though the play may have been available in the Italian original, I have yet to find any word about an English translation.

3. *The Mandragola* has a sizable performance history in Italy both before and after Machiavelli's appearance on the Index Librorum Prohibitorum of 1559, instituted by Pope Paul IV; but performances were probably limited to Italy. The play "enjoyed prestigious early performances (Rome 1520, Venice 1522, and Florence 1526) and was anthologized [with other drama] both early (1525) and later (1554) in the century; it was performed by the Accademia Olimpica in Vicenza in 1564" (Martinez "Comedian, Tragedian" 206). For more on the performance history of this play, especially in Venice, see Giorgio Padoan.

Yet the play's print history is another story. The manuscript, dated 1519, was only discovered in the 1960s when it was found among the papers of Lorenzo the Magnificent. Mario Monaco has meticulously outlined the many editions and difficulties in dating. The first three Italian editions likely date around 1524, though the first dated editions include one in Venice (1531) and another in Florence (1533). This play may not have circulated as widely because the comedies were often excluded from collected editions of Machiavelli's work, such as that by *Bladiana* (1549) published in Florence. The first Italian edition of *all* of Machiavelli's works together probably appeared sometime between 1609 and 1619, and this was not reissued until later in the century (1629, 1645, and 1650). More editions of this play were available throughout the eighteenth and nineteenth centuries, although there was still little literary criticism. *La mandragola* was translated into French, however, because Machiavelli's condemnation by the Catholic Church was far less felt in France than Italy; Jean de la Fontaine did an adaptation for one of his tales, but the play may not have been translated in full until Briencourt (1664) and then by Rousseau in 1725. Nonetheless, for every one French translation of *La mandragola* there were six of the *Il Principe* (Beecher "Introduction to *The Mandragola* " 103–14).

4. See Melissa Walter's *Italian Novella and Shakespeare's Comic Heroines* (19–22) for extensive discussion of the various ways in which Shakespeare might have had access to translations of Boccaccio in either English or French through commissions by de Navarre, imitators of Boccaccio, or direct access to the Italian since it is not as simple as just Painter's *Palace of Pleasure* (1566). Moreover, even if one can make a case instead for Painter's translation (1.38) of Boccaccio's tale over the original, frankly the two versions differ very little (K. Muir *Sources* 172). Kirsten Inglis argues that the problematic nature of *All's Well* might be the fact that Shakespeare was not only inspired by Giletta but also the stories that surround it in Painter's translation (81). Shakespeare's changes to the Boccaccian source are well-documented. Aside from the Parolles subplot, the play adds a starker class difference between Bertram and Helena. Boccaccio's Giletta comes from a wealthy family and has several attendant suitors fighting for her hand since being a wealthy physician's daughter was practically the equivalent of a knighthood (Cassell 78–79). The King's related speech on nobility in birth or virtue is another Shakespearean addition.

5. Scholars, strangely, have not cited Giletta's tale specifically as a source for Machiavelli's play, even though both bed tricks have a child as their desired outcome. The exception to this claim is the master's thesis of Inglis, where she names this tale and (3.6) as the common sources for Machiavelli's play, Marston's *The Malcontent*, and Shakespeare's *All's Well*. However, rather than focus on the bed trick's biblical cast, she cites the cuckold figure as the common thread, leading to rather different results. The seminal work by Douglas Radcliff-Ulmstead repeatedly cites various novelle from Boccaccio's *Decameron* as sources, and Charles Singleton cites Boccaccio's influence but without pointing to any specific novella. Gay Bardin's more recent work cites three novelle as fundamental: Lodovico and Madonna Beatrice (7.7), Ambrogiuolo and Ginevra (2.9), and Ricciardo Minutolo (3.6) (5). Di Maria's "Machiavelli On Stage" makes a strong case for the last of these three. However, one should note that it is a different kind of bed trick: Catella goes into the baths believing the man to be her philandering husband, but she instead finds Ricciardo who blackmails her into keeping it secret and continuing their affair.

6. Citations from Boccaccio in the original Italian as well as the translation come from Brown University's Decameron Web (and thus are without page numbers). The English translation posted there is by J. M. Rigg, originally published in London in 1903.

7. See Carl Dennis, Claire McEachern, Helen Wilcox, and Alison Findlay. See also Joshua Avery's "Faith in the Unseen," which highlights Helena's actions as motivated by a sacramental vision.

8. W. L. Godshalk calls attention to the lack of God in Helena's motivation for the cure despite the talk of God as the agent of the king's restoration to health (64). Michele Osherow, however, claims that God's absence in *All's Well* is due to a parallel absence in Tamar's story (165). I suppose I take issue with two aspects of Osherow's account. First, the divine sanction of Tamar's bed trick is already a foregone conclusion due to its context, regardless of whether it is specifically referenced in this story. Second, given that Boccaccio includes a very explicit divine sanction via the narrator Neifile, one would expect its presence in Shakespeare unless there were a specific reason *not* to include it.

9. Shakespeare likely composed *All's Well* with this title some time between 1590 and

1592. It was only revised as *All's Well That Ends Well* around ten years later, c. 1602 (see Claire McEachern's introduction to *All's Well* in *The Complete Pelican Shakespeare*).

10. This reading resembles that of Kaara Peterson who writes on the bed trick as the crafty virgin's way to cure her own greensickness (380).

11. The very quick change of heart on Bertram's part, for example, is much easier to stomach in the novella genre. Beltramo's change of heart happens in one clause in which Beltramo "laid aside his harsh obduracy" ("pose giú la sua ostinata gravezza"). Many Shakespeare critics have commented on *All's Well's* experimentation with genre: Steven Mentz and Pamela D. Stewart (the "Doctor She" plot of the Italian novella and "Idolator" plot of the Cupid/Psyche myth), Gerard Gross (comedy under the lens of realism), Ian Donaldson (the "paradox of comedy"), Alexander Leggatt (conflict between romantic and realist genres), Bergeron's "'The Credit of Your Father,'" and Michael Shurgot (a fairy-tale genre requiring the miraculous healing of both the King and Bertram), Robert Turner and W. David Kay (a failed attempt at a "prodigal son" play or a successful one), Godshalk (the morality play), Regina Buccola (popular beliefs about fairies), Patricia Parker's "*All's Well That Ends Well*" (the early modern tradition of verbal *copia*), and Clubb's *Italian Drama*, and Marrapodi's "From Narrative" (mixture of a variety of Italian topoi or "theatergrams").

12. Older criticism of this play often turned on how scholars interpreted the two protagonists and with whom they sympathized. R. B. Parker shows how both characters are unlikable because they must ultimately compromise their respective masculine and feminine ideals (masculine honor of prowess in war and female honor as chastity in love) in order to reach a union. But more often, the debate focuses on Helena as either a saint (see M. C. Bradbrook and Harold Wilson) or a crafty, scheming, social climber (see Richard Levin). The latter of these readings are aided by the more obviously scheming character of Boccaccio's Giletta, who clearly sees the King's disease as an opportunity to procure Beltramo as her husband. And recently, Eric Nicholson has demonstrated Helena's roots in female character types from the *commedia dell'arte* (especially the *prima donna innamorata*) and possible similarity to actual Italian traveling actresses, to explain the more troubling aspects of her personality. Most contemporary critics, however, move beyond this dichotomy by placing Helena in the historical context of gender relations, though still accounting for discomfort or dislike of her character (see Marliss Desens, Carolyn Asp, Ellen Belton, Eileen Cohen, Kathryn Schwarz, and David McCandless). Barbara Traister, Catherine Field, and Kent Lehnhoff all underscore that the problem is Helena's engagement with medicine, making her akin to either a female mountebank or a prostitute, which would have been unnerving to an early modern audience. Lastly, David Strong places Helena in a legal context, simply acting in a way that protects her natural rights (as outlined by Hugh Grotius) and reinforcing the community's stability via her marriage.

13. Elizabeth Steinway likewise reads the play as one that interrogates discourses of kinship. However, her reading focuses not on blood discourses but rather Helena's indeterminate pregnancy. She maintains, "Helen[a]'s unconventional use of motherhood (i.e., a mother without children) allows us to read for an alternative kinship formation that is not bound to linear time, is not dependent on the figure of the child, and is not contingent on a shared physical likeness between members. Indeed, kinship here becomes something that Helen[a] 'does'—and her enactment of kinship asks that we accept what might unsettle our expectations about both pregnancy and motherhood" (289). While I agree with the general thrust of Steinway's argument, this reading seems somewhat complicated by the "if" clauses

of the conclusion, or the question of what would happen to Helena and her marriage if she does not pony up a child. Moreover, we ultimately disagree due to our readings of Helena. I believe that Helena's views on family/kinship are quite conservative in nature; it is merely that Helena has found a loophole in this ideology.

14. Helen Wilcox makes this comment about the multivalence of the word *grace* in the play, despite arguing for its staunchly spiritual nature. But given the play's discomfiting tone, it would seem that we should gloss *grace* as the former rather than the latter.

15. See especially Janet Adelman's *Suffocating Mothers*, Michele Osherow, Carol Thomas Neely's *Broken Nuptials*, Alexander Welsh, John Adam, and W. Speed Hill.

16. For Garrett Sullivan, blood discourses are an obstacle to the process of self-forgetting, a forgetting of one's dictated place, in order to reconstitute a new identity in relation to one's own desires. He claims that Helena forgets herself (symbolized by her father's death) and translates her desire for Bertram into a socially acceptable perpetuation of society, that is, marriage, which would allow her to bridge social difference (59–60, 61).

17. Hopkins's work on Shakespearean marriages accords with this reading, as she aligns Shakespeare with an aristocratic ethos (*Shakespearean Marriage* 9–11). Ann Jennalie Cook reads *All's Well* similarly for she believes that the original audience may have actually sympathized with Bertram (65). Berkeley and Keesee further use this framework of blood to attack Helena's character, arguing that she accords with the stereotypes of baser blood in the period, therefore giving credence to Bertram's beliefs. Berkeley and Keesee insist on Helena's similarity to Parolles regarding her self-interest and ambition. They likewise maintain that Helena conforms to the class stereotype of "egocentric and destitute of love toward family members," as well as being more melancholic (253).

18. John Love makes this observation when explaining the comedy's unsatisfactory conclusion. He notes that while in most comedies the source of darkness is usually embodied in one alien character (e.g., Jacques or Malvolio), who can be exiled at the play's end in service of the romantic conclusion, in *All's Well*, there is no one character to exile.

19. Kirsten Inglis's study of *All's Well* contains a similar reading of the play as a critique of naturalized hierarchies. But even though the thrust of the argument is similar, she reaches a very different conclusion, commenting in particular on the power of the King and the source of divine authority.

20. In this respect, my reading is friendly to certain feminist interpretations of the play that outline how Helena must negotiate existing codes of patriarchy. See especially Marliss Desens, Carolyn Asp, Caroline Bicks, Sullivan, Sachini Seneviratne, and Melissa Walter, but also Schwarz's "'My Intents,'" Theodora Jankowski, McCandless, and (in a certain light) Adelman's *Suffocating Mothers*. I emphasize this point about their similarity because most critics have read Helena and Bertram as diametrically opposed. This particularly applies to feminist readings of *All's Well* that identify a "sisterhood" of female characters all working together for Helena's goal (see Asp, Findlay, Osherow, and, to a degree, Adelman's *Suffocating Mothers*). Yet I believe that the different viewpoints divide more cleanly along generational lines (Helena/Bertram versus King/Countess), rather than coalescing around gendered camps. As I hope to show, Helena and Bertram, in fact, share assumptions about blood's role in determining identity.

21. Moreover, Bertram would not require reminding from his mother if he had inherited this virtue. Regarding the analysis of this line, many scholars claim that blood is actually in contention with virtue, a seemingly natural reading of the sentence, though it does not

entirely make sense in the context of Bertram's inheritance to which the Countess refers. It thus remains ambiguous with what the blood and virtue are contending. In my mind, it may be his baser nature that the Countess does not wish to acknowledge.

22. Bergeron's "'The Credit of Your Father'" likewise glosses Bertram's failure to resemble his father, though he does not explain it as a potential failure of blood. For him, this play turns around absent fathers whose memories the characters must sustain and uphold; in this, Helena succeeds, and Bertram fails.

23. This gloss shows how strange *All's Well's* plot stands in relation to traditional comedy. In Roman New Comedy, or its later Italian iteration the *commedia erudita,* the father figure advocates for the socially appropriate marriage alliance whereas the younger generation (the noble youth and a peasant girl, sometimes a prostitute) scheme for a love match.

24. Sullivan argues that Bertram's investment in blood can be explained by an explicitly masculinist worldview that "figures aristocratic descent in terms of blood's passage through women from one generation of men to another, blood that can be contaminated by women but which in its purity signifies only as a marker of male identity" (63). While this is true, I hope to show that Helena sees blood as equally relevant to her own identity.

25. Bruce Smith approaches this reading by noting that Bertram's "decampment to Florence has less to do with personal repugnance toward Helena than with avoiding legal consummation of a marriage to a social inferior" (171). Smith sees this contamination in legal rather than bodily terms.

26. In Boccaccio's version, the narrative acknowledges that Beltramo finds Giletta beautiful, likewise making the sudden conversion at the close more palatable. Moreover, the text highlights Giletta's desirability, having already turned down many suitors.

27. R. B. Parker, by contrast, glosses Bertram's refusal to bed Helena as a fear of losing his "manly marrow" in sex and marriage, as Parolles mentions. But I think it is important to note that Parolles has voiced this concern and not Bertram. For more discussion of this play as a parody of "a certain kind of military masculinity," see Melissa Walter's chapter on *All's Well* in her *Italian Novella.*

28. This talk of being a "true husband" is unique to Shakespeare. For Boccaccio, Beltramo employs terms of cohabitation: "Io per me vi tornero allora ad esser con lei che elle questo anello avrà in dito, e in braccio figliuol di me acquistato" ("For my part I will go home and live with her, when she has this ring on her finger and a son gotten of me upon her arm").

29. Schwarz likewise sees this moment as pointing toward the bed trick ("'My Intents'" 207).

30. Adelman also views the ring as a symbolic placeholder for either patrilineal descent or the child begotten by Helena's body of his descent (*Suffocating Mothers* 81).

31. Interesting too, in this moment, is the specification of Bertram's noble lineage going back four or five generations. This number was considered the minimum for qualifying as nobility in the French court of this period. This small moment then might acknowledge how nobility is *fashioned* over time and not simply innate. But given the serious tone with which Helena treats concerns for blood, I would read this detail as the playwright's moment to take Bertram down a peg rather than reflecting Helena's opinion.

32. I am not the first to notice this logical inconsistency. Anthony Cassell also notes that one issue for the class consciousness of Boccaccio's Beltramo (as well as Shakespeare's Bertram) is that he is willing to fornicate with a penniless girl of even baser station than his own wife (73). Cassell, however, might be misreading Boccaccio since even though the Florentine

girl is penniless, she is nevertheless a gentlewoman. The text narrates, "Il piú innamorato uom del mondo d'una nostra vicina, la quale è gentil femina, ma è povera" ("He is in the last degree enamoured of one of our neighbors here, who is a gentlewoman, but in poor circumstances"). Bertram, therefore, still appears to be more hypocritical than Beltramo.

33. As countless scholars have shown, the addition of Parolles to the *All's Well* plot means that he must serve as foil to Bertram. Parolles's scene of shaming, being exposed as an empty man of hot air, *should* act as the means for Bertram's learning or conversion. However, as many readers have found, though it *should* work this way, it clearly does not.

34. For a performance history of Helena's pregnant belly, see Kathryn M. Moncrief's "'Show Me a Child Begotten of Thy Body That I Am Father To'" in *Performing Maternity in Early Modern England*, ed. Kathryn M. Moncrief and Kathryn R. McPherson and (London: Routledge, 2007).

35. This couplet is the cause for much spilt ink since the reading of the play's conclusion hinges upon it. Cohen and Donaldson note that the "if" is a problem, but they try to gloss over it by pointing to the unknown future of the play. Osherow, Gross, Howard C. Cole, and Love, however, all agree that the "if" contributes to the unromantic ending of *All's Well.* Howard glosses the uncomfortable ending somewhat differently, underlining that the goal of this marriage in the play is social reproduction, which has very little to do with personal happiness (57).

36. The comments of the Countess, in particular, depart from the original source since she plays no role in the *Decameron.* Robert Miola convincingly shows that the older generation scheming to help a young and unorthodox love inverts the conventions of Terence's New Comedy ("New Comedy" 27). But this point might require a qualification. Solomon maintains that "it is important to view the older generation's destabilization of distinctions between nature and culture as an effort to extend the life of the social order with which they are familiar" ("Morality" 158).

37. Erin Ellerbeck calls attention to the Countess's use of a grafting metaphor in order to suggest a new connection, an adoptive one, between herself and Helena, thereby competing with the nature metaphors that describe consanguineous connections. While I agree with Ellerbeck's reading of the Countess's comments, she goes on to equate the Countess's views with Helena's: "*All's Well* features female grafters and excludes men, to a degree, from family generation and formation, thereby moderating heteronormative and biological methods of procreation" (308). I contend that Helena still feels the need to join the family stock, to become of one *blood,* mating the hare and lion, and she may not share the Countess's opinions.

38. Boccaccio's rendition of the King's attempt to convince Beltramo, by contrast, is not so forward-thinking. He acknowledges that status may matter, or at least it exists, but he maintains that Beltramo would undoubtedly be happier with a woman who loves him very deeply: "'Sí sarete,' disse il re 'per ciò che la damigella è bella e savia e ammavi molto; per che speriamo che molto piú lieta vita con lei avrete che con una donna piú alto legnaggio non avreste'" ("'Nay, but you will,' replied the King; 'for the damsel is fair and discreet, and loves you well; wherefore we anticipate that you will live far more happily with her than with a dame of much higher lineage'").

39. Solomon remarks that the King is often read as fickle and arbitrary. He adamantly refuses Helena's newfangled medicine, convinced by his conservative doctors, but quickly insists upon his ability to "create" nobility only to contradict himself later when he says that

he is merely validating Helena's innate virtue ("Morality" 157–58). This vacillation makes it "difficult to assess the ontological status of nobility or of 'honor'" (158).

40. Love likewise maintains that Lavatch is the greatest social upstart of the play, exposing the darker implications of class that will not matter for the "narrow gate" to heaven (526–27). One could read a similar point into Lavatch's riddle with a the "bountiful answer," one that will answer all questions. He compares the "bountiful answer" to a chair that will seat anyone, "It is like a barber's chair that fits all buttocks—the pin buttock, the quatch buttock, the brawn buttock, or any buttock" (2.2.16–18). The metaphor may delineate differences in butts, but they are, nevertheless, all butts.

41. The precise date of the play's first performance is unknown. Radcliff-Ulmstead and others place the date of composition around 1518 given its reference to the Ottomans in act 3 (118). However, many scholars debate whether it was written for festivities honoring Lorenzo de Medici Jr.'s return from France with his wife, Madeleine de la Tour d'Auvergne, in September, or, alternatively, if it was composed for the season of carnival earlier that year alongside two other comedies including *Il Frate* by Lorenzo Strozzi, as reported in a letter by Alfonsina Orsini (Bertelli and Innocenti xxi–xxii). For an extended discussion of the dating controversy, see Antonio Sorella (9–17) and Robert Black (186–89). Black settles on some time between 1515 and 1517, during Machiavelli's time at the Ruccellai gardens.

42. See Julia Ivanova's "Machiavelli's Soteriology" for more evidence to support these allusions.

43. The translation of *La mandragola* is by Laura Giannetti and Guido Ruggiero in *Five Comedies of the Italian Renaissance*. The in-text citations first show the act and scene numbers, followed by the page number of this translation.

44. See Carnes Lord, A. Parronchi, Robert Black, and, for a more general allegorical reading, Joseph Mazzeo. For a possible Christian allegory of the play, see Giovanni Aquilecchia. For an overview of the many possible allegorical readings of the play, see Antonio Sorella (17–43). Political theorist Boris Litvin has recently deemed this whole project misguided, suggesting that we struggle to find a prince in the play because there is not one, thereby satirizing any one character's ability to wield power, for all dupe and manipulate one another (18–19).

45. Augustus Mastri reads Ligurio as Machiavelli and revises the allegorical reading, demonstrating that the play is what would have taken place if Lorenzo de Medici had taken Machiavelli's advice (*The Prince*), namely a more successful marriage between himself and Florence, where he was quite unpopular (5). Joseph Francese likewise sees the play as the revenge of the meritocrat (Ligurio), one of minor station who forces his own will on others and who succeeds in a fictional world where Machiavelli failed. Patricia Vilches sees the *savio* in all of the characters, though especially Lucrezia who like "the *segretario* seeks accommodation in Florentine society. Bending his own republican ideals and ingratiating himself to the Medici who will allow him to achieve that goal" (108). Paul Corey sees Machiavelli in Ligurio as his practical advisory role to politicians and in Nicia who must "procreate indirectly through his writings" (200–201).

46. For arguments on behalf of Callimaco, see Di Maria's "Ethical Premises," Theodore A. Sumberg, and Guido Ruggiero's *Machiavelli in Love*. For arguments for Ligurio, see John Bernard, Franco Fido, Timothy J. Lukes, George Thomas, Jane Tylus, and Gay Bardin.

For supporters of Lucrezia as the *donna di virtù*, see Joseph A. Barber's "Irony of Lucrezia," Jack D'Amico, Giulio Ferroni, Arlene W. Saxonhouse, Heather Wright, Alexandra Coller, and Yael Manes. Lastly, and most surprisingly, for supporters of Nicia as deceptively cunning, see Harvey Mansfield, Valeria Finucci, Guido Ruggiero, Michael Palmer, James F. Pontuso, and Angus Fletcher. Wayne A. Rebhorn, by contrast, identifies two species of men in the play, confidence men (Callimaco, Ligurio, and Fra Timoteo) and dupes (Nicia and Sostrata). For him, *La mandragola* is a tale of Lucrezia's transformation into the "perfect confidence woman" (45). Francesca Sparacio refuses to identify any single character as the *savio*, but she suggests that each character displays the quality of *virtù*, the ability to take advantage of Fortune to serve their own desires in turn, which is part of the play's grand joke on the audience who is searching for the singular ruler.

47. For this reading of a societal rebirth, see also William Kennedy and Salvino Bizzarro. This rebirth is also cast as the triumph of youthful lust and desire over obstructions as is common with erudite comedy, as suggested by Robert Williams (387), or as the power of desire to disrupt the status quo, as Martin Fleisher holds (380). Ivanova, too, looks at the play's engagement with the fourteenth-century humanist dialogue genre to show the play's production of a different "ethical world" by "carnivalizing" the New Testament's evangelical message (71). Alternatively, for an interesting reading of the prince's role as scapegoat in the revitalization of society, see Carlo Celli's reading of the carnivalesque in Machiavelli. Finally, Aria Dal Molin points to the French medieval hagiographical tradition of *L'Aveugle et le Boiteux* to show how *La mandragola* retells the story of two disabled men, one blind (the love-blind Callimaco) and one lame (the impotent Nicia), who might cooperate to become the perfect, 'whole' husband for Lucrezia: "Messer Nicia and Callimaco serve as an emblem for cooperation and mutual assistance" (61).

48. Ronald Martinez posits that if Lucretia is the sacrificial victim for the rebirth of a new and better state (the Roman republic), then Lucrezia's rape, figured as the mandrake root's simultaneous poison and cure, "is itself an etiology of a new community, albeit one founded not on historical sacrifice but on the rational calculation of private advantage, on acquisitiveness or *guadagno*" ("Pharmacy of Machiavelli" 10). For responses to Martinez and the play's engagement with Livy, see Melissa Matthes, Catherine Conners, Jack D'Amico's "Virtù of Women," Paul Corey, and Mera J. Flaumenhaft.

49. George Thomas similarly depicts the play's moral as essentially "all's well that ends well" (an ironic coincidence, of course), wherein the end fulfills a public good even if the means are immoral, deceitful, and hypocritical (193). See also Marrapodi's "From Narrative to Drama" and Flaumenhaft on the corrupt, pragmatic ethic of *La mandragola*. For a history of the critical interpretation of the play as moralistic satire, see Carlo Dionisotti.

50. Dal Molin, building on the work of Harvey Mansfield and Gerry Milligan, puts it in this way: "What is more important, in a world of Machiavelli, is that Messer Nicia should *appear* fertile and able-bodied enough to engender an offspring who will eventually preserve his house and name" (53). One could gloss this moment as literary theorist Stephen Guy-Bray does in *Against Reproduction*, whereby "to produce a child is to reproduce a particular vision of society" (15).

51. Mansfield suggests that "the aging *messere* is a shrewd individual who uses others to achieve his goal, namely, to have an heir" (quoted in Di Maria "From Prose to Stage" 144). Michael Palmer and James Pontuso likewise suggest that Nicia is simply "playing the fool" to achieve his ends. For another rendition of this argument, see Paul Corey. In *Comic Democracies*, Angus Fletcher casts Nicia as the prince-like figure who embraces low strategies and help from more impetuous and low persons in order to be rid of his own impotence here

both literal and figurative as a lack of social power (35–36). Lastly, Finucci characterizes Callimaco as Nicia's "sperm donor," but she takes these observations in a psychoanalytic direction, looking at the play's sexual stakes in order to paint a picture of male anxiety.

52. Snyder describes dissimulation: "Early modern dissimulation involved first and foremost the exercise of strict self-control over the expression of thoughts, emotions or passions. As a practice of self-censorship, dissimulation assisted those who sought not to reveal or disclose anything of their own interiority, but were at the same time intent on not uttering any untruth to others" (6).

53. See Renzo Sereno for his note on the etymology of Nicia's name. He suggests this name could simply be ironic, but it might be genuine since Nicia seems to succeed after all.

54. Both Ruggiero's *Machiavelli in Love* and Joseph Francese suggest that Nicia's frustration likely mirrors Machiavelli's own in his state of exile (Ruggiero 143–44; Francese 167).

55. See especially Mansfield and Finucci.

56. Nicia himself insists upon seeing the plot all the way through to the end, including "testing the merchandise." Nicia feels the need to subject the *garzonaccio* (the "big stud") to scrutiny, supposedly checking his health but also ensuring that the coupling takes place. Nicia recounts, "Volli toccare con mano come la cosa andava" ("before I left, I decided I should feel to make sure that everything was going as it should") (5.2.111). And he continues that he only left after "Tocco e sentito che io ebbi ogni cosa" ("I had touched and felt everything") (5.2.112). Nicia, when explaining himself to Ligurio, insists, "Poi volsi vedere s'egli era sano; s'egli avessi avuto le bolle, dove mi travavo io?" ("I wanted to see if he was healthy. If he had syphilis, where would that have left me?") (5.2). Yet, as he describes the young stud's genitalia, "ma tu non vedesti mai le più belle carni: bianco, morbido, pastoso [smooth]! E dell'altre cose non ne domandate" ("but you've never seen more handsome flesh—white, soft, comely! And about the rest, don't ask"), it is clear that Nicia has touched his member in a fashion not purely for the purposes of a medical examination (5.2). Di Maria hints that Nicia might be gay, originally posited by Alonge ("From Prose to Stage" 142–43). While the moment certainly flirts with homoeroticism, it might likewise demonstrate his concern, his willingness to take a "hands on" approach to check for potency and fertility.

57. While "who's your daddy" apparently never presents an issue, the child does need to be a boy in order to satisfy the social demands of patriarchy. Callimaco promises Nicia, "Se, oggi ad un anno, la vostra donna non ha un suo figlio in braccio, io voglio avervi a donare duemila ducati" ("if a year from today your wife is not carrying her little baby boy in her arms, I will give you two thousand ducats") (2.6.86). Later, Nicia responds to the good news that Fra Timoteo has agreed to help convince Lurezia, asking, "Fia egli maschio?" ("Will the child be a boy?") (3.8.95). The child's masculinity seems to be Nicia's only requirement: the one way that the body does matter is that it needs a penis. A girl would produce heirs for another family, but a boy could inherit his wealth and carry his name. Fra Timoteo, after having "convinced" Lucrezia to agree to the plan, congratulates Nicia, "Voi vi beccherete un fanciullo maschio" ("You're going to have a son") (3.12.99).

58. Intellectuals like Stephano Guazzo (see Delille 129) and John Ferne provide two such examples of this tradition; see the introduction.

59. Nonetheless, it is true that this spatial metaphor quickly comes to designate biological connection, as "of the same house" does in English, so does "casata" (family lineage) in Italian.

60. To be fair, from a moral and theological standpoint, it was never as simple as "no soul,

no sin." Historian Nicholas Terpstra explains, "Preachers, theologians, and churchmen never accepted this simple correspondence and aimed to close what laymen might take as a loophole. The scholastic philosopher Thomas Aquinas argued that terminating any pregnancy was still 'a serious sin,' even if only the removal of an animated fetus was murder" (89). Additionally, as the early modern period progressed, the toleration of abortion waned, possibly due to decreasing theological tolerance or political motivations to promote population growth. For instance, the ordinances of Bamberg in 1507 ruled that any woman who brings an abortion on herself is guilty of a crime and to be severely punished, even if it was not equivalent to homicide (J. Riddle 126–27). And later, Catholic theologian Paul Laymann (1574–1635) asserted that "one who took or gave something 'maliciously' to prevent contraception or ensoulment [what we would call early-term abortifacients] 'commits not a true but a quasi-homicide and sins mortally'" (J. Riddle 162). Moreover, arts related to birth control or abortion became associated with witchcraft (J. Riddle 137). What information was available was somewhat hidden in herbals (i.e., manuals for homeopathic medicine).

61. Ligurio's timing, describing the woman as four months pregnant, is *either* a mistake *or* it might be too late on purpose! The prevailing theory was that the fetus was "ensouled" approximately forty to ninety days after conception (Terpstra 88).

62. My translation here is more literal than that of Giannetti and Ruggiero.

63. In a more conventional reading of the bed trick's divine sanction, Osherow explains the Lot story as follows, "while the elder daughter's acts are despicable, her motives are good. Her son becomes the forebearer of kings" (162).

64. Maristella de Panizza Lorch, for example, views Lucrezia's concession as a willing choice to embrace sensual enjoyment so long as it is in the service of procreation (265–66). Ruggiero, in a similar vein, credits Callimaco with convincing Lucrezia to bed him (*Machiavelli in Love* 152). But I would first like to trouble the assumption that we can trust Callimaco's word for what the audience has not witnessed. Moreover, it would seem that Lucrezia's presence in the bed in the first place means that she has already conceded to this scheme and whatever "convincing" that Callimaco employs is of minimal import.

65. As Melissa Walter notes, by staging Callimaco's entrance to Lucrezia's room, and her body, this play "can model in physical space the abstract issue of individual integrity. The room provides an intermediate, concrete model for the problems of bodily and personal integrity and interpenetration that are at stake in these dramas of national and personal identity" ("Dramatic Bodies" 68).

66. Di Maria agrees that Sostrata's role in convincing Lucrezia is about her pragmatic concern for her daughter's well-being, not moral shortcomings or her colorful past ("Machiavelli on Stage" 185–86). Manes alternatively reads this moment as Sostrata's demonstration of *virtù* because she dismisses conventional virtue, looking to practical effects and taking advantage of the occasion. Moreover, Sostrata is the one that Lucrezia finally listens to (21–23).

67. I am by no means the first to view this play in terms of a sociopolitical critique of nobility and hereditary power; see especially Francese. Moreover, as Inglis suggests, this play is "a scathing denunciation of a political system in which the foolish and corrupt become rich and powerful by virtue of their birth rather than their wits while true political thinkers are relegated to positions as base functionaries" (28). Yet I believe that this play is not only a critique of hereditary nobility and power structures (or as I would term them "blood logics") but it additionally exposes that inheritance/biology beneath this structure is baseless, for nobility does not actually function in this way.

Chapter 4

1. Citations of both plays come from *The Complete Pelican Shakespeare* unless noted otherwise.

2. Of course, the nobility's desire to maintain visible differences between the classes likely explains the popularity of sumptuary laws throughout early modern Europe.

3. Many could argue that there is, nevertheless, a disanalogy between the two characters in question. Though Othello and Jessica would both be considered different in kind in the older ideology of blood, Othello's relationship must fail because the lovers would be considered, or at least perceived as, *more* different and taboo (especially given the emphasis on Othello's color, which we do not see to the same degree with Jessica, or, if we do, it has a whitening effect). This comparison also runs the risk of disanalogy due to the gendered difference between the couples. Namely, Jessica's female body could be seen as more pliable and passive in an Aristotelian framework such that her body would not "contaminate" Lorenzo's Christian bloodline (see Lindsay Kaplan). By contrast, Othello would be perceived as more contaminator than contaminated within this same framework (Britton *Becoming Christian* 34). To all of these points I concede.

4. I choose not to put race in quotation marks, unless I am designating word usage or definition, following the lead of Ania Loomba in *Shakespeare, Race, and Colonialism* and Peter Erickson so as not to indicate skepticism or neuter the term such that it belittles the force it has in people's lives (even if we want to denaturalize it).

5. See Michael Omi and Howard Winant, *Racial Formation in the United States: From the 1960s to the 1990s* (New York: Routledge, 1986); Anthony Appiah, "Race," in *Critical Terms for Literary Study*, ed. Frank Lentricchia and Thomas McLaughlin (University of Chicago Press, 1990); Kimberlé Crenshaw, Neil Gotanda, Gary Peller, and Kendell Thomas, eds., *Critical Race Theory: The Key Writings That Formed a Movement* (New York: New Press, 1996). This discussion also involves some quibbling about how many (if any) racialized persons were present in England in Shakespeare's lifetime.

6. For a discussion of the intersection between racial and religious discourses, specifically as they occurred within the creation of the Black legend (the name for Spain's colonial brutality in the Americas in the sixteenth and seventeenth centuries, deemed "unique" by other European powers), see Margaret R. Greer, Walter D. Mignolo, and Maureen Quilligan, eds., *Rereading the Black Legend: The Discourses of Religious and Racial Difference in the Renaissance Empires* (Chicago: University of Chicago Press, 2007). Arthur Little's "Re-Historicizing Race" critiques this scholarly trend, maintaining that New Historicism has the effect of whitewashing history, making Shakespeare the property of whiteness as well as presuming that whiteness indicates the human or universal. See also R. M. Christofides, "Hamlet versus Othello: Or, Why the White Boy Keeps Winning," *Shakespeare* 17, no.1 (2021): 6–14, which marks the disciplinary partiality for *Hamlet* over *Othello*.

7. The geohumoral framework, dating back to the classical period, placed ideal, moderate, or temperate complexions in the middle-temperate region or, unsurprisingly, the Mediterranean. The two extreme climates of the north (including England) and south had a certain affinity as nonideal, though they were inversions of each other: "Scythians garnered praise for their physical strength and martial prowess, while Ethiopians were associated with natural wisdom and spirituality" (Floyd-Wilson *English Ethnicity* 2). In the early modern period, however, the English aimed to revise this picture due to its unflattering implications. Thus, drama of the time reflects a tension between this older framework and the emergent racial

discourse that re-created Blackness as innately inferior. This new project is evident in the retelling of Ham's lineage associated with Africa (wherein in previous tellings he was associated with Britain), or John Pory's English translation of Leo Africanus (Floyd-Wilson *English Ethnicity* 10–11, 140–41).

8. In *Race and Rhetoric*, Smith returns our attention to the classical means of establishing cultural difference and hierarchy, which is language (access to either Greek or Latin). Smith argues that the designation of barbarity ultimately becomes a "racial awareness" since England aimed to distance itself and its language from its own "barbaric" past (from the classical perspective) and then projected this identity onto early modern Africans, in part, because they could claim whiteness as ethnic continuity with the ancients (6). Moreover, this project for him explains Elizabethan humanism as a "collective enterprise toward English rhetorical eloquence . . . fueled by the old anxiety that the native tongue was inefficient, unsophisticated, unsuited to grand ideas" (17).

9. In *Things of Darkness*, Hall outlines how the vocabulary of "blackness" in aesthetic discourses established clear hierarchies in the realms of both gender and class; this vocabulary then easily transitioned to the realm of colonialism, playing an important role in the formation of English identity.

10. See Smith's fascinating work on textiles in "Textile Black Body" and "Othello's Black Handkerchief."

11. Boose offers evidence of an evolution wherein skin color slowly becomes more important than religio-ethnic differences, pointing to discussions of "black" and "white" Moors in early modern texts (38). She goes on to examine the particularly threatening figure of the Black woman for White patriarchy because "through her [the Black woman], all free-floating anxieties about 'the mother's dark place' contaminating the father's designs for perfect self-replication become vividly literal" (46). See also Loomba's study of the Black Bride figure from the Song of Songs in "Identities and Bodies," which traces the evolution of Blackness from associations with earthly contact to sin and faithlessness.

12. Karim-Cooper maintains that anti-Black racist discourses are not only an inheritance from the classical and medieval periods but that they were amplified during the Renaissance, citing royal portraits, the whitening of both Christ and the Virgin in art, and the blackening of the fool, malcontent, or revenger in drama (including evidence for both white and black face in early modern stage practice).

13. Dadabhoy's "Barbarian Moors" examines George Peele's *The Battle of Alcazar* (1594) to show the construction of English identity through various registers of difference: "The play offers evidence of early modern race in its dramatization of technologies of empire that facilitate the plunder of Africa and naturalize the bondage and enslavement of black bodies" (33).

14. Spiller argues that race as a visible marker is partly the by-product of the printing press and the circulation of texts and images articulating social identities (especially in the genre of romance). She maintains that racial identity's central assumption is the same as that of romance: "The narrative assumption that otherwise ineffable qualities are physically manifest in those that possess them" (*Reading and the History of Race* 11). In other words, the romance genre is part of the story about the shift from race as blood lineage/kinship to that of chromatic difference.

15. This commentary was printed in Hakluyt's Principal Navigations (1598–1600) with the following title: "A True Discourse of the Three Voyages of Discoverie. . . . Penned by Master George Best, a Gentleman Employed in the Same Voyages."

16. Dennis Britton, too, notes Best's *A True Discourse of the Late Voyages of Discoverie* (1578) as an important text that established "a causal relationship between the color of the flesh and the condition of the blood; he implies that one can know blood and race by looking at the skin" ("Flesh and Blood" 114).

17. Feerick traces the changing dynamic in which differences in class or social station determined by blood started to wane and differences in phenotype or chromatic difference replaced them. She argues, for instance, that in plays like *Othello* and *Titus*:

> differences of colour emerge, as it were, in dialectical relation to social rank, allowing social tensions originating with difference in rank to be resolved, mitigated, or exploited with reference to this emerging difference of colour. As such, they expose with uncanny similarities of representational strategy, the social dynamics that enable colour to accrue value as a cultural marker of difference, revealing how its emergence is predicated on and entangled with the decline of a deeply established system of difference that places a metaphysical value on bloodline independent of colour, complexion or culture. As such, these plays depict the processes that enable disparities of social power to be actively rechannelled into emerging modes of difference. (*Strangers in Blood* 4–5)

18. In "Delicious Traffick," Loomba argues that racialization in the period occurs in both the contexts of Jewishness (where a lack of visible difference causes anxiety) and for Moors (where there is more clear color differentiation); moreover, blood is the mechanism by which this happens because "purity of faith and purity of blood were necessarily yoked together," and there existed a persistent tension between Black skin and Christian interiority (209–11). Ultimately, Loomba's overarching project asks that we layer our examination of various discourses of otherness in the period in the examination of race including skin color, class, religion, community, but with a special focus on questions of gender, sexuality, and miscegenation.

19. Hendricks cites the changing, multifarious concept of race from "difference born of a class-based concept of genealogy" to other group typologies as the result of both colonial practices and changing class dynamics in Europe wherein money grants greater social mobility (20). She states, "It's not a coincidence, then, that the word 'race' and its conceptual semantics require lexical elucidation at the same time that the English capitalist/mercantile class is just beginning to solidify its presence as a political, cultural, and social power as a consequence of its economic hegemony" (18).

20. Bovilsky highlights just how contingent and complicated early modern race is since it activates so many different vectors of identity. It includes not only the geohumoral framework that helped to create differences around national, religious, geographic, or climactic origin but also issues around class or degree "with beliefs about heritable differences between the two groups in skin color, feature and body type, psychology, and quality of blood" (14).

21. Loomba makes this point explicitly when she underlines that "poorer people were widely perceived to be darker or more barbaric than their masters. Conversely, those whom the Europeans colonized were portrayed in terms that had already been applied to poor people—rude, uncultured, dirty, unrefined, and unintelligent" (*Shakespeare, Race, and Colonialism* 33–34). Akhimie makes the point as follows: "Race is not so much distinguishable from as contiguous with other systems of differentiation in its reliance on a strategy of

naturalizing social difference through biological narratives such as blood, lineage, phenotype, or geohumoralism" (13).

22. In this project, I find excellent company in Patricia Akhimie, whose work examines the role of conduct literature in the creation of race. Akhimie's thesis is as follows:

> Blackness, or any stigmatized somatic mark, may invite or justify heavy scrutiny and harsh judgement of one's conduct even as it comes to signify detrimental attributes such as an inability to improve one's conduct due to a lack of self-knowledge or a lack of imaginative capacity. I seek out such forms of oppression and prejudice and demonstrate that they are the by-products of a conduct system in which social identity is understood as both fixed and fluid. Conduct is essential to our understanding of race and class in the early modern period, a period in which a privileged group was characterized by its capacity and desire to improve, while others were marked by a devastating lack: an inability to be better and even to know better—that is, to know that they *should* be better. (5)

Akhimie further argues that Shakespeare's drama stages the failures of conduct for social climbers and earnest strivers in order to critique these discourses, though the characters do not offer an alternative: "These characters present accounts of how inferior status is not inherited but assigned, stamped on, inflicted, and enforced, and then later presented as the dictate of nature" (32).

23. For a recent comparison of these plays and their respective clandestine marriages, see Cleland's *Irregular Unions*. In it, she argues, "Shakespeare compellingly explores how clandestine marriage creates opportunities for racial and/or religious outsiders to enter into white, Christian society, underscoring the importance of the marriage ritual to identity in the early modern period" (16).

24. I am in fact skeptical that this is a live possibility. The couple faces so many obstacles from racial prejudice that it is an unlikely counterfactual. Nonetheless, the couple's acceptance of certain ideological confines does not help, and this point is the thrust of my argument here.

25. Whigham glosses the whole play in light of performing status and belonging, thus highlighting that these aristocratic and religious discourses are not only similar but are functionally the same.

26. As Lawrence Normand notes, the Prince of Morocco "invokes the human body as a place where certain disputed questions can be tested and decided" (55). Notably too, Portia is most similar to the prince, and not Bassanio, on this measure.

27. See Albert Gerard and Laurence Lerner.

28. See Elliot H. Tokson for an early version of this reading. Feminist scholarship of the play (e.g., Karen Newman's "Portia's Ring," Patricia Parker's "Fantasies of 'Race,'" and Ania Loomba's *Shakespeare, Race, and Colonialism*) has extended this reading to show that Othello assumes misogynist or patriarchal discourses in an effort to assimilate to Venetian society.

29. Anthony Barthelemy argues that the few virtuous Black characters like Othello "offer further validation of the more common, harmful, and denigrating representations of black Moors because they prove that it is possible to resist the call of evil, though most unusual" (147). The ending in particular might render Othello's individual character sympathetic, but Othello feels compelled to kill the "villainous Moor" within himself (159–60).

30. Peter Erickson argues that, though *Othello* stages "the fashioning of a discourse of racial whiteness," the play ultimately reimposes racial stereotypes in which the titular character "confirms the link between black skin and a predisposition to moral dislocation and depravity" (36). In Ian Smith's reading, Othello's *black* handkerchief "reinforces his cultural barbarity already implied in his blackness" (21). And Kim Hall likewise underlines the constant calls for light to push away the darkness and expose the truth, which indicate "the wish to destroy and exclude evil is expressed in terms of blackness and cultural difference" ("*Othello* and The Problem of Blackness" 370). Adelman's "Iago's Alter Ego," informed by Kleinian psychoanalytic theory, argues that while Othello's Blackness is not meant to indicate his innate evil or sin, for Iago, Othello's visual Blackness becomes the perfect vehicle/receptacle for him to project his own darkness within (caused by self-obsession and a feeling of injury, fractured identity, and thus envy of Othello's own self-assurance and "fullness"). And Floyd-Wilson maintains that, though Othello's jealousy would have been more properly associated by a contemporary audience with the Italian Iago, "the Moor's violent metamorphosis—his transformation from an extraordinarily calm state to passionate rage—severs external blackness from its longstanding (though now obscured) geohumoral associations with dispassion and constancy. And it is the legacy of Shakespeare's play that this portrait of 'Moorish behavior' established many of the strains of modern racial discourse" (*English Ethnicity* 142).

31. See G. K. Hunter and Eldred Jones (87–88). Carol Thomas Neely's reading is adjacent, placing the play in the geohumoral context of "hot blood." She holds that "although *Othello* sees hot love as an anxiety and threat, it does not deny Mediterranean bodies, even African ones, the capacity for self-control, idealized love, and audience sympathy" ("Hot Blood" 61). In the most extreme of reparative readings I have encountered, Elaine L. Robinson argues that the play actively critiques bigotry. In *Shakespeare Attacks Bigotry*, she makes the rather contentious claim that Shakespeare was both gay and a Catholic. Moreover, his faith, "real Christianity" as articulated by the theologian Saint Bernard of Clairvaux, informed his work; his oeuvre thus attacks all bigotry, but especially the African slave trade, and the mechanisms that support it, especially Renaissance humanism.

32. For instance, Martin Orkin argues that, while the visual of a Black man strangling a White woman would play into racial prejudices, the play repeatedly flags our inability to trust our own vision or perception (176). Similarly, David Shalkwyk argues that *Othello* stages an epistemological problem central to love and desire, namely the inability to fully "know" or see into another person. The play's tragedy stems from Othello's desperate project to fully know Desdemona, which is doomed to failure.

33. Michael Neill underscores that, while the play does not directly "call out" racism, it does illuminate how racism as "visceral superstitions" emerges in culture ("Unproper Beds"). Ian Smith outlines the two racial tactics of the play: one of overt racial hostility and the other wherein Othello assumes this stigmatized identity, "reinforcing the dynamic of power between white mastery and racialized blackness that degrades black persons at the entry level of social interaction" ("We Are Othello" 110). Finally, Akhimie notes that in the play racial difference "emerges as the product of marking, a form of scrutiny or social judgement within a system that advocates universal access to strategies of self-improvement while denying certain groups access to these strategies. . . . Blackness is imagined as opposite to cultivation and Othello as a seeming exception whose true nature will out. In fact, Othello's plight is shared by all who strive to rise" (49–50).

34. In Britton's *Becoming Christian*, he argues that early modern English authors grappled

with the infidel-conversion motif typical of the romance genre at a time when the notion of a universal Christian community was becoming less desirable for an English Protestant country, looking to distance itself from the miraculous power of the sacraments. In this context, *Othello* shows that "if romantic love can cement a religious identity, Iago shows that dissolving the relationship between Othello and Desdemona can undo Othello's Christian identity, transforming him into a 'turban'd Turk' at the end of the play" (33). Julia Reinhard Lupton's *Citizen-Saints* argues that the religious difference would have been perceived in the period as more "other" and dangerous than the racial difference. She maintains that the play's central issue is not whether Othello can convert but which religion he has converted *from*, therein dictating whether his conversion is successful. Namely, is he a Black Gentile who is a blank slate to be easily converted, or is he previously Muslim in which case he would be more resistant to Christian revelation?

35. Loomba argues that "despite his seeming difference from other Moors, Othello ultimately embodies the stereotype of Moorish lust and violence—a jealous, murderous husband of a Christian lady," which was a common setpiece for stories about Ottomans in the period (*Shakespeare, Race, and Colonialism* 95). For Daniel Vitkus, Othello represents the dark threat of the Muslim at the edge of Christendom and the fear that one could "turn Turk," "dramatized as a fall into a bestial, sex-obsessed condition" ("Turning Turk" 156). Similarly, Dadabhoy cites the image of the *devşirme* (children that converted to Islam and then assumed elite positions in the Ottoman army) to highlight contemporary anxieties about conversion and the possibility of multiple, layered identities. *Othello's* narrative then "exposes a hierarchy that favours the natural over and above the assimilated" ("Two-Faced" 133).

36. For Emily Bartels, Othello is not a Moor *in* Venice but a Moor *of* Venice, that is, typical of Venetian society and well assimilated, maintaining "erase Othello's identity as a Moor, and the play might go on" (167). Virginia Vaughan likewise sees the play as a profoundly political drama, one that negotiates the tension between the myth of Venice—an ideal republic whose tolerance, justice, and peace accounted for its mercantile success—and a place on the edge of Christian civilization threatened by barbaric, evil forces.

37. For instance, Edward Berry maintains that Othello has imbibed an ideology typical of colonial attitudes toward the Other; either they are not fully human or they must be saved by Christianity, and he "comes to see himself as his own stereotype" (318–19). Jack D'Amico likewise casts Othello's tragedy as "accept[ing] a perspective that deprives him of his humanity" (*Moor in English Renaissance Drama* 177).

38. Brabantio's language is unsurprising in the context of Venetian culture and policy. As Vaughan points out, the city fathers actually promoted prostitution so that their own daughters (and their bloodlines) would not be at risk to contamination from its many strange visitors (16).

39. See Akhimie's *Shakespeare and the Cultivation of Difference* (70–72). See also Bovilsky's reading in *Barbarous Play* (60–63). Her argument, however, proceeds to trace Desdemona's darkening throughout the play as indicative of her violations of normative femininity, especially her sexual agency perceived as promiscuity.

40. Langley's "Plagued by Kindness" also examines the vocabulary of likeness in the play, albeit in the context of a Paracelsan view of contagion and disease. Langley's reading proffers Othello as a pharmakon: the poison that will be the cure for the Ottoman invasion due to his similarity to this foreign "other." I extend this discussion by looking at the other dangers of forcing similarity or likeness at the cost of individuality or difference.

41. See, for instance, those of Robert Crofts in *The Lover, or Nuptiall Love* (1638), Ludovico Dolce's *Della institutione delle donne* (1545), and John Ferne's *The Blazon of Gentrie* (1586) in the introduction.

42. Dennis F. Bormann argues that Othello's change in character reflects a change in his physiology, whereby his blood has been literally infected by the poison of Iago's words (81). Bormann's argument primarily focuses on Iago's villainy as indicative of his base blood.

43. David Hume's dictum that "the minds of men are mirrors to one another" implies this kind of parity as well as Adam Smith's imaginative perspective-taking, or, as we might say, putting ourselves in their shoes to imagine how we might feel in their position (Stanford Encyclopedia of Philosophy, "Empathy," https://plato.stanford.edu/).

44. Here, I disagree with Britton's reading of pity, which presumes that Aristotle's pity "requires audience members to identify some form of likeness between the hero and themselves" for a proper functioning of catharsis ("*Contaminatio*, Race, and Pity " 47). He thus contends that "Desdemona's pity . . . responds to Othello's role as defender of Christian Venice (and its colonial interests)" (57). However, I am unsure that Aristotle's definition of pity requires such a similarity between the subject and the object but rather a recognition that a *similar evil* could happen to the subject: "Let pity then be a kind of pain excited by the sight of evil, deadly or painful, which befalls one who does not deserve it; an evil which one might expect to come upon himself or one of his friends, and when it seems near" (Aristotle 2.8.2) Additionally, if we are called to "pity" what we find similar, it would be odd to couple Desdemona's pity with repetitions of Othello's story as "strange."

45. While potentially an unfair criticism since often the language is meant to underline that the couple's love is reciprocal, Senator 1 provides another example. He asks Othello, just accused of witchcraft, whether Desdemona's affection rather "came . . . by request and such fair question / As soul to soul affordeth" (1.3.127–28). This language, not incorrectly mind you, points to the level of the soul rather than the body in order to cast this love relationship as one of parity and similitude.

46. Evidence of their discord further increases as phrases are repeated (e.g., heaven, false, etc.) but out of confusion rather than understanding (4.2.31–92).

47. Arguably, Desdemona attempts to find another point of connection or similarity by ventriloquizing Barbary's song.

48. The difficulty of assimilating Jessica is, interestingly, mimicked by her name, which is thought to have been the first use in English. Nominally, she is the first of her kind.

49. There is no daughter in either of the commonly cited sources for *Merchant*: *Il Pecorone* or *The Ballad of Geruntus*. Rather, Jessica's predecessor is likely Abigail from Marlowe's *The Jew of Malta* (Drakakis 145–46). For an extended comparison of Jessica and Abigail, see Maurice Charney and Joan Holmer. Beatrice D. Brown cites literary predecessors in popular tales from medieval exempla of the thirteenth to fifteenth centuries. And, as John Hale reminds us, she is not only the largest addition to the bondholder plot but her scenes outnumber those of her father. In fact, her role contains many of the "quintessential attributes of heroines in romance comedy—male disguise, giggling about it, participation in a masque, abandoning family for a lover" (191). And yet many productions elect to omit a great number of Jessica's scenes.

50. Most famously, Sigurd Burkhardt insists that Jessica and Lorenzo offer a perfect counterpoint to Portia and Bassanio, creating an effect of chiaroscuro to further elevate the central pair (253). Similarly, John Drakakis believes that Jessica's character takes on some of the negative aspects of the mercantile marriage contract that would otherwise be applied to

Portia (157). Nicoleta Cinpoeş, however, draws a close comparison between the heroines who demonstrate agency by defrauding their fathers. For more Portia criticism, especially pertaining to her role in the homosocial relationship between Antonio and Bassanio, see Lawrence Hyman, Shalkwyk, Newman's "Portia's Ring," and Lars Engle. For more discussion of the casket test, see Harry Berger. For more on Portia and the trial scene, see Rebecca Lemon and Burkhardt.

51. Camille Slights presents this as the key issue because Jessica's conversion and Christian ethics cannot fully justify her immoral actions. Furthermore, one cannot deny the pathos of Shylock's moment of anguish over the loss of Leah's ring (359–60). Even if one can find evidence to counteract accusations of greed and frivolity, "she is undeniably disloyal" (364). Although in an effort to salvage her character, Slights insists that "the design of the play directs us to be glad that Jessica loves rather than hates, to approve her decision to reject her father's values and to ally herself through Lorenzo with those of Portia, but to regret the pain and loss that accompany her choices" (360).

52. I purposefully set aside Shylock's character and the "Jewish question" criticism by and large for Jessica's "Jewishness" here is the impetus for asking a different question about the workings of blood's ideology. Thus, the demonization or rescue of Shylock's Jewishness is too much to take on here. For some wonderful scholarship on this issue, see James Shapiro, Lawrence Normand, Kenneth Gross, Jerome Friedman, Bernard Glassman, and Edna Nashon and Michael Shapiro's recent edited collection, *Wrestling with Shylock*.

53. Hall in "Guess Who's Coming to Dinner?," for instance, reads the relationship in accordance with the "like likes like" logic of blood's ideology, maintaining that because Jessica loves a Christian she cannot possibly share her father's blood. Irene Middleton underlines that most productions of the play favor this gloss of Jessica's character whereby she is always already Christian, having these traits prior to her conversion and elopement (298). See also Loomba's "'Delicious Traffick.'"

54. According to some scholars, for example Mary Metzger, early modern persons held that Jessica could convert while Shylock cannot because his body carries a visible mark of Jewishness, that is, his circumcision. Moreover, Jessica can easily be incorporated because she can take her husband's name and thus be socially integrated into Christian society (59). See Lindsay Kaplan and Britton's *Becoming Christian* for similar arguments. See also Holmer, who claims that Jessica's successful conversion to Christianity symbolizes the promise of the second coming (124).

55. Brett Hirsch claims that despite linguistic attempts to make Jessica alike, these attempts are unsuccessful ("Counterfeit Professions"). Shapiro says of Jessica's character, which sounds quite reminiscent of Brabantio's warning to Othello: "Lorenzo's exchange with Jessica . . . raises the possibility that the obedient and converted Jewess who had to disobey one man and one set of religious principles to embrace another might revert to her true Jewish nature" (159). Adelman insists upon Jessica's persistent, bodily Jewishness as it activates Christian anxieties about their own Jewish heritage, saying, "Even as she herself would convert to Christianity, then, Jessica's name carries the potential reminder that the fleshly lineage of Jesus comes from his Jewish mother. But if the Jewish womb is the bearer of the Jewish lineage that must authenticate Jesus's status as the Messiah, it is also the symbolic repository of the fleshy remains that should be left behind; and this double valence reiterates the double valence of Judaism within Christianity" (*Blood Relations* 69). See also Michelle Ephraim (135) and Britton's revised thesis in "Flesh and Blood."

56. As Adelman suggests, "the relation between the Christian and the Jew in *Merchant* turns crucially on blood: on the Jewish blood lineage of Christianity, on Jessica's inability to rid herself of her father's blood, and . . . the spectacle of the Jewish convert to Christianity requires desperate remedies of a proto-racism and contradictory desire that the Jew continue at all costs to play that bloody role assigned to him by the Christian" (*Blood Relations* 37). Additionally, Hall argues that *Merchant* attends to England's anxiety about national identity being contaminated on account of foreign difference present in economies like that of Venice ("Guess Who's Coming to Dinner?" 95).

57. Cleland argues instead that Jessica's claim to Christianity is not bodily but that the wedding vows constitute conversion; she maintains that the problem with the marriage ceremony is not that it fails to convert Jessica, it is that the community has not seen it for themselves due to its clandestine nature, and this is why she remains an outsider (102–3). The only issue here is that marriage is likewise a bodily affair. I choose rather to interpret the discomfort here as due to the obvious rhetoricity of blood, especially given the emphasis on this vocabulary throughout.

58. Middleton notes that this conflict over Jessica's signification can either be played up or resolved in costume choices, for instance in the "continuance of Jewish elements," once she reaches Belmont (303–5). Similarly, many productions choose to visually isolate Jessica from the action of Belmont in staging (Drakakis 145).

59. A. R. Braunmuller in Shakespeare's *Complete Pelican Shakespeare* opts to maintain three distinct characters (Salarino, Solanio, and Salerio) following M. M. Mahood's edition, rather than combine Salarino and Salerio as is customary. I have decided to keep the three characters as well since it provides a nice foil for my argument. Namely, Jessica's uniqueness and individuality stands out even more in contrast with these minor Christian characters who can be, and clearly have been, easily conflated.

60. Kaplan assumes that this bodily difference is meant in earnest, namely that Jews were believed to have darker skin, but, in Jessica's case, her fairness implies "the female convert's willingness implicitly codes her as authentically Christian and white" (19). I rather believe that it speaks to an anxiety about bodily difference and a discursive overcompensation due to this anxiety. Nonetheless, the easy, ideologically conservative equivocation between pallor and virtue is likewise seen in the Prince of Morocco's failure of the casket test. For more discussion of this scene in *The Merchant of Venice*, see Hall's "Guess Who's Coming to Dinner?," Spiller's "From Imagination to Miscegenation," and Boose.

61. Adelman likewise reads this moment as "a gentile and no Jew," which establishes, in her mind, an impossible method of translation/conversion. She explains that even in Jessica's attempt to become a Christian, she could never first be a Gentile and thus could never be a Christian (*Blood Relations* 75).

62. Certainly, Shylock's characterization might be one way to vilify a Jew who became too big for his britches. Some scholars have suggested that these bodily differences in *Merchant* have very little to do with religion and everything to do with money and status. Both main plots are, after all, set in motion by Shylock's loan to Antonio, a loan predicated upon Shylock's difference. Shylock's body dictates his inability to own property, and thus, he loans with interest to make money. Furthermore, flesh and money are seemingly translatable: Shylock conflates his daughter with his ducats; Portia's body and her dowry are similarly confused; and the promise of Antonio's flesh can be deemed as collateral for a loan. For more, see René Girard, Newman's "Portia's Ring," and Thomas Moisan.

63. For an argument in favor of importing Galen's assumptions about psychophysiology over Aristotle's, see Hirsch's "Counterfeit Professions."

64. When Lorenzo arrives on the scene shortly thereafter and Jessica relays the conversation, Lorenzo does not exactly contradict Lancelot or promise Jessica's salvation. Instead, he simply retorts that Lancelot has himself gotten a Moor/Black woman pregnant (3.5.33–35).

65. Whether this means that Jessica does not love Lorenzo is unclear. I certainly would not commit myself to that claim, but the play could be convincingly staged this way, and certainly, many scholars hold this viewpoint.

66. For a reading of the play through the lens of generational conflict, as well as early modern debates about the nature-nurture question, see Stephanie F. Gearhart's *Drama and the Politics of Generational Conflict.*

67. Lancelot's rhyme could easily be played in a teasing manner, or alternatively with a kind of earnestness for Jessica's future well-being, since he previously deemed her a "sweet Jew" (2.3.11).

68. The skepticism voiced by most critics pertains to the potential *economic* motivations for Lorenzo and Jessica's elopement; see Hymen and Hirsch's "Gentle and No Jew." Cinpoeş actively reads Jessica's elopement as a double fraud and the relationship between the pair as "transaction-like," though "Jessica is equally clear that her interest is in salvation by becoming a Christian('s wife)" (141–42).

69. For instance, when Portia asks him to safeguard her estate in the name of his love for her, he responds, "Madam with all my heart; / I shall obey you in all fair commands" (3.4.35–36). Or, for example, when looking to bait Jessica for a compliment by praising himself, Lorenzo compares himself to Portia: "Even such a husband / Hast thou of me as she is for a wife" (3.5.76–77). Lorenzo and Portia seem ripe for comparison, seemingly more alike than the two lovers.

70. Hirsch reads Lorenzo's profession of love for Jessica as tainted since it is tempered extensively by conditional statements ("Counterfeit Professions").

71. Frank Whigham explains that Bassanio's success in wooing Portia is due to his demonstration of stylistic affinity and thus a similarity in rank: "The scene is set with a series of allusions to artistic signs of harmony. They conduct a witty duet, cleverly playing variations on the Petrarchan theme of love torture, creating an effect not dissimilar to the sonnet spoken by Romeo and Juliet. As they test and reveal their verbal affinity, they establish social congruence and foreshadow a decorous love match" (100).

72. Hale glosses that the litany of lovers do not apply (195); Holmer claims that "these tragic endings may be better understood as ominously reflexive, suggesting what might have been" (130); Spiller likewise reads Jessica and Lorenzo as a successful cross-cultural marriage in contrast to the romances cited, which they revise as an "alternative, happy ending" by domesticating Jessica's otherness ("From Imagination to Miscegenation" 155); and Lawrence Danson maintains that "it is precisely the movement out of legend into the realm of living flesh and blood that averts the tragic" (176).

73. See Adelman's *Blood Relations*, Metzger, and Britton's *Becoming Christian*. David Goldstein says that this litany "challenges the very possibility of absorbing aliens within the body politic, and of equalizing power relations more broadly" (331).

74. Shapiro reads this moment in a negative light, highlighting Lorenzo's anxiety about the possibility of Jessica's future treachery turned against himself, especially in light of the earlier allusion to Medea (158–59).

75. Metzger suggests, on the contrary, that this debate illustrates their own potential future tragedy. Lorenzo's reading emphasizes thieving and betrayal, while Jessica's pertains to "the difficulty of establishing trust between persons of different religions, colors, classes, and especially genders that is played up in the rest of act 5" (60).

76. Adelman's *Blood Relations* and Hirsch's "Counterfeit Professions" go so far as to suggest Jessica's persistent Jewishness. They claim that Jessica's disposition isolates her from Belmont and aligns her temperament with that of Shylock earlier in the play, hearkening back to his own displeasure with mirth (2.5).

Epilogue

1. The play, however, supports the notion that once struck by love for a specific object the love cannot be projected elsewhere. Once captured by Diana, Cupid is given the punishment of attempting to untie love knots. He complains to her that this is an impossible task: "If they be true love-knots, 'tis unpossible to unknit them; if false, I never tied them" (4.2.23–24). The conclusion likewise supports this point for both Galatea and Phillida insist that they cannot change their love objects despite the inconvenience that both of them are women (5.3.135–38). Neptune calls it a "constant faith" even if a "strange and foolish" one since there is "no cause of affection" (glossed as "no instigation for sexual desire") (5.3.139–41). The solution then is that one must agree to undergo a metamorphosis (or sex change). Yet the fact that this does not occur within the space of the play might suggest that sex does not matter either, despite the fact that both parties swear up and down that they would not have fallen in love had they known the other's true identity.

2. See Ultsch ("Maddalena Campiglia"). See also Virginia Cox, "Single Self," 548–50. Ultsch remarks on the difficulty around using this term for Campiglia since she likely was not a member of the official order of *dimesse* in Vicenza even though she left property to them upon her death. Moreover, many critics argue that associating Campiglia with this religious order of sorts was an attempt to retroactively cloister or ideologically contain her in a morally acceptable frame for a woman's social identity. Yet Ultsch argues that viewing her in this framework is both useful and productive. For the hypothesis that Campiglia may have had same-sex desires given the prominence of sapphic desire in her work, see Carlachiara Perrone.

3. Virginia Cox and Lisa Sampson argue that Campiglia's success suggests that "provincial cities may have provided more supportive environments for female writers than larger and more prestigious cultural centers, precisely because of the more intimate scale of their literary circles and the institutions that sustained them" (7). For more on Campiglia's autobiography and her navigation of the literary scene, see Philiep Bossier and Sampson's chapter "Imitations and Innovations after Tasso's *Aminta*" in *Pastoral Drama*.

4. See Maggie Gunsberg for a more extensive discussion of women in Italian comedy.

5. For more on this and other basics of Italian pastoral drama, see Sampson, Richard Andrews's "Pastoral Drama," and Clubb's *Italian Drama in Shakespeare's Time*.

6. Both the Italian and the English translation come from the University of Chicago bilingual edition by Cox and Sampson.

BIBLIOGRAPHY

Adair, Richard. *Courtship, Illegitimacy and Marriage in Early Modern England.* Manchester: Manchester University Press, 1996.

Adam, John F. "*All's Well That Ends Well*: The Paradox of Procreation." *Shakespeare Quarterly* 12, no.3 (1961): 261–70.

Adamson, Sylvia. "Questions of Identity in Renaissance Drama: New Historicism Meets Old Philology." *Shakespeare Quarterly* 61, no.1 (2010): 56–77.

Adelman, Janet. *Blood Relations: Christian and Jew in* The Merchant of Venice. Chicago: University of Chicago Press, 2008.

———. "Her Father's Blood: Race, Conversion and Nation in *The Merchant of Venice*." *Representations* 81, no.1 (2003): 4–30.

———. "Iago's Alter Ego: Race as Projection in *Othello*." *Shakespeare Quarterly* 48, no.2 (1997): 125–44.

———. *Suffocating Mothers: Fantasies of Maternal Origin in Shakespeare's Plays,* Hamlet *to* The Tempest. New York: Routledge, 1992.

Akhimie, Patricia. *Shakespeare and the Cultivation of Difference: Race and Conduct in the Early Modern World.* New York: Routledge, 2018.

Albanese, Ralph. "Masculin/féminin: Discours et pouvoir dan *Le Cid*." *Romantic Review* (1994): 361–70.

Amtower, Laurel. "This Idol Thou Ador'st: The Iconography of *'Tis Pity She's a Whore*." *Papers on Language and Literature* 34, no.2 (1998): 179–206.

Anderson, Donald K., Jr. "The Heart and the Banquet: Imagery in Ford's *'Tis Pity* and *The Broken Heart*." *Studies in English Literature, 1500–1900* 2, no.2 (1962): 209–17.

Andrews, Richard. "Anti-feminism in Commedia Erudita." In *Contexts of Renaissance Comedy,* edited by Janet Clare and Roy Eriksen, 11–31. Oslo: Novus Press, 1997.

———. "Pastoral Drama." In *The Cambridge History of Italian Literature,* edited by Peter Brand and Lino Pertile, 292–98. Cambridge: Cambridge University Press, 1996.

Appelbaum, Robert. "Flowing or Pumping? The Blood of the Body Politic in Burton, Harvey, and Hobbes." In *The Cultural Politics of Blood, 1500–1900,* edited by Kimberly Anne Coles, Ralph Bauer, Zita Nunes, and Carla L. Peterson, 171–90. New York: Palgrave Macmillan, 2015.

———. "'Standing to the Wall': The Pressures of Masculinity in *Romeo and Juliet*." *Shakespeare Quarterly* 48, no.3 (1997): 251–72.

Aquilecchia, Giovanni. "La favola *Mandragola* si chiama." In *Collected Essays on Italian Language and Literature Presented to Kathleen Speight,* edited by Giovanni Aquilecchia, Stephen N. Cristea, and Sheila Ralphs, 74–100. Manchester: Manchester University Press, 1971.

Aristotle. *Rhetoric. Aristotle in 23 Volumes.* Vol. 22. Translated by J. H. Freese. Cambridge, MA: Harvard University Press; London: William Heinemann, 1926.

Asp, Carolyn. "Subjectivity, Desire and Female Friendship in *All's Well That Ends Well*." *Literature and Psychology* 32, no.4 (1986): 48–63.

Atkinson, James B. "An Essay on Machiavelli and Comedy." In *The Comedies of Machiavelli*. Translated and edited by David Sices and James B. Atkinson, 1–37. Indianapolis: Hackett Publishing Company, 1985.

Ault, H. C. "The Tragic Genius of Corneille." *Modern Language Review* 45, no.2 (1950): 164–76.

Avery, Joshua. "Faith in the Unseen: Helena's Sacramental Vision in *All's Well That Ends Well*." *Renascence* 69, no.1 (2017): 33–48.

Bakas, Adrianna E., ed. *Politics, Ideology and the Law in Early Modern Europe: Essays in Honor of J. H. M. Salmon*. Rochester, NY: University of Rochester Press, 1994.

Balizet, Ariane M. *Blood and Home in Early Modern Drama*. New York: Routledge, 2014.

———. "'Enamoured of Thy Parts': Dismemberment and Domesticity in *Romeo and Juliet*." *Early Modern Literary Studies* 12, no.19 (2009): 1–31. Academic One File.

Bandello, Matteo. *Le Novelle*. Vol. 2. Edited by Gioachino Brognoligo. Bari: Gius. Laterza and Figli, 1928.

Banerjee, Pompa. "Gift: Economies of Kinship and Sacrificial Desire in *'Tis Pity She's a Whore*." *Studies in the Humanities* 29, no.2 (2002): 137–49.

Barber, Joseph A. "The Irony of Lucrezia: Machiavelli's Donna di virtù." *Studies in Philology* 82, no.4 (1985): 450–59.

———. "La Strategia Linguistica di Ligurio nella 'Mandragola' di Machaivelli." *Italianistica: Rivista di letteratura italiana* 13, no.3 (1984): 387–95.

Barclay, Katie. "Emotional Lineages: Blood, Property, Family and Affection in Early Modern Scotland." In *Historicising Heritage and Emotions: The Affective Histories of Blood, Stone and Land*, edited by Alicia Marchant, 84–98. New York: Routledge, 2019.

Bardin, Gay. "Machiavelli Reads Boccaccio: *Mandragola* between *Decameron* and *Corbaccio*." *Italian Quarterly* 38, no.149–50 (2001): 5–26.

Barker, Simon. *'Tis Pity She's a Whore*. Florence: Routledge, 1997.

Barnes, Elizabeth. *Incest and the Literary Imagination*. Gainesville: University of Florida Press, 2002.

Barroll, J. Leeds. *Artificial Persons: The Formation of Character in the Tragedies of Shakespeare*. Columbia: University of South Carolina Press, 1974.

Bartels, Emily. *Speaking of the Moor: From Alcazar to Othello*. Philadelphia: University of Pennsylvania, 2008.

Barthelemy, Anthony Gerard. "Ethiops Washed White: Moors of the Nonvillainous Type." In *Black Face, Maligned Race: The Representation of Blacks in English Drama from Shakespeare to Southerne*, 147–81. Baton Rouge: Louisiana State University Press, 1987.

Bauman, Kentston D. *"Incestuous Sheets" and "Adulterate Beasts": Incest and Miscegenation in Early Modern Drama*. PhD diss., University of Michigan, 2011. Ann Arbor: University of Michigan Press, 2011.

Bauman, Zygmunt. *Liquid Love: On the Frailty of Human Bonds*. Cambridge: Polity Press, 2003.

Beecher, Donald. "Introduction to *The Mandragola* by Niccolò Machiavelli." In *Renaissance Comedy: The Italian Masters*. Vol. 2, edited by Donald Beecher, 103–14. Toronto: University of Toronto Press, 2009.

———. "Introduction to *The Sister* by Giambattista Della Porta." In *Renaissance Comedy: The*

Italian Masters. Vol. 2, edited by Donald Beecher, 375–84. Toronto: University of Toronto Press, 2009.

———. "Machiavelli's *La Mandragola* and the Emerging Animateur." *Quaderni d'italianistica* 5, no.2 (1984): 171–89.

———. "Quatrocento Views on the Eroticization of the Imagination." In *Eros and Anteros: The Medical Traditions of Love in the Renaissance*, edited by Donald A. Beecher and Massimo Ciavolella, 49–73. Ottawa: Dovehouse Editions, 1992.

Behuniak-Long, Susan. "The Significance of Lucrezia in Machiavelli's *La Mandragola*." *Review of Politics* 51, no.2 (1989): 264–80.

Bell, Rudolph M. *How to Do It: Guides to Good Living for Renaissance Italians*. Chicago: University of Chicago Press, 1999.

Belling, Catherine. "Acts of Flesh and Blood: Anatomy and Physiology in English Renaissance Drama." PhD diss., State University of New York, Stony Brook, 2000.

———. "Infectious Rape, Therapeutic Revenge: Bloodletting and the Health of Rome's Body." In *Disease, Diagnosis, and Cure on the Early Modern Stage*, edited by Stephanie Moss and Kaara L. Peterson, 113–32. Burlington, VT: Ashgate, 2004.

Belsey, Catherine. "The Elephants' Graveyard Revisited: Shakespeare at Work in *Antony and Cleopatra, Romeo and Juliet* and *All's Well That Ends Well*." *Shakespeare Survey* 68 (2015): 62–72.

———. "Love in Venice." *Shakespeare Survey* 44 (1992): 41–53.

———. *Shakespeare and the Loss of Eden: The Construction of Family Values in Early Modern Culture*. New Brunswick, NJ: Rutgers University Press, 1999.

Belton, Ellen. "'To Make the "Not" Eternal': Female Eloquence and Patriarchal Authority in *All's Well That Ends Well*." In *New Critical Essays*, edited by Gary Waller, 125–39. New York: Routledge, 2007.

Bénichou, Paul. *Morales du grand siècle*. Paris: Éditions Gallimard, 1948.

Berger, Harry, Jr. "Marriage and Mercifixion in *The Merchant of Venice*: The Casket Scene Revisited." *Shakespeare Quarterly* 32, no.2 (1991): 155–62.

Bergeron, David M. "'The Credit of Your Father': Absent Fathers in *All's Well That Ends Well*." In *New Critical Essays*, edited by Gary Waller, 169–82. New York: Routledge, 2007.

———. "Middleton's *No Wit, No Help* and Civic Pageantry." In *Pageantry in the Shakespearean Theater*, edited by David M. Bergeron, 65–80. Athens: University of Georgia Press, 1985.

Bernard, John. "Writing the Paradox of the Self: Machiavelli's Literary Vocation." *Renaissance Quarterly* 59, no.1 (2006): 59–89.

Berkeley, David S. "Shakespeare's Severall Degrees in Bloud." In *Shakespeare's Theories of Blood, Character, and Class: A* Festschrift *in Honor of David Shelley Berkeley*, edited by Peter C. Rollins and Alan Smith, 7–18. New York: Peter Lang, 2001.

Berkeley, David S., and Donald Keesee. "Bertram's Blood-Consciousness in *All's Well That Ends Well*." *Studies in English Literature, 1500–1900* 31, no.2 (1991): 247–58.

Berry, Edward. "Othello's Alienation." *Studies in English Literature, 1500–1900* 30, no.2 (1990): 315–33.

Bertelli, Sergio, and Piero Innocenti. *Bibliografia Machiavelliana*. Verona: Edizioni Valdonega, 1979.

Bicks, Caroline. "Planned Parenthood: Minding the Quick Woman in *All's Well*." *Modern Philology* 103, no. 3 (2006): 299–331.

Biet, Christian. *Théâtre de la cruauté et récits sanglants en France (xvie-xviie siècle)*. Paris: Éditions Robert Laffront, 2006.

Bigliazzi, Silvia. "Romeo before Romeo: Notes on Shakespeare Source Study." *Memoria di Shakespeare. A Journal of Shakespeare Studies* 5 (2018): 13–39.

Billing, Christian. "Modelling the Anatomy Theatre and the Indoor Hall Theatre: Dissection on the Stages of Early Modern London." *Early Modern Literary Studies* Special Issue 13 (2004): 1–17.

Bizzarro, Salvino. "Debauchery, Mayhem, and Sex in Machiavelli's *Mandragola*." In *Seeking Real Truths: Multidisciplinary Perspectives on Machiavelli*, edited by Patricia Vilches and Gerald Seaman, 199–217. Boston: Brill, 2007.

Black, Cordell W. *Corneille's Denouements: Text and Conversion*. Potomac: Studia Humanitatis, 1984.

Black, Robert. *Machiavelli*. New York: Routledge, 2013.

Blackburn, Simon. *The Oxford Dictionary of Philosophy*. New York: Oxford University Press, 1994.

Boaistuau, Pierre. *Historiques Tragiques*. Edited by Richard A. Carr. Paris: Librarie Honoré Champion, 1977.

Boccaccio, Giovanni. *Il Decamerone*. Italian text based on V. Braca. Translated by J. M. Rigg. In Decameron Web, Brown University, Department of Italian Studies. www.brown.edu.

Boehrer, Bruce. *Monarchy and Incest in Renaissance England: Literature, Culture, Kinship and Kingship*. Philadelphia: University of Pennsylvania Press, 1992.

———. "'Nice Philosophy': *'Tis Pity She's a Whore* and the Two Books of God." *Studies in English Literature, 1500–1900* 24, no.2 (1984): 355–71.

Bolam, Robyn. "John Ford, Mary Wroth, and the Final Scene of *'Tis Pity She's a Whore*." In *A New Companion to English Renaissance Literature and Culture*, edited by Michael Hattaway, 276–83. Blackwell Reference Online, 2010. www.onlinelibrary.wiley.com.

Boose, Lynda E. "'The Getting of a Lawful Race': Racial Discourse in Early Modern England and the Unrepresentable Black Woman." In *Women, "Race," and Writing in the Early Modern Period*, edited by Margo Hendricks and Patricia Parker, 35–54. New York: Routledge, 1994.

Bormann, Dennis F. "'Thou Art a Villain': From the Ensign to Iago—Blood Changes in *Othello*." In *Shakespeare's Theories of Blood, Character, and Class: A* Festschrift *in Honor of David Shelley Berkeley*, edited by Peter C. Rollins and Alan Smith, 77–94. New York: Peter Lang, 2001.

Bornedal, Peter. "The Law of the Name. The Imaginary Recipient in Corneille's *Le Cid*." *Orbis Litterarum* 52 (1997): 157–77.

Bossier, Philiep. "Female Writing and the Use of Literary Byways. Pastoral Drama by Maddalena Campiglia (1553–1595)." In *Women Writing Back/Writing Women Back*, edited by Anke Gilleir, Alicia Montoya, and Suzan van Dijk, 115–33. Leiden, Netherlands: Brill, 2010.

Bouchard, Constance Brittain. *"Those of My Blood": Constructing Noble Families in Medieval Francia*. Philadelphia: University of Pennsylvania Press, 2001.

Bovilsky, Lara. *Barbarous Play: Race on the English Renaissance Stage*. Minneapolis: University of Minnesota Press, 2008.

Bradbrook, M. C. "Virtue Is the True Nobility: A Study of Structure in *All's Well That Ends Well*." *Review of English Studies* 1, no.4 (1950): 289–301.

Brian, Peter. *Galen on Bloodletting: A Study of the Origins, Development and Validity of His Opinions, with a Translation of the Three Works*. Cambridge: Cambridge University Press, 1986.

Briggs, Julia. "Shakespeare's Bed Tricks." *Essays in Criticism* 44, no.4 (1994): 293–314.

Brissenden, Alan. "Impediments to Love: A Theme in John Ford." *Renaissance Drama* 7, no.1 (1964): 95–102.

Britton, Dennis. *Becoming Christian*. New York: Fordham University Press, 2014.

———. "*Contaminatio*, Race, and Pity in *Othello*." In Britton and Walter, *Rethinking Shakespeare Source Study*, 46–64.

———. "Flesh and Blood: Race and Religion in *The Merchant of Venice*." In Thompson, *Cambridge Companion to Shakespeare and Race*, 108–22.

Britton, Dennis Austin, and Melissa Walter, eds. *Rethinking Shakespeare Source Study: Audiences, Authors, and Digital Technologies*. New York: Routledge, 2018.

Brown, Beatrice D. "Medieval Prototypes of Lorenzo and Jessica." *Modern Language Notes* 44, no.4 (1929): 227–32.

Brown, Carolyn E. "Juliet's Taming of Romeo." *Studies in English Literature, 1500–1900* 36, no.2 (1996): 333–55.

Brown, Pamela. "Othello and Italophobia." In Marrapodi, *Shakespeare and Intertextuality*, 179–92.

Brundage, James A. *Sex, Law and Marriage in the Middle Ages*. Brookfield: Variorum Ashgate Publishing Company, 1993.

Bruster, Douglas. *Quoting Shakespeare: Form and Culture in Early Modern Drama*. Lincoln: University of Nebraska Press, 2000.

Buccola, Regina. "'As Sweet as Sharp': Helena and the Fairy Bride Tradition." In *New Critical Essays*, edited by Gary Waller, 71–84. New York: Routledge, 2007.

Bueler, Lois E. "The Structural Uses of Incest in English Renaissance Drama." *Renaissance Drama* 15 (1984): 115–45.

Bürger, Peter. "*Le Cid* de Corneille et le matériau de la tragicomédie." *Papers on French Seventeenth Century Literature* 21, no.1 (1984): 426–45.

Burkhardt, Sigurd. "*The Merchant of Venice*: The Gentle Bond." *ELH* 29, no.3 (1962): 239–62.

Bynum, Caroline Walker. *Wonderful Blood: Theology and Practice in Late Medieval Northern Germany and Beyond*. Philadelphia: University of Pennsylvania Press, 2007.

Calbi, Maurizio. *Approximate Bodies: Gender and Power in Early Modern Drama and Anatomy*. New York: Routledge, 2005.

Callaghan, Dympna C. *The Weyward Sisters: Shakespeare and Feminist Politics*. Oxford: Blackwell Publishers, 1994.

Campiglia, Maddalena. *Flori, A Pastoral Drama. A Bilingual Edition*. Edited by Virginia Cox and Lisa Sampson. Chicago: University of Chicago Press, 2004.

Camporesi, Piero. *Juice of Life: The Symbolic and Magic Significance of Blood*. Translated by Robert R. Barr. New York: Continuum, 1995.

Canning, Patricia. *Style in the Renaissance: Language and Ideology in Early Modern England*. London: Bloomsbury Publishing, 2014.

Carlin, Claire L. *Women Reading Corneille: Feminist Psychocriticisms of* Le Cid. New York: Peter Lang, 2000.

Carroll, Stuart. *Blood and Violence in Early Modern France*. New York: Oxford University Press, 2006.

Carroll, William C. "'We Were Born to Die': 'Romeo and Juliet.'" *Comparative Drama* 15, no.1 (1981): 54–71.

Cartwright, Kent. "Diachronic and Synchronic: Two Problems of Textual Relations in *The Comedy of Errors*." In Britton and Walter, *Rethinking Shakespeare Source Study*, 184–205.

Cassell, Anthony. "Pilgrim Wombs, Physicke and Bed Tricks: Intellectual Brilliance, Attenuation and Elision in *Decameron* III:9." *MLN* 121, no.1 (2006): 53–101.

Castiglione, Baldesar. *The Book of the Courtier*. Translated by George Bull. New York: Penguin Books, 1967.

———. *Il Cortegiano*. Edited by Vittorio Cian. Firenze: G. C. Sansoni, 1894.

Céard, Jean. "The Devil and Lovesickness: Views of 16th Century Physicians and Demonologists." Translated by Faith Wallis. In *Eros and Anteros: The Medical Traditions of Love in the Renaissance*, edited by Donald A. Beecher and Massimo Ciavolella, 33–47. Ottawa: Dovehouse Editions, 1992.

Celli, Carlo. *Il Carnevale di Machiavelli*. Florence: Leo S. Olschki, 2009.

Chakravarty, Uvrashi. "More Than Kin, Less Than Kind: Similitude, Strangeness, and Early Modern English Homonationalisms." *Shakespeare Quarterly* 67, no.1 (2016): 14–29.

———. "Race, Natality, and the Biopolitics of Early Modern Political Theology." *Journal of Early Modern Cultural Studies* 18, no.2 (2018): 140–66.

Chakravorty, Swapan. *Society and Politics in the Plays of Thomas Middleton*. Oxford: Clarendon Press, 1996.

Champion, Larry S. F. "Ford's *'Tis Pity She's a Whore* and the Jacobean Tragic Perspective." *PMLA* 90, no.1 (1975): 78–87.

Chaplin, Gregory. "'One Flesh, One Heart, One Soul': Renaissance Friendship and Miltonic Marriage." *Modern Philology* 99, no.2 (2001): 266–92.

Charney, Maurice. "Jessica's Turquoise Ring and Abigail's Poisoned Porridge: Shakespeare ad Marlowe as Rivals and Imitators." *Renaissance Drama* 10, no.1 (1979): 33–44.

Cherpack, Clifton. *The Call of Blood in French Classical Tragedy*. Baltimore: Johns Hopkins University Press, 1958.

Cherry, Caroline Lockett. *The Most Unvaluedst Purchase: Women in the Plays of Thomas Middleton*. Edited by Erwin A. Stürzl. Salzburg: Institute for Salzburg Studies in English Literature, 1973.

Chevlowe, Susan. "Christian Iconography and Jewish Accommodation in Maurycy Gottlieb's Painting, *Shylock and Jessica*." In Nashon and Shapiro, *Wrestling with Shylock*, 359–80.

Ciavolella, Massimo. *La "Malattia D'Amore" dall'Antichità al Medioevo*. Rome: Bulzoni editore, 1976.

Cinpoeş, Nicoleta. "Defrauding Daughters Turning Deviant Wives?: Reading Female Agency in *The Merchant of Venice*." *Sederi* 21 (2011): 133–46.

Clare, Janet. *Writers and Their Work: Revenge Tragedies of the Renaissance*. Tavistock: GBR Writers and Their Work, 2003.

Clark, Glen. "The Civil Mutinies of *Romeo and Juliet*." *English Literary Renaissance* 41, no.2 (2011): 280–300.

Clarke, David. *Pierre Corneille: Poetics and Political Drama under Louis XIII*. Cambridge: Cambridge University Press, 1992.

Cleland, Katherine. *Irregular Unions: Clandestine Marriage in Early Modern English Literature*. Ithaca, NY: Cornell University Press, 2021.

Clerico, Terri. "The Politics of Blood: John Ford's *'Tis Pity She's a Whore*." *English Literary Renaissance* 22, no.3 (1992): 405–34.

Clubb, Louise George. *Giambattista Della Porta Dramatist*. Princeton, NJ: Princeton University Press, 1965.

———. *Italian Drama in Shakespeare's Time*. New Haven, CT: Yale University Press, 1989.

———. "Looking Back on Shakespeare and Italian Theater." *Renaissance Drama* 36 (2010): 3–19.

Cohen, Eileen Z. "'Virtue is Bold': The Bed-Trick and Characterization in *All's Well That Ends Well* and *Measure for Measure*." *Philological Quarterly* 65, no.2 (1986): 171–86.

Cohen, Walter. "Eurasian Literature." In *Comparative Early Modernities, 1100–1800*, edited by David Porter, 47–72. New York: Palgrave Macmillan, 2012.

Cole, Howard C. *The* All's Well *Story from Boccaccio to Shakespeare*. Urbana: University of Illinois Press, 1981.

Colie, Rosalie L. *The Resources of Kind: Genre-Theory in the Renaissance*. Edited by Barbara K. Lewalski. Berkeley: University of California Press, 1973.

———. *Shakespeare's Living Art*. Princeton, NJ: Princeton University Press, 1974.

Coller, Alexandra. "*Bella Creanza* and Female *Destrezza:* Women in Italian Renaissance Comedy." PhD diss., New York University, 2005.

Conners, Catherine. "Love and Piety in Machiavelli's *Mandragola*." In *The Pious Sex: Essays on Women and Religion in the History of Political Thought*, edited by Andrea Radasanu, 74–91. Lanham, MD: Lexington Books, 2012.

Conticelli, Valentina. "*Sanguis Suavis*: Blood between Microcosm and Macrocosm." Translated by Jane Carroll and Beryl Stockman. In *Blood: Art, Power, Politics and Pathology*, edited by James M. Bradburne, 55–63. Munich: Prestel Verlag, 2001.

Cook, Ann Jennalie. *Making a Match: Courtship in Shakespeare and his Society*. Princeton, NJ: Princeton University Press, 1991.

Cope, Jackson I. *Secret Sharers in Italian Comedy: From Machiavelli to Goldoni*. Durham, NC: Duke University Press, 1996.

Corey, Paul. *Messiahs and Machiavellians: Depicting Evil in the Modern Theatre*. Notre Dame, IN: University of Notre Dame Press, 2008.

Corneille, Pierre. *Agesilas Tragedy*. In *Works of P. Corneille*. Vol. 7, edited by Ch. Marty-Laveaux. Paris: Library of L. Hachette and C., 1862. https://www.gutenberg.org.

———. *Corneille's Le Cid*. Translated by Roscoe Mongan. New York: Hinds and Noble Publishers, 1896. https://www.gutenberg.org/.

———. *Le Cid Tragi-Comédie*. Edited by Milorad R. Margitić. Philadelphia: John Benjamins Publishing Company, 1989.

Couton, Georges. *Corneille et la tragédie politique*. Paris: Presses Universitaires de France, 1984.

Cox, Virginia. "The Single Self: Feminist Thought and the Marriage Market in Early Modern Venice." *Renaissance Quarterly* 48, no.3 (1995): 513–81.

Cox, Virginia, and Lisa Sampson. Introduction to *Flori, A Pastoral Drama. A Bilingual Edition*. By Maddalena Campiglia. Edited by Virginia Cox and Lisa Sampson, 1–35. Chicago: University of Chicago Press, 2004.

Craik, Katharine A. "Sorting Pistol's Blood: Social Class and the Circulation of Character in Shakespeare's *2 Henry IV* and *Henry V*." In Johnson and Decamp, *Blood Matters*, 43–58.

Crane, Mary Thomas. "Linguistic Change, Theatrical Practice, and the Ideologies of Status in *As You Like It*." *English Literary Renaissance* 27 (1997): 361–92.

Crawford, Patricia M. *Blood, Bodies, and Families in Early Modern England*. Harlow, UK: Longman, 2004.

Cressy, David. "Kinship and Kin Interaction in Early Modern England." *Past and Present* 113 (1986): 38–69.

Crouch, David. *The Birth of Nobility: Constructing Aristocracy in England and France 900–1300*. Harlow, UK: Pearson Education, 2005.

Dadabhoy, Ambereen. "Barbarian Moors: Documenting Racial Formation in Early Modern England." In Thompson, *Cambridge Companion to Shakespeare and Race*, 30–46.

———. "Two-Faced: The Problem of Othello's Visage." In *Othello: The State of Play*, edited by Lena Cowen Orlin, 121–47. London: Bloomsbury Publishing, 2014.

Dal Molin, Aria. "Deceiving Disability: The Blind Man and the *Boiteux* in Machiavelli's *Mandragola*." *MLN* 134, no.1 (2019): 42–64.

D'Amico, Jack. *The Moor in English Renaissance Drama*. Tampa: University of South Florida Press, 1991.

———. "The Virtù of Women: Machiavelli's *Mandragola* and *Clizia*." *Interpretation* 12, no. 2/3 (1984): 261–73.

Danson, Lawrence. *The Harmonies of* The Merchant of Venice. New Haven, CT: Yale University Press, 1978.

Davey, Kibrina. "Ungovernable Passion in John Ford's *'Tis Pity She's a Whore*." *Early Modern Literary Studies* Special Issue 26: John Ford (2017): 1–19.

Dawson, Lesel. "'In Every Wound There Is a Bloody Tongue': Cruentation in Early Modern Literature and Psychology." In Johnson and Decamp, *Blood Matters*, 151–66.

———. *Lovesickness and Gender in Early Modern English Literature*. Oxford: Oxford University Press, 2008.

Decameron Web. Brown University, Department of Italian Studies. Updated January 3, 2014. https://www.brown.edu/.

Dear, Peter. "What Is the History of Science the History Of?: Early Modern Roots of the Ideology of Modern Science." *Isis* 96, no.3 (2005): 390–406.

Defaye, Claudine. "Annabella's Unborn Baby: The Heart in the Womb in *'Tis Pity She's a Whore*." *Cahiers élisabéthains* 15 (1979): 35–41.

D'Elia, Anthony F. *The Renaissance of Marriage in Fifteenth-Century Italy*. Cambridge, MA: Harvard University Press, 2004.

Delille, Gerard. "The Shed Blood of Christ: From Blood as Metaphor to Blood as Bearer of Identity." In Johnson et al., *Blood and Kinship*, 125–143.

Della Porta, Giambattista. *Le Commedie: A Cura di Vincenzo Spampanato*. Vol. 1. Bari: Gius. Laterza and Figli, 1911. www.gutenburg.org.

———. *The Sister (La Sorella)*. Translated by Bruno Ferraro and Donald Beecher. In *Renaissance Comedy: The Italian Masters*. Vol. 1, edited by Donald Beecher, 385–460. Toronto: University of Toronto Press, 2008.

Demeter, Tamás. "Values, Norms and Ideologies in Early Modern Inquiry: An Introduction." In *Conflicting Values of Inquiry: Ideologies of Epistemology in Early Modern Europe*, edited by Tamás Demeter, Kathryn Murphy, and Claus Zittel, 1–9. Leiden, Netherlands: Brill, 2014.

De Navarre, Margueritte. *L'Heptaméron*. Paris: Éditions Garnier Frères, 1967.

Dennis, Carl. "*All's Well That Ends Well* and the Meaning of *Agape*." *Philological Quarterly* 50, no.1 (1971): 75–84.

Dente, Carla. "Reading Symptoms of Decadence in Ford's *'Tis Pity She's a Whore*." In *Romancing Decay: Ideas of Decadence in European Culture*, edited by Michael St. John, 27–38. Brookfield, VT: Ashgate, 1999.

Descimon, Robert. "The Birth of the Nobility of the Robe: Dignity versus Privilege in the Parlement of Paris, 1500–1700." In *Changing Identities in Early Modern France*, edited by Michael Wolfe, 95–123. Durham, NC: Duke University Press, 1997.

Desens, Marliss S. *The Bed-Trick in English Renaissance Drama*. Newark: University of Delaware Press, 1994.

Dessen, Alan C. "The Logic of Elizabethan Stage Violence: Some Alarms and Excursions for Modern Critics, Editors, and Directors." *Renaissance Drama, New Series* 9 (1978): 39–69.

Devyver, André. *Le Sang Épuré: Les préjugés de race chez les gentilshommes français de l'Ancien Régime (1560–1720)*. Bruxelles: Editions de l'Universite de Bruxelles, 1973.

DiGangi, Mario. *The Homoerotics of Early Modern Drama*. Cambridge: Cambridge University Press, 1997.

———. "John Ford." In *A Companion to Renaissance Drama*, edited by Arthur F. Kinney, 567–83. Blackwell Reference Online. www.onlinelibrary.wiley.com.

Di Maria, Salvatore. "The Ethical Premises for the Mandragola's New Society." *Italian Culture* 7 (1986–9): 17–33.

———. "From Prose to Stage: Machiavelli's *Mandragola*." *MLN* 121, no.1 (2006):130– 53.

———. "Machiavelli on Stage: *Mandragola* and *Clizia*." In *Seeking Real Truths: Multidisciplinary Perspectives on Machiavelli*, edited by Patricia Vilches and Gerald Seaman, 173–98. Boston: Brill, 2007.

Dionisotti, Carlo. "Appunti sulla *Mandragola*." *Belfagor* 39 (1984): 621–44.

Dolan, Frances E. *Marriage and Violence: The Early Modern Legacy*. Philadelphia: University of Pennsylvania Press, 2008.

———. *Whores of Babylon: Catholicism, Gender, and Seventeenth-Century Print Culture*. Ithaca, NY: Cornell University Press, 1999.

Dollimore, John. *Sexual Dissidence: Augustine to Wilde, Freud to Foucault*. Oxford: Clarendon Press, 1991.

Donaldson, Ian. "*All's Well That Ends Well*: Shakespeare's Play of Endings." *Essays in Criticism* 27, no.1 (1977): 34–55.

Doniger, Wendy. *The Bed Trick: Tales of Sex and Masquerade*. Chicago: University of Chicago Press, 2000.

Donini, Pierluigi. "Psychology." Translated by R. J. Hankinson. In *The Cambridge Companion to Galen*, edited by R. J. Hankinson, 184–209. Cambridge: Cambridge University Press, 2015.

Doran, John. "Shakespeare in France." *Littell's Living Age (1844–1896)* 136, 1759 (March 2, 1878): 533.

Doubrovsky, Serge. *Corneille et la dialectique du heros*. Paris: Gallimard, 1968.

Drakakis, John. "Jessica." In *The Merchant of Venice: New Critical Essays*, edited by John W. Mahon and Ellen Macleod Mahon, 145–65. New York: Routledge, 2002.

Du Laurens, André. "Of Another Kinde of Melancholie Which Commeth by the Extremitie of Love." In *A Discourse of the Preservation of the Sight: Of Melancholike Diseases*, translated by R. Surphlet, 1597.

Duncan-Jones, Katherine. "'O Happy Dagger': The Autonomy of Shakespeare's Juliet." *Notes and Queries* September 1998: 314–16.

Dye, Fredd Arthur. "Love and Lust in the Dramatic Works of Thomas Middleton." PhD diss., Indiana University, 1969.

Edelman, Lee. *No Future: Queer Theory and the Death Drive.* Durham, NC: Duke University Press, 2004.

Ehland, Christoph. "Happiness against All Odds: Incestuous Desires in John Ford's *'Tis Pity She's a Whore.*" *Critical Survey* 32, no.3 (2020): 94–104.

Elam, Keir. "'Tis Pity She's Italian: Performing the Courtesan on the Early Seventeenth-Century English Stage." In *Shakespeare and Renaissance Literary Theories: Anglo-Italian Transactions,* edited by Michele Marrapodi, 235–46. Farnham, UK: Ashgate Publishing, 2011.

Ellerbeck, Erin. "Adoption and the Language of Horticulture in *All's Well That Ends Well.*" *Studies in English Literature* 51, no.2 (2011): 305–26.

Engle, Lars. "'Thrift Is a Blessing': Exchange and Explanation in The Merchant of Venice." *Shakespeare Quarterly* 37, no.1 (1986): 20–37.

Ephraim, Michelle. *Women and Gender in the Early Modern World: Reading the Jewish Woman on the Elizabethan Stage.* Abingdon, UK: Ashgate, 2013.

Erasmus. *The Colloquies of Erasmus.* Translated by N. Bailey. London: Reeves and Turner, 1878.

Erickson, Peter. "The Moment of Race in Renaissance Studies." *Shakespeare Studies* 26 (1998): 27–36.

Farr, Dorothy M. *John Ford and the Caroline Theatre.* New York: Macmillan Press, 1979.

Feerick, Jean E. "'Rude Uncivill Blood': The Pastoral Challenge to Hereditary Race in Fletcher and Milton." In *The Cultural Politics of Blood, 1500–1900,* edited by Kimberly Anne Coles, Ralph Bauer, et. al., 65–83. New York: Palgrave Macmillan, 2015.

———. *Strangers in Blood: Relocating Race in the Renaissance.* Toronto: University of Toronto Press, 2010.

Ferne, John. *The Blazon of Gentrie.* London: John Windet for Andrew Maunfell, 1586.

Ferroni, Giulio. "Transformation and Adaptation in Machiavelli's *La Mandragola.*" Translated by Ronald L. Martinez. In *Machiavelli and the Discourse of Literature,* edited by Albert Russell Ascoli and Victoria Kahn, 81–116. Ithaca, NY: Cornell University Press, 1993.

Ficino, Marsilio. *Commentary on Plato's Symposium on Love.* Translated by Sears Jayne. Dallas: Spring Publications, 1985.

Fido, Franco. "Politica e teatro nel badalucco di Messer Nicia." *Italica* 46, no.4 (1969): 359–75.

Field, Catherine. "'Sweet Practicer, Thy Physic I Will Try': Helena and Her 'Good Reciept' in *All's Well, That Ends Well.*" In *New Critical Essays,* edited by Gary Waller, 194–208. New York: Routledge, 2007.

Findlay, Alison. "'One That's Dead Is Quick': Virgin Re-birth in *All's Well That Ends Well.*" In *Marian Moments in Early Modern Drama,* edited by Regina Buccola and Lisa Hopkins, 35–45. Abingdon, UK: Ashgate Publishing, 2007.

Fineman, Joel. *Shakespeare's Perjured Eye: The Invention of Poetic Subjectivity in the Sonnets.* Berkeley: University of California Press, 1986.

———. *The Subjectivity Effect in Western Literary Tradition: Essays toward the Release of Shakespeare's Will.* Cambridge: MIT Press, 1991.

Finucci, Valeria. *The Manly Masquerade: Masculinity, Paternity and Castration in the Italian Renaissance.* Durham, NC: Duke University Press, 2003.

Fishelov, David. "Genre Theory and Family Resemblance—Revisited." *Poetics* 20 (1991): 123–38.

Fitter, Chris. "'Mock Not Flesh and Blood with Solemn Reverence': Recovering Radical Shakespeare." *Literature Compass* 9, no.6 (2012): 420–30.

———. "'The Quarrel Is between Our Masters and Us Their Men': *Romeo and Juliet*, Dearth, and the London Riots." *English Literary Renaissance* 30, no.2 (2000): 154–83.

Flaumenhaft, Mera J. "The Comic Remedy: Machiavelli's Mandragola." *Interpretation: A Journal of Political Philosophy* 7, no.2 (1978): 33–74.

Fleisher, Martin. "Trust and Deceit in Machiavelli's Comedies." *Journal of the History of Ideas* 27, no.3 (1966): 365–80.

Fletcher, Angus. *Comic Democracies: From Ancient Athens to the American Republic.* Baltimore: Johns Hopkins University Press, 2016.

Florio, John. *A Worlde of Wordes.* Edited byHermann W. Haller. Toronto: University of Toronto Press, 2013.

Floyd-Wilson, Mary. "'Angry Mab with Blisters Plague': The Pre-Modern Science of Contagion in *Romeo and Juliet.*" In *The Palgrave Handbook of Early Modern Literature and Science,* edited by H. Marchitello and E. Tribble, 401–22. New York: Palgrave Macmillan, 2017.

———. *English Ethnicity and Race in Early Modern Drama.* Cambridge: Cambridge University Press, 2003.

———. *Occult Knowledge: Science and Gender on the Shakespearean Stage.* Cambridge: Cambridge University Press, 2013.

Foakes, R. A. *Shakespeare and Violence.* Cambridge: Cambridge University Press, 2003.

Ford, John. *'Tis Pity She's a Whore.* In *English Renaissance Drama: A Norton Anthology,* edited by David Bevington, Lars Engle, Katharine Eisaman Maus, and Eric Rasmussen. New York: W. W. Norton and Company, 2002.

Forker, Charles R. "'A Little More Than Kin and Less Than Kind': Incest, Intimacy, Narcissism and Identity in Elizabethan and Stuart Drama." *Medieval and Renaissance Drama in England* 4 (1989): 13–51.

Forsyth, Elliott. *La Tragédie Francaise de Jodelle à Corneille (1553–1640): Le Thème de la Vengeance.* Paris: A. G. Nizet, 1962.

Foster, Verna. "*'Tis Pity She's a Whore* as City Tragedy." In *John Ford: Critical Re-Visions,* edited by Michael Neill, 181–200. Cambridge: Cambridge University Press, 1988.

Foucault, Michel. *The History of Sexuality* Vol. 1. Translated by Robert Hurley. New York: Vintage Books, 1978.

Fowler, Elizabeth. *Literary Character: The Human Figure in Early English Writing.* Ithaca, NY: Cornell University Press, 2003.

Francese, Joseph. "La meritocrazia di Machiavelli: Dagli scritti politici alla *Mandragola.*" *Italica* 71, no.2 (1994): 153–75.

Freccero, Carla. *Father Figures: Genealogy and Narrative Structure in Rabelais.* Ithaca, NY: Cornell University Press, 1991.

Friedman, Jerome. "Jewish Conversion, the Spanish Pure Blood Laws and Reformation: A Revisionist View of Racial and Religious Antisemitism." *Sixteenth Century Journal* 18, no.1 (1987): 3–30.

Gager, Kristen Elizabeth. *Blood Ties and Fictive Ties.* Princeton, NJ: Princeton University Press, 1996.

Galen. *The Capacities of the Soul Depend on the Mixtures of the Body*. Translated by P. N. Singer. In Singer, *Galen*, 374–424.

———. *Character Traits*. Translated by Daniel Davies. In Singer, *Galen*, 135–201.

———. *On the Natural Faculties*. Translated by Arthur John Brock. Cambridge: Loeb Classical Library, Harvard University Press, 1916.

Gauer, Denis. "Heart and Blood: Nature and Culture in *'Tis Pity She's a Whore*." *Cahiers élizabéthains* 31 (1987): 45–57.

Gearhart, Stephanie S. *Drama and the Politics of Generational Conflict in Shakespeare's England*. New York: Routledge, 2018.

Geevers, Liesbeth, and Mirella Marini, eds. *Dynastic Identity in Early Modern Europe: Rulers, Aristocrats and the Formation of Identities*. New York: Routledge, 2016.

Gerard, Albert. "'Egregiously an Ass': The Dark Side of the Moor. A View of Othello's Mind." In *Aspects of Othello*, edited by Kenneth Muir and Philip Edwards. Cambridge: Cambridge University Press, 1977.

Giannetti, Laura. *Lelia's Kiss: Imagining Gender, Sex, and Marriage in Italian Renaissance Comedy*. Toronto: University of Toronto Press, 2009.

Giannetti, Laura, and Guido Ruggiero, ed. and trans. *Five Comedies of the Italian Renaissance*. Baltimore: Johns Hopkins University Press, 2003.

Giese, Loreen L. *Courtships, Marriage, Customs, and Shakespeare's Comedies*. New York: Palgrave Macmillan, 2006.

Gil Harris, Jonathan. "Shakespeare and Race." In *The New Cambridge Companion to Shakespeare*, edited by Margreta De Grazia and Stanley Wells, 201–15. 2nd ed. Cambridge University Press, 2010.

Girard, René. "'To Entrap the Wisest.'" In *William Shakespeare's* The Merchant of Venice, edited by Harold Bloom, 91–105. New York: Chelsea House Publishers, 1986.

Giuliani, Pierre. "D'un XVIIe siècle à l'autre: La question du sang sur scène. Une mise en perspective." *Revue d'histoire littéraire de la France* 104, no.2 (2004): 305–32.

Glassman, Bernard. *Anti-Semitic Stereotypes without Jews: Images of Jews in England, 1290–1700*. Detroit: Wayne State University Press, 1975.

Glimp, David. *Increase and Multiply: Governing Cultural Reproduction in Early Modern England*. Minneapolis: University of Minnesota Press, 2003.

Godshalk, W. L. "*All's Well That Ends Well* and the Morality Play." *Shakespeare Quarterly* 25, no.1 (1974): 61–70.

Goldberg, Jonathan, and Madhavi Menon. "Queering History." *PMLA* 120, no.5 (2005): 1608–17.

Goldstein, David B. "Jews, Scots, and Pigs in *The Merchant of Venice*." *Studies in English Literature 1500–1900* 54, no.2 (2014): 315–48.

Goode, William O. "Hand, Heart, and Mind: The Complexity of the Heroic Quest in *Le Cid*." *PMLA* 91, no.1 (1976): 44–53.

Goodkin, Richard E. "Neoclassical Dramatic Theory in Seventeenth-Century France." In *A Companion to Tragedy*, edited by Rebecca Bushnell, 373–91. Malden, MA: Blackwell Publishing, 2005.

Gordon, D. J. "Middleton's *No Wit, No Help Like a Woman's* and Della Porta's *La Sorella*." *Review of English Studies* 17, no.68 (1941): 400–14.

Gossip, C. J. "The Denouement of *Le Cid*, Yet Again." *Modern Language Review* 75, no.2 (1980): 275–81.

Greenberg, Mitchell. *Corneille, Classicism, and the Ruses of Symmetry*. Cambridge: Cambridge University Press, 1986.

———. "Stage Whispers: Corneille's Absolutism and the Voice of the M/other." *Biblio* 17, no.30. Edited by M. Bareau, J. Barchilon, D. Stanton, and J. Alter, 167–84. Paris: Papers on French Seventeenth Century Literature, 1987.

Greenblatt, Stephen. *Renaissance Self-Fashioning: From More to Shakespeare*. Chicago: University of Chicago Press, 1980.

Greene, Roland. *Five Words: Critical Semantics in the Age of Shakespeare and Cervantes*. Chicago: University of Chicago Press, 2013.

Greene, Thomas M. *The Light in Troy: Imitation and Discovery in Renaissance Poetry*. New Haven, CT: Yale University Press, 1982.

Gross, Gerard J. "The Conclusion to *All's Well That Ends Well*." *Studies in English Literature, 1500–1900* 23, no.2 (1983): 257–76.

Gross, Kenneth. *Shylock is Shakespeare*. Chicago: University of Chicago Press, 2006.

Guerreau-Jalabert, Anita. "Flesh and Blood in Medieval Language about Kinship." In Johnson et al. *Blood and Kinship*, 61–82.

Guillén, Claudio. *Literature as System: Essays toward the Theory of Literary History*. Princeton, NJ: Princeton University Press, 1971.

Gunsberg, Maggie. *Gender and the Italian Stage: From the Renaissance to the Present Day*. Cambridge: Cambridge University Press, 1997.

Gurney, Tessa C. "Echoes of Wartime in Late Sixteenth-Century Italian Comedy." *Journal of Iberian and Latin American Studies* 24, no.1 (2018): 155–74.

Guy-Bray, Stephen. *Against Reproduction: Where Renaissance Texts Come From*. Toronto: University of Toronto Press, 2009.

Haddad, Elie. "The Question of the Imprescriptibility of Nobility in Early Modern France." In Romaniello and Lipp, *Contested Spaces of Nobility*, 147–66.

Hale, John K. "Does Source Criticism Illuminate the Problems of Interpreting *The Merchant* as a Soured Comedy?" In *The Merchant of Venice: New Critical Essays*, edited by John W. Mahon and Ellen Macleod Mahon, 187–97. New York: Routledge, 2002.

Haley, David. *Shakespeare's Courtly Mirror: Reflexivity and Prudence in "All's Well That Ends Well."* Newark: University of Delaware Press, 1993.

Hall, Kim F. "Guess Who's Coming to Dinner? Colonization and Miscegenation in *The Merchant of Venice*." *Renaissance Drama* 23 (1992): 87–111.

———. "*Othello* and the Problem of Blackness." In *A Companion to Shakespeare's Works*. Vol. 1: *The Tragedies*, edited by Richard Dutton and Jean E. Howard, 357–74. Hoboken, NJ: Wiley-Blackwell, 2005.

———. *Things of Darkness: Economies of Race and Gender in Early Modern England*. Ithaca, NY: Cornell University Press, 1995.

Hamlin, William M. *Tragedy and Skepticism in Shakespeare's England*. New York: Palgrave Macmillan, 2005.

Hankinson, R. J. "Philosophy of Nature." In *The Cambridge Companion to Galen*, edited by R. J. Hankinson, 210–41. Cambridge: Cambridge University Press, 2009.

Harris, Jonathan Gil. "All Swell That End Swell: Dropsy, Phantom Pregnancy, and the Sound of Deconception in *All's Well That Ends Well*." *Renaissance Drama* 35 (2006): 169–89.

Harrison, Helen L. "Payer or Récompenser: Royal Gratitude in *Le Cid*." *French Review* 72, no.2 (1998): 238–49.

Harvey, William. *On the Motion of the Heart and Blood in Animals*. Translated and edited by Alex Bowie. Chicago: Gateway Editions, 1962.

Hauvette, Henri. "Une variante française de la légende de Roméo et Juliette." *Revue de littérature comparée* 1, no.3 (1921): 329–37.

Helmbrecht, Johannes. "Politeness Distinctions in Personal Pronouns—A Case Study in Competing Motivations." In *Competing Motivations in Grammar and Usage*, edited by MacWhinney, Brian, Andrej Malchukov, and Edith Moravcsik, 315–32. Oxford: Oxford University Press, 2014.

Hendricks, Margo. "Surveying 'Race' in Shakespeare." In *Shakespeare and Race*, edited by Catherine M. S. Alexander and Stanley Wells, 1–22. Cambridge: Cambridge University Press, 2000.

Hill, Ruth. "The Blood of Others: Breeding Plants, Animals, and White People in the Spanish Atlantic." In *The Cultural Politics of Blood, 1500–1900*, edited by Kimberly Anne Coles, Ralph Bauer, Zita Nunes, and Carla L. Peterson, 45–64. New York: Palgrave Macmillan, 2015.

Hill, W. Speed. "Marriage as Destiny: An Essay on *All's Well That Ends Well*." *English Literary Renaissance* 5 (1975): 344–59.

Hillman, David, and Carla Mazzio, eds. *The Body in Parts: Fantasies of Corporeality in Early Modern Europe*. New York: Routledge, 1997.

Hirsch, Brett D. "Counterfeit Professions: Jewish Daughters and the Drama of Failed Conversion in Marlowe's *The Jew of Malta* and Shakespeare's *The Merchant of Venice*." *Early Modern Literary Studies* 12, no.19 (2009): 1–37. Academic OneFile. www.purl.oclc.org.

———. "'A Gentle and No Jew': The Difference Marriage Makes in *The Merchant of Venice*." *Parergon* 23, no.1 (2006): 119–29.

Holbrook, Peter. *Shakespeare's Individualism*. Cambridge: Cambridge University Press, 2010.

Holmer, Joan Ozark. "Jewish Daughters: The Question of Philo-Semitism in Elizabethan Drama." In *The Merchant of Venice: New Critical Essays*, edited by John W. Mahon and Ellen Macleod Mahon, 107–43. New York: Routledge, 2002.

Hopkins, Lisa. "John Ford: Silence and Suffering in *Perkin Warbeck* and *'Tis Pity She's a Whore*." In *The Cambridge Companion to Shakespeare and Contemporary Dramatists*, edited by Ton Hoenselaars, 197–211. Cambridge: Cambridge University Press, 2012.

———, ed. *'Tis Pity She's a Whore: A Critical Guide*. London: Continuum, 2010.

———. "Knowing Their Loves: Knowledge, Ignorance, and Blindness in *'Tis Pity She's a Whore*." *Renaissance Forum* 3, no.1 (1998). www.web.archive.org.

———. *The Shakespearean Marriage: Merry Wives and Heavy Husbands*. New York: St. Martin's Press, 1998.

———. "A Source for John Ford's *'Tis Pity She's a Whore*." *Notes and Queries* 41, no.4 (1994): 520.

———. "Speaking Sweat: Emblems in the Plays of John Ford." *Comparative Drama* 29, no.1 (1995): 133–46.

———. "Venice in *'Tis Pity She's a Whore*." *Early Theatre* 13, no.2 (2010): 79–88.

———. "Wooing Scenes in *'Tis Pity She's a Whore* and *Richard III*." *Notes and Queries* 40, no.2 (1993): 227–28.

Howard, Jean E. "Female Agency in *All's Well That Ends Well*." *AUMLA: Journal of the Australian Universities Modern Language Association* 106 (2006): 43–60.

Hunter, G. K. "Othello and Colour Prejudice." In *Proceedings of the British Academy* 53. London: Oxford University Press, 1968.

Hunter, Lynette. "Cankers in Romeo and Juliet: Sixteenth-Century Medicine at a Figural/Literal Cusp." In *Disease, Diagnosis, and Cure on the Early Modern Stage*, edited by Stephanie Moss and Kaara L. Peterson, 171 85. Burlington, VT: Ashgate, 2004.

Hutson, Lorna. "'Che indizio, che prova . . . ?': Ariosto's Legal Conjectures and the English Renaissance Stage." *Renaissance Drama* 36 (2010): 179–205.

———. "Fortunate Travelers: Reading for the Plot in Sixteenth-Century England." *Representations* 41 (1993): 83–103.

Hyman, Lawrence W. "The Rival Lovers in The Merchant of Venice." *Shakespeare Quarterly* 21, no.2 (1970): 109–16.

Ibbett, Katherine. *The Style of the State in French Theater, 1630–1660: Neoclassicism and Government*. Burlington, VT: Ashgate, 2009.

Inglis, Kirsten. *Reading "Circe's Court": Politics and the Bed-Trick from Boccaccio to Shakespeare*. Master's thesis, Dalhousie University, 2007. University of Michigan Dissertation Publishing, 2007.

Ingram, Martin. *Church Courts, Sex and Marriage in England, 1570–1640*. Cambridge: Cambridge University Press, 1987.

Irvine, Christopher. *Medicina Magnetica; or, The Rare and Wonderful Art of Curing By Sympathy*. Edinburgh: C. Higgins, 1656.

Ivanova, Julia V. "Machiavelli's Soteriology and the Humanist Quattrocento Dialogue." In *History and Drama: The Pan-European Tradition*, edited by Joachim Küpper, Jan Mosch, and Elena Penskaya, 60–71. Boston: De Gruyter, 2019.

Iyengar, Sujata. *Shades of Difference: Mythologies of Skin Color in Early Modern England*. Philadelphia: University of Pennsylvania Press, 2009.

James, Heather. "The Ovidian Girlhood of Shakespeare's Boy Actors: Q2 Juliet." In *Shakespeare Survey*, edited by Peter Holland, 106–22. Cambridge: Cambridge University Press, 2016.

Jankowski, Theodora A. "Hymeneal Blood, Interchangeable Women, and the Early Modern Marriage Economy in *Measure for Measure* and *All's Well That Ends Well*." In *A Companion to Shakespeare's Works*. Vol. 4, *The Poems, Problem Comedies, Late Plays*, edited by Richard Dutton and Jean E. Howard, 89–105. Malden, MA: Blackwell Publishing, 2005.

Johnson, Bonnie Lander. "Blood, Milk, Poison: Romeo and Juliet's Tragedy of 'Greeen' Desire and Corrupted Blood." In Johnson and Decamp, *Blood Matters*, 134–48.

Johnson, Bonnie Lander, and Eleanor Decamp, eds. *Blood Matters: Studies in European Literature and Thought, 1400–1700*. Philadelphia: University of Pennsylvania Press, 2018.

Johnson, Christopher H., Bernhard Jussen, David Warren Sabean, and Simon Teuscher, eds. *Blood and Kinship: Matter for Metaphor from Ancient Rome to the Present*. New York: Berghahn Books, 2013.

Jones, Eldred D. *Othello's Countrymen: The African in English Renaissance Drama*. London: Oxford University Press, 1965.

Jouanna, Arlette. *L'idée de race en France au XVIe et au debut du XVIIe siècle*. Paris: Librairie Honoré Champion, 1976.

———. *The Saint Bartholomew's Day Massacre: The Mysteries of a Crime of State*. Translated by Joseph Bergin. Manchester: Manchester University Press, 2015.

Jowett, John. "Introduction: *No Wit/Help Like a Woman's*; or, *The Almanac*." In *Thomas Middleton*. Vol 1., *The Collected Works*. Edited by Gary Taylor, John Lavagnino, MacDonald P. Jackson, John Jewett, Valerie Wayne, and Adrian Weiss, 779–83. Oxford: Oxford University Press, 2007.

———. "Middleton's *No Wit* at the Fortune." *Renaissance Drama* 22 (1991): 191–208.

Kahn, Coppélia. "Coming of Age in Verona." *Modern Language Studies* 8, no.1 (1977): 5–22.

———. *Man's Estate: Masculine Identity in Shakespeare*. Berkeley: University of California Press, 1981.

Kahn, Victoria. "Machiavelli's Afterlife and Reputation to the Eighteenth Century." In *The Cambridge Companion to Machiavelli*, edited by John M. Najemy, 239–55. Cambridge: Cambridge University Press, 2010.

Kaplan, Lindsay M. "Jessica's Mother: Medieval Constructions of Jewish Race and Gender in *The Merchant of Venice*." *Shakespeare Quarterly* 58, no.1 (2007): 1–30.

Karim-Cooper, Farah. "The Materials of Race: Staging the Black and White Binary in the Early Modern Theatre." In Thompson, *Cambridge Companion to Shakespeare and Race*, 17–29.

Kay, W. David. "Shakespeare's Transformative Art: Theatrical Paradigms as Sources in *All's Well That Ends Well* and *Macbeth*." In Britton and Walter, *Rethinking Shakespeare Source Study*, 159–79.

Keller, Eve. "'That Sublimest Juyce in Our Body': Bloodletting and Ideas of the Individual in Early Modern England." *Philological Quarterly* 86 (2007): 97–122.

Keller, Marcus. *Figurations of France: Literary Nation-Building in Times of Crisis (1550–1650)*. Newark: University of Delaware, 2011.

Kennedy, William J. "Comic Audiences and Rhetorical Strategies in Machiavelli, Shakespeare, and Molière." *Comparative Literature Studies* 21, no.4 (1984): 363–82.

Kenny, Amy. *Humoral Wombs on the Shakespearean Stage*. New York: Palgrave Macmillan, 2019.

Kenny, Neil. *Born to Write: Literary Families and Social Hierarchy in Early Modern France*. Oxford: Oxford University Press, 2020.

Kerwin, William. *Beyond the Body: The Boundaries of Medicine and English Renaissance Drama*. Amherst: University of Massachusetts Press, 2005.

Kibler, William. "Objective and Subjective *Gloire* as seen in *Le Cid, Horace*, and *Cinna* of Corneille." *Romance Notes* 7 (1966): 171–75.

Kirschbaum, Leo. "Shakespeare's Stage Blood and Its Critical Significance." *PMLA* 64, no.3 (1949): 517–29.

Kistner, Arthur L., and M. K. Kistner. "The Dramatic Functions of Love in the Tragedies of John Ford." *Studies in Philology* 70, no.1 (1973): 62–76.

Klapsich-Zuber, Christiane. *Women, Family, and Ritual in Renaissance Italy*. Translated by Lydia Cochrane. Chicago: University of Chicago Press, 1985.

Knight, R. C. *Corneille's Tragedies: The Role of the Unexpected*. Washington, DC: Rowman and Littlefield Publishers, 1991.

Knippenberg, Joseph M. "Virtue, Honor, and Reputation: Machiavelli's Appropriation of Christianity in the 'Rape' of Lucrezia." In *Poets, Princes, and Private Citizens*, edited by Joseph M. Knippenberg and Peter Augustine Lawler, 21–38. Lanham, MD: Rowman and Littlefield Publishers, 1996.

Knowles, Ronald. "Carnival and Death in *Romeo and Juliet*: A Bakhtinian Reading." *Shakespeare Survey* 49 (1996): 69–84.

Koch, Erec R. *The Aesthetic Body: Passion, Sensibility and Corporeality in Seventeenth-Century France.* Newark: University of Delaware Press, 2008.

Kodera, Sergius. "The Laboratory as Stage: Giovan Battista della Porta's Experiments." *Journal of Early Modern Studies* 3, no.1 (2014): 15–38.

Kottman, Paul A. "Defying the Stars: Tragic Love as the Struggle for Freedom in *Romeo and Juliet.*" *Shakespeare Quarterly* 63, no.1 (2012): 1–38.

Kowsar, Mohammad. "In Defense of Desire: Chimène's Role in *Le Cid* Reconsidered." *Theatre Journal* 343 (1982): 289–301.

Kriegel, Jill. "A Case against Natural Magic: Shakespeare's Friar Laurence as *Romeo and Juliet's* Near-Tragic Hero." *Logos: A Journal of Catholic Thought and Culture* 13, no.1 (2010): 132–45.

Kugler, Georg. "'The Predestination of Our Hapsburg Dynasty': Hapsburg Marriage Policies and the Phenomenon of 'Pure Blood' in European Dynasties." Translated by Paul Aston and Ishbel Flett. In *Blood: Art, Power, Politics and Pathology,* edited by James M. Bradburne, 120–35. Munich: Prestel Verlag, 2001.

Kuzner, James. "*All's Well That Ends Well* and the Art of Love." *Shakespeare Quarterly* 68, no.3 (2017): 215–40.

Langley, Eric. "Plagued by Kindness: Contagious Sympathy in Shakespearean Drama." *Medical Humanities* 37 (2011): 103–9.

———. *Shakespeare's Contagious Sympathies: Ill Communications.* Oxford: Oxford University Press, 2018.

Laroque, François. "The 'Science' of Astrology in Shakespeare's Sonnets, *Romeo and Juliet* and *King Lear.*" In *Spectacular Science, Technology and Superstition in the Age of Shakespeare,* edited by Sophie Chiari and Mickaël Popelard, 29–42. Edinburgh: Edinburgh University Press, 2017.

Leggatt, Alexander. "*All's Well That Ends Well*: The Testing of Romance." *Modern Language Quarterly* 32, no.1 (1971): 21–41.

Lehnhoff, Kent R. "Performing Woman: Female Theatricality in *All's Well That Ends Well.*" In *New Critical Essays,* edited by Gary Waller, 111–24. New York: Routledge, 2007.

Lemnius, Levinus. *The Touchstone of Complexions Generallye Appliable, Expedient and Profitable for All Such, as Be Desirous and Carefull of Their Bodylye Health* [...]. 1576.

Lemon, Rebecca. "Shakespeare and Law." In *Oxford Guide to Shakespeare,* edited by Arthur F. Kinney, 548–64. Oxford: Oxford University Press, 2011.

LePage, Raymond. "A Study in Dramatic Transposition and Invention: Della Porta's *La Sorella,* Rotroi's *La Soeur,* and Middleton's *No Wit, No Help Like A Woman's.*" *Comparative Literature Studies* 24, no.4 (1987): 335–52.

Lerner, Laurence. "The Machiavel and the Moor." *Essays in Criticism* 9, no.4 (1959): 339–60.

Levenson, Jill. "Romeo and Juliet before Shakespeare." *Studies in Philology* 81, no.3 (1984): 325–47.

Levin, Harru. "Form and Formality in *Romeo and Juliet.*" *Shakespeare Quarterly* 11, no.1 (1960): 3–11.

Levin, Richard A. "*All's Well That Ends Well* and 'All Seems Well.'" *Shakespeare Studies* 13 (1980): 131–44.

Lezra, Jacques. *Unspeakable Subjects: The Genealogy of the Event in Early Modern Europe.* Stanford, CA: Stanford University Press, 1997.

Linke, Uli. *Blood and Nation: The European Aesthetics of Race*. Philadelphia: University of Pennsylvania Press, 1999.

Lipscomb, Suzannah. "Subjection and Companionship: The French Reformed Marriage." *Reformation and Renaissance Review* 6, no.3 (2004): 349–60.

Little, Arthur L., Jr. "Re-Historicizing Race, White Melancholia, and the Shakespearean Property." *Shakespeare Quarterly* 67, no.1 (2016): 84–103.

———. "The Rites of Queer Marriage in *The Merchant of Venice*." In *Shakesqueer: A Queer Companion to the Complete Works of Shakespeare*, edited by Madhavi Menon, 216–24. Durham, NC: Duke University Press, 2011.

Litvin, Boris. "Mapping Rule and Subversion: Perspective and the Democratic Turn in Machiavelli Scholarship." *European Journal of Political Theory* 18, no.1 (2019): 3–25.

Loomba, Ania. "'Delicious Traffick': Racial and Religious Difference on Early Modern Stages." In *Shakespeare and Race*, edited by Catherine M. S. Alexander and Stanley Wells, 203–24. Cambridge: Cambridge University Press, 2000.

———. "Identities and Bodies in Early Modern Studies." In *The Oxford Handbook of Shakespeare and Embodiment: Gender, Sexuality, and Race*, edited by Valerie Traub, 228–45. Oxford: Oxford University Press, 2016.

———. *Shakespeare, Race, and Colonialism*. Oxford: Oxford University Press, 2002.

Loomba, Ania, and Jonathan Burton, eds. *Race in Early Modern England: A Documentary Companion*. New York: Palgrave Macmillan, 2007.

Lorch, Maristella de Panizza. "Women in the Context of Machiavelli's *Mandragola*." In *Donna: Women in Italian Culture*, edited by Ada Testaferri, 253–72. Ottawa: Dovehouse Editions, 1989.

Lord, Carnes. "Allegory in Machiavelli's *Mandragola*." In *Political Philosophy and the Human Soul: Essays in Memory of Allan Bloom*, edited by Michael Palmer and Thomas L. Pangle, 149–74. London: Rowman and Littlefield Publishers, 1995.

Love, John M. "'Though Many of the Rich Are Damn'd: Dark Comedy and Social Class in *All's Well That Ends Well*." *Texas Studies in Literature and Language* 18, no.4 (1977): 517–27.

Low, Jennifer A. "'Bodied Forth': Spectator, Stage, and Actor in the Early Modern Theater." *Comparative Drama* 39, no.1 (2005): 1–29.

Luis-Martínez, Zenón. *In Words and Deeds: The Spectacle of Incest in English Renaissance Tragedy*. New York: Rodopi, 2002.

Lukes, Timothy J. "Fortune Comes of Age (in Machiavelli's Literary Works)." *Sixteenth Century Journal* 11, no.4 (1980): 33–50.

Lupton, Julia Reinhard. "Character Criticism and Psychoanalytic Criticism." In *The Cambridge Guide to the World of Shakespeare*, edited by Bruce Smith, 1761–68. Cambridge: Cambridge University Press, 2016.

———. *Citizen-Saints: Shakespeare and Political Theology*. Chicago: University of Chicago Press, 2005.

———. "Making Room, Affording Hospitality: Environments of Entertainment in *Romeo and Juliet*." *Journal of Medieval and Early Modern Studies* 43, no.1 (2013): 145–72.

———. "Response to Paul A. Kottman 'Defying the Stars: Tragic Love as the Struggle for Freedom in *Romeo and Juliet*.'" *Shakespeare Quarterly* 63, no.1 (2012): 39–45.

Lydgate, John. *Troy Book*. Edited by Robert R. Edwards. TEAMS Middle English Texts Series, Robbins Library Digital Projects, University of Rochester, 1998.

Lyly, John. *Galatea*. Edited by Leah Scragg. Manchester: Manchester University Press, 2012.

Lyons, John D. *Kingdom of Disorder: The Theory of Tragedy in Classical France*. West Lafayette, IN: Purdue University Press, 1999.

———. *The Tragedy of Origins: Pierre Corneille and Historical Perspective*. Stanford, CA: Stanford University Press, 1996.

Macfarlane, Alan. *Marriage and Love in England: Modes of Reproduction 1300–1840*. New York: Basil Blackwell, 1986.

Machiavelli, Niccolò. *La mandragola*. In *The Comedies of Machiavelli: The Woman from Andros, The Mandrake, Clizia*. Bilingual Edition, edited and translated by David Sices and James B. Atkinson, 153–275. Indianapolis: Hackett Publishing, 2007.

———. *La mandragola (The Mandrake Root)*. In *Five Comedies from the Italian Renaissance*, translated and edited by Laura Giannetti and Guido Ruggiero, 71–116. Baltimore: Johns Hopkins University Press, 2003.

MacKenzie, Clayton G. "Love, Sex and Death in *Romeo and Juliet*." *English Studies* 88, no.1 (2007): 22–42.

Manes, Yael. *Motherhood and Patriarchal Masculinities in Sixteenth Century Italian Comedy*. New York: Routledge, 2011.

Mansfield, Harvey C. "The Cuckold in Machiavelli's *Mandragola*." In *The Comedy and Tragedy of Machiavelli: Essays on the Literary Works*, edited by Vickie B. Sullivan, 1–29. New Haven, CT: Yale University Press, 2000.

Marchitello, Howard. *The Machine in the Text: Science and Literature in the Age of Shakespeare and Galileo*. Oxford: Oxford University Press, 2011.

Margitic, Milorad R. *Essai sur la mythologie du "Cid."* Chapel Hill: Romance Monographs, 1976.

Markham, Francis. *The Booke of Honour: Or Five Decades of Epistles of Honour*. London, 1625.

Marrapodi, Michele. "From Narrative to Drama: The Erotic Tale and the Theater." In *The Italian World of English Renaissance Drama: Cultural Exchange and Intertextuality*, edited by Michele Marrapodi, 41–70. Newark: University of Delaware Press, 1998.

———, ed. *Shakespeare and Intertextuality: The Transition of Cultures between Italy and England in the Early Modern Period*. Rome: Bulzoni, 2000.

Martin, John Jeffries. *Myths of Renaissance Individualism*. New York: Palgrave Macmillan, 2004.

Martinez, Ronald L. "Comedian, Tragedian: Machiavelli and Traditions of Renaissance Theater." In *The Cambridge Companion to Machiavelli*, edited by John M. Najemy, 206–22. Cambridge: Cambridge University Press, 2010.

———. "The Pharmacy of Machiavelli: Roman Lucretia in *Mandragola*." *Renaissance Drama* 14 (1983): 1–43.

Masten, Jeffrey. *Textual Intercourse: Collaboration, Authorship, and Sexualities in Renaissance Drama*. Cambridge: Cambridge University Press, 1997.

Mastri, Augustus A. "Machiavelli's *La Mandragola*: A Political and Personal Statement." *Forum* 28, no.3 (1987): 3–15.

Matthes, Melissa M. "The Seriously Comedic, or Why Machiavelli's Lucrezia Is Not Livy's Virtuous Roman." In *Feminist Interpretations of Niccolò Machiavelli*, edited by Maria J. Falco, 247–66. University Park: Pennsylvania State University Press, 2004.

Maus, Katherine Eisaman. *Inwardness and Theater in the English Renaissance*. Chicago: University of Chicago Press, 1995.

May, Georges. *Tragédie Cornelienne, tragédie racinienne*. Urbana: University of Illinois Press, 1948.

Mazzeo, Joseph Anthony. *Renaissance and Revolution: The Remaking of European Thought*. New York: Pantheon Books, 1965.

McCabe, Richard A. *Incest, Drama and Nature's Law 1550–1700*. Cambridge: Cambridge University Press, 1993.

McCandless, David. "Helena's Bed Trick: Gender and Performance in *All's Well That Ends Well*." *Shakespeare Quarterly* 45, no.4 (1994): 449–68.

McEachern, Claire. "Introduction: *All's Well That Ends Well*." In *William Shakespeare The Complete Works*, edited by Stephen Orgel and A. R. Braunmuller, 565–69. New York: Penguin Books, 2002.

Melzer, Sara E. *Colonizer or Colonized: The Hidden Stories of Early Modern French Culture*. Philadelphia: University of Pennsylvania Press, 2012.

Mentz, Steven. "Revising the Sources: Novella, Romance and the Meanings of Fiction in *All's Well That Ends Well*." In *New Critical Essays*, edited by Gary Waller, 57–70. New York: Routledge, 2007.

Merlin-Kajman, Hélène. *L'absolutisme dans les lettres et la théorie des deux corps: Passions et politique*. Paris: H. Champion, 2000.

Metzger, Mary Janell. "'Now by My Hood, a Gentle and No Jew': Jessica, *The Merchant of Venice*, and the Discourse of Early Modern English Identity." *PMLA* 113, no.1 (1998): 52–63.

Middleton, Irene. "A Jew's Daughter and a Christian's Wife: Performing Jessica's Multiplicity in *The Merchant of Venice*." *Shakespeare Bulletin* 33, no.2 (2015): 293–317.

Middleton, Thomas. *No Wit, No Help Like a Woman's*. Edited by Michael Taylor. Oxford: Oxford University Press, 1995.

Miller, Jonathan. "The Pump: Harvey and the Circulation of the Blood." In *Blood: Art, Power, Politics and Pathology*, edited by James M. Bradburne, 100–107. Munich: Prestel Verlag, 2001.

Milstein, Jouanna. *The Gondi: Family Strategy and Survival in Early Modern France*. Farnham, UK: Routledge, 2014.

Minel, Emmanuel. *Pierre Corneille, Le Héros et le Roi: Stratégies d'héroïsation dans le théâtre cornélien*. Paris: Eurédit, 2010.

Mintz, Susannah B. "The Power of 'Parity' in Ford's *'Tis Pity She's a Whore*." *Journal of English and Germanic Philology* 102, no.2 (2003): 269–91.

Miola, Robert S. "New Comedy in *All's Well That Ends Well*." *Renaissance Quarterly* 46, no.1 (1993): 23–43.

———. "Seven Types of Intertextuality." In Marrapodi, *Shakespeare and Intertextuality*, 23–38.

Moisan, Thomas. "'Which Is the Merchant Here? And Which the Jew?': Subversion and Recuperation in *The Merchant of Venice*." In *Shakespeare Reproduced: The Text in History and Ideology*, edited by Jean E. Howard and Marion F. O'Connor, 188–206. New York: Routledge, 1987.

Monaco, Mario. "La fortuna della Mandragola." *Bibliofilia* 39, no.3 (1937): 113–21.

Monga, Luigi. "Romeo and Juliet Revisited: More *novelle* from the Italian Renaissance." *Manuscripta* 33, no.1 (1989): 47–53.

Monsarrat, Gilles. "The Unity of John Ford: *'Tis Pity She's a Whore* and *Christ's Bloody Sweat*." *Studies in Philology* 77, no.3 (1980): 247–70.

Montgomery, Marianne. "Listening to the Emissary in Middleton's *No Wit, No Help Like a Woman's.*" In *Emissaries in Early Modern Literature and Culture,* edited by Brinda Charry and Gitanjali Shahani, 193–203. Abingdon, UK: Ashgate Publishing, 2009.

Moore, Olin H. *The Legend of Romeo and Juliet.* Columbus: Ohio State University Press, 1950.

Moss, Stephanie. "Transformation and Degeneration: The Paracelsan/Galenic Body in Othello." In *Disease, Diagnosis, and Cure on the Early Modern Stage,* edited by Stephanie Moss and Kaara L. Peterson, 151–70. Burlington, VT: Ashgate, 2004.

Moulton, Ian Frederick. "The Italian Encounter with Tudor England: A Cultural Poetics of Translation." *Shakespeare Studies* 35 (2007): 300–305.

Mousley, Andy. *Renaissance Drama and Contemporary Literary Theory.* New York: Palgrave Macmillan, 2000.

Muir, Edward. *Mad Blood Stirring: Vendetta and Factions in Friuli during the Renaissance.* Baltimore: Johns Hopkins University Press, 1993.

Muir, Kenneth. "The Case of John Ford." *Sewanee Review* 84, no.4 (1976): 614–29.

———. *The Sources of Shakespeare's Plays.* New Haven, CT: Yale University Press, 1978.

———. "Two Plays Reconsidered: *More Dissemblers besides Women* and *No Wit, No Help Like A Woman's.*" In *"Accompaninge the players": Essays Celebrating Thomas Middleton, 1580–1980,* edited by Kenneth Friedenreich, 147–59. New York: AMS Press, 1983.

Mullaney, Steven. *The Reformation of Emotions in the Age of Shakespeare.* Chicago: University of Chicago Press, 2015.

Munro, Lucy. "'They Eat Each Others Arms': Stage Blood and Body Parts." In *Shakespeare's Theaters and the Effects of Performance,* edited by Farah Karim-Cooper and Tiffany Stern, 73–93. London: Arden Shakespeare Library, 2013.

Muratore, Mary Jo. *The Evolution of the Cornelian Heroine.* Madrid: Studia Humanitatis, 1982.

Murphy, Andrew. *But the Irish Sea Betwixt Us: Ireland, Colonialism, and Renaissance Literature.* Lexington: University of Kentucky Press, 1999.

Nadal, Octave. *Le sentiment de l'amour dans l'oeuvre de Pierre Corneille.* Paris: Gallimard, 1948.

Nashon, Edna, and Michael Shapiro, eds. *Wrestling with Shylock: Jewish Responses to* The Merchant of Venice. Cambridge: Cambridge University Press, 2017.

Neely, Carol Thomas. *Broken Nuptials in Shakespeare's Plays.* Urbana: University of Illinois Press, 1985.

———. "Hot Blood: Estranging Mediterranean Bodies in Early Modern Medical and Dramatic Texts." In *Disease, Diagnosis, and Cure on the Early Modern Stage,* edited by Stephanie Moss and Kaara L. Peterson, 55–68. Burlington, VT: Ashgate, 2004.

Neill, Michael. "Unproper Beds: Race, Adultery, and the Hideous in *Othello.*" *Shakespeare Quarterly* 40, no.4 (1989): 383–412.

———. "'What Strange Riddle Is This?': Deciphering *'Tis Pity She's a Whore.*" In *John Ford: Critical Re-Visions,* edited by Michael Neill, 153–79. Cambridge: Cambridge University Press, 1988.

Nelson, Robert J. *Corneille: His Heroes and Their Worlds.* Philadelphia: University of Pennsylvania Press, 1963.

Neuschel, Kristen B. *Word of Honor: Interpreting Noble Culture in Sixteenth-Century France.* Ithaca, NY: Cornell University Press, 1989.

Nevo, Ruth. "Tragic Form in Romeo and Juliet." *Studies in English Literature, 1500–1900* 9, no.2 (1969): 241–58.

Newcomb, Lori Humphrey. "Toward a Sustainable Source Study." In Britton and Walter, *Rethinking Shakespeare Source Study*, 19–45.

Newman, Karen. "Corneille's City Comedy: Courtship and Consumption in Early Modern Paris." *Renaissance Drama* 27 (1996): 105–22.

———. *Fashioning Femininity and English Renaissance Drama*. Chicago: University of Chicago Press, 1991.

———. "Portia's Ring: Unruly Women and Structures of Exchange in *The Merchant of Venice*." *Shakespeare Quarterly* 38, no.1 (1987): 19–33.

Nicholson, Eric. "Helen, the Italianate Theatrical Wayfarer of *All's Well That Ends Well*." In *Shakespeare and the Italian Renaissance: Appropriation, Transformation, Opposition*, edited by Michele Marrapodi, 163–80. Burlington, VT: Ashgate, 2014.

Normand, Lawrence. "Reading the Body in *The Merchant of Venice*." *Textual Practice* 5, no.1 (1991): 55–73.

Novy, Marianne. *Love's Argument: Gender Relations in Shakespeare*. Chapel Hill: University of North Carolina Press, 1984.

———. "Violence, Love, and Gender in *Romeo and Juliet*." In *Romeo and Juliet: Critical Essays*, edited by John F. Andrews, 359–370. New York: Routledge, 1993.

Nunn, Hillary M. *Staging Anatomies: Dissection and Spectacle in Early Stuart Tragedy*. Burlington, VT: Ashgate, 2005.

O'Brian, Mary. "The Root of the Mandrake: Machiavelli and Manliness." In *Feminist Interpretations of Niccolò Machiavelli*, edited by Maria J. Falco, 173–96. University Park: Pennsylvania State University Press, 2004.

O'Callaghan, Michelle. *Thomas Middleton, Renaissance Dramatist*. Edinburgh: Edinburgh University Press, 2009.

O'Hara, Diana. *Courtship and Constraint: Rethinking the Making of Marriage in Tudor England*. Manchester: Manchester University Press, 2000.

Oliver, Kelly, eds. *The Portable Kristeva: Updated Edition*. New York: Columbia University Press, 2002.

Orkin, Martin. "Othello and the 'Plain Face' of Racism." *Shakespeare Quarterly* 38, no.2 (1987): 166–88.

Ornstein, Robert. *The Moral Vision of Jacobean Tragedy*. Madison: University of Wisconsin Press, 1960.

Osherow, Michele. "She Is in the Right: Biblical Maternity and *All's Well, That Ends Well*." In *New Critical Essays*, edited by Gary Waller, 155–68. New York: Routledge, 2007.

Padgett, John F. "Open Elite? Social Mobility, Marriage, and Family in Florence, 1282–1494." *Renaissance Quarterly* 63, no.2 (2010): 357–411.

Padoan, Giorgio. "*La Mandragola* del Machiavelli nella Venezia cinquecentesca." *Lettere Italiane* 22, no.2 (1970): 161–86.

Palmer, Michael, and James F. Pontuso. "The Master Fool: The Conspiracy of Machiavelli's *Mandragola*." *Perspectives on Political Science* 25 (1996): 124–32.

Park, Katherine. "Holy Autopsies: Saintly Bodies and Medical Expertise, 1300–1600." In *The Body in Early Modern Italy*, edited by Julia L. Hairston and Walter Stephens, 61–73. Baltimore: Johns Hopkins University Press, 2010.

Parker, John. "Holy Adultery: Marriage in *The Comedy of Errors*, *The Merchant of Venice*, and *The Merry Wives of Windsor*." In *The Oxford Handbook of Shakespearean Comedy*, edited by Heather Hirschfeld, 491–504. Oxford: Oxford University Press, 2018.

Parker, Patricia. "*All's Well That Ends Well*: Increase and Multiply." In *Creative Imitation: New Essays on Renaissance Literature In Honor of Thomas M. Greene*, edited by David Quint, Margaret W. Ferguson, Wayne A. Rebhorn, and G. W. Pigman, 355–90. Binghamton: Medieval and Early Renaissance Studies, 1992.

———. "Fantasies of 'Race' and 'Gender': Africa, Othello and Bringing to Light." In *Women, "Race," and Writing in the Early Modern Period*, edited by Margo Hendricks and Patricia Parker, 84–100. New York: Routledge, 1994.

Parker, R. B. "War and Sex in *All's Well That Ends Well*." *Shakespeare Survey* 37 (1948): 99–113.

Parronchi, A. "La prima rappresentazione della *Mandragola*. Il modello dell'apparato, L'allegoria." *La Bibliofilia* 64 (1962): 37–86.

Paster, Gail Kern. "The Humor of It: Bodies, Fluids, and Social Discipline in Shakespearean Comedy." In *A Companion to Shakespeare's Works*. Vol. 3, *The Comedies*, edited by Richard Dutton and Jean E. Howard, 47–66. Malden, MA: Blackwell Publishing, 2003.

———. *Humoring the Body: Emotions and the Shakespearean Stage*. Chicago: University of Chicago Press, 2004.

Paster, Gail Kern, Katherine Rowe, and Mary Floyd-Wilson, eds. *Reading the Early Modern Passions: Essays in the Cultural History of Emotion*. Philadelphia: University of Pennsylvania Press, 2004.

Pearce, Spencer. "Dante: Order, Justice and the Society of Orders." In *Orders and Hierarchies in Late Medieval and Renaissance Europe*, edited by Jeffrey Denton, 33–55. Toronto: University of Toronto Press, 1998.

Perkins, Marina. "A Message from the Margins: The Function of the Infante in Corneille's *Le Cid*." *French Studies* 74, no. 4 (2020): 519–35.

Perocco, Daria. "The First Variations of a Legendary Short Story: *Romeo and Juliet*." In *The Italian Short Story through the Centuries: The Met(A)morphoses of the Novella*, edited by Roberto Nicosia, 42–54. Cambridge: Cambridge Scholars Publishing, 2018.

Perrone, Carlachiara. *"So che Donna Amo Donna"*: La Calisa *di Maddalena Campiglia*. Galatina, Italy: Congedo Editore, 1996.

Peters, Christine. "Gender, Sacrament and Ritual: The Making and Meaning of Marriage in Late Medieval and Early Modern England." *Past and Present* 169 (2000): 63–96.

Peterson, Kaara L. "Medical Discourses of Virginity and the Bed-Trick in Shakespearean Drama." In *The Palgrave Handbook of Early Modern Literature and Science*, edited by H. Marchitello and E. Tribble, 377–99. New York: Palgrave MacMillan, 2017.

Pettigrew, Todd H. J. *Shakespeare and the Practice of Physic: Medical Narratives on the Early Modern Stage*. Newark: University of Delaware Press, 2007.

Pichichero, Christy L. "Pierre Corneille and Military Drama: Power, Potlach, *Mérite*." *MLN* 132 (2017): 1090–1117.

———. "Words of the 'Wise Captain': Pierre Corneille, *Le Cid*, and Fidelity." *Renaissance Drama* 43, no.1 (2015): 27–52.

Piesse, A. J. "Identity." In *A New Companion to English Renaissance Literature and Culture*, edited by Michael Hattaway, 461–73. Malden, MA: Blackwell Publishing, 2010.

Planinc, Zdravko. "Politics, Religion and Love's Transgression: The Political Philosophy of *Romeo and Juliet*." *Philosophy and Literature* 43, no.1 (2019): 11–37.

Poulsen, Rachel E. "Incest and Inflection in Della Porta's *La Sorella*." In *Sex Acts in Early*

Modern Italy: Practice, Performance, Perversion, Punishment, edited by Allison Levy, 181–94. Burlington, VT: Ashgate Publishing, 2010.

Powell, Raymond. "The Adaptation of a Shakespearean Genre: *Othello* and Ford's *'Tis Pity She's a Whore.*" *Renaissance Quarterly* 48, no.3 (1995): 582–92.

Prigent, Michel. *Le héros et l'état dans la tragedie de Corneille.* Paris: Presses Universitaires de France, 1986.

"Proceedings and Conclusions: Stage Blood, A Roundtable." *Shakespeare's Globe Theater History Seminar.* Edited by Farah Karim-Cooper and Ryan Nelson. July 13, 2006, 1–37.

Prunster, Nicole. "Theory and Practice in the Theatre of Giambattista Della Porta." PhD diss., University of Toronto, 1979.

Pullan, Brian. "Three Orders of Inhabitants': Social Hierarchies in the Republic of Venice." In *Orders and Hierarchies in Late Medieval and Renaissance Europe,* edited by Jeffrey Denton, 147–68. Toronto: University of Toronto Press, 1998.

Pullini, G. "Stile di transizione nel teatro di Giambattista Della Porta." *Lettere Italiane* 8, no.3 (1956): 299–310.

Quilligan, Maureen. *Incest and Agency in Elizabeth's England.* Philadelphia: University of Pennsylvania Press, 2005.

Radcliff-Umstead, Douglas. *The Birth of Modern Comedy in Renaissance Italy.* Chicago: University of Chicago Press, 1969.

Räisänen, Kalervo. "Le dilemme et la métaphore étendue dans *Le Cid.*" *PFSCL* 40, no.78 (2013): 81–98.

Rebhorn, Wayne A. *Foxes and Lions: Machiavelli's Confidence Men.* Ithaca, NY: Cornell University Press, 1988.

Reilly, Terry. "'Doctor She!' Helena and Sisterhood in William Shakespeare's *All's Well That Ends Well* (ca. 1602–1603)." In *Women in Literature: Looking through the Lens of Gender,* edited by Jerily Fisher and Ellen S. Silber, 8–11. Westport, CT: Greenwood Press, 2003.

Riddle, John M. *Eve's Herbs: A History of Contraception and Abortion in the West.* Cambridge, MA: Harvard University Press, 1997.

Riddle, Lawrence Melville. *The Genesis and Sources of Pierre Corneille's Tragedies from Medée to Pertharite.* Baltimore: Johns Hopkins University Press, 1926.

Rigolot, Francois. "Review: Father Figures: Genealogy and Narrative Structure in Rabelais by Carla Freccero." *Modern Philology* 91, no.4 (1994): 480–82.

Roberts, Penny. "The Kingdom's Two Bodies?: Corporeal Rhetoric and Royal Authority during the Religious Wars." *French Historical Studies* 21 (2007): 147–64.

Robinson, Elaine L. *Shakespeare Attacks Bigotry: A Close Reading of Six Plays.* Jefferson, NC: McFarland, 2009.

Rogers, Daniel. *Matrimoniall Honour; Or, The Mutuall Crowne and Comfort of Godly, Loyall, and Chaste Marriage.* London: Thomas Harper, 1642.

Romaniello, Matthew P., and Charles Lipp, eds. *Contested Spaces of Nobility in Early Modern Europe.* New York: Routledge, 2011.

Rosen, Carol C. "The Language of Cruelty in Ford's *'Tis Pity She's a Whore.*" *Comparative Drama* 8, no.4 (1974): 356–68.

Row, Jennifer E. "Queer Time on the Early Modern Stage: France and the Drama of Biopower." *Exemplaria* 29, no.1 (2017): 58–81.

Rowe, George E. *Thomas Middleton and the New Comedy Tradition.* Lincoln: University of Nebraska Press, 1979.

Rozett, Martha Tuck. "The Comic Structures of Tragic Endings: The Suicide Scenes in *Romeo and Juliet* and *Anthony and Cleopatra.*" *Shakespeare Quarterly* 3, no.2 (1985): 152–64.

Rubright, Marjorie. *Doppelgänger Dilemmas: Anglo-Dutch Relations in Early Modern Literature and Culture.* Philadelphia: University of Pennsylvania Press, 2014.

Ruggiero, Guido. *Binding Passions: Tales of Magic, Marriage, and Power at the End of the Renaissance.* Oxford: Oxford University Press, 1993.

———. *Machiavelli in Love: Sex, Self and Society in the Italian Renaissance.* Baltimore: Johns Hopkins University Press, 2007.

———. *Violence in Early Renaissance Venice.* New Brunswick, NJ: Rutgers University Press, 1980.

Russell, Nicholas. *Like Engend'ring Like: Heredity and Animal Breeding in Early Modern England.* Cambridge: Cambridge University Press, 1986.

Rzepka, Adam. "Chamber, Tomb, and Theater. Living in Romeo and Juliet's Spaces of the Dead." In *Critical Insights* Romeo and Juliet, edited by Robert C. Evans, 181–200. Ipswich, MA: Salem Press, 2017.

Sabean, David Warren. "Descent and Alliance: Cultural Meanings of Blood in the Baroque." In Johnson et al., *Blood and Kinship,* 144–174.

Sabean, David, and Simon Teuscher. "Introduction." In Johnson et al., *Blood and Kinship,* 1–17.]

Sampson, Lisa. *Pastoral Drama in Early Modern Italy: The Making of a New Genre.* Ann Arbor: University of Michigan Press, 2006.

Santing, Catrien. "'For the Life of a Creature Is in the Blood' (Leviticus 17:11). Some Considerations on Blood as the Source of Life in Sixteenth-Century Religion and Medicine and Their Interconnections." In *Blood, Sweat and Tears—The Changing Concepts of Physiology from Antiquity into Early Modern Europe,* edited by Manfred Horstmanshoff, Helen King, and Claus Zittel, 415–41. Boston: Brill, 2012.

Sawday, Jonathan. *The Body Emblazoned: Dissection and the Human Body in Renaissance Culture.* New York: Routledge, 1995.

Saxonhouse, Arlene W. "Niccolò Machiavelli: Women as Men, Men as Women, and the Ambiguity of Sex." In *Feminist Interpretations of Niccolò Machiavelli,* edited by Maria J. Falco, 93–116. University Park: Pennsylvania State University Press, 2004.

Schalk, Ellery. *From Valor to Pedigree: Ideas of Nobility in France in the Sixteenth and Seventeenth Centuries.* Princeton, NJ: Princeton University Press, 1986.

Schalkwyk, David. "Othello's Consummation." In *Othello: The State of Play,* edited by Lena Cowen Orlin, 203–34. London: Bloomsbury Arden Shakespeare, 2014.

Schoenfeldt, Michael C. *Bodies and Selves in Early Modern England: Physiology and Inwardness in Spenser, Shakespeare, Herbert and Milton.* Cambridge: Cambridge University Press, 1999.

Schwartz-Gastine, Isabelle. "*Romeo and Juliet* on the French Stage: From the Early Versions to the English Production at the Odéon Theatre in 1827." In Romeo and Juliet *in European Culture,* edited by Juan F. Cerdá, Dirk Delabastita, and Keith Gregor, 77–99. Philadelphia: John Benjamins Publishing, 2017.

Schwarz, Kathryn. "Held in Common: *Romeo and Juliet* and the Promiscuous Seductions of Plague." In *Queer Shakespeare: Desire and Sexuality,* edited by Goric Stanivukovic, 245–61. London: Bloomsbury Arden Shakespeare, 2017.

———. "'My Intents Are Fix'd': Constant Will in *All's Well That Ends Well*." *Shakespeare Quarterly* 58, no. 2 (2007): 200–227.

———. *Tough Love: Amazon Encounters in the English Renaissance*. Durham, NC: Duke University Press, 2000.

Scott, Paul. "'Ma Force Est Trop Petite': Authority and Kingship in *Le Cid*." *Modern Language Studies* 45, no.3 (2009): 292–304.

Seneviratne, Sachini K. "Rethinking the 'Good Receipt' in *All's Well That Ends Well*." *Essays in Criticism* 70, no. 2 (2020): 160–77.

Sensabaugh, G. F. "John Ford and Platonic Love in the Court." *Studies in Philology* 36, no.2 (1939): 206–26.

Sereno, Renzo. "A Note on the Names of the Personages of Machiavelli's *Mandragola*." *Italica* 26, no.1 (1949): 56.

Seuntjens, Wolter. "Vapours and Fumes, Damps and Qualms: Windy Passions in the Early Modern Age (1600–1800)." *English Studies* 87, no.1 (2006): 35–52.

Shalkwyk, David. "The Impossible Gift of Love in *The Merchant of Venice* and the Sonnets." *Shakespeare* 7, no.2 (2011): 142–55.

"Shakespeare and Embodiment: An E-Conversation." *Literature Compass* 2 (2005) SH 180: 1–13.

Shakespeare, William. *An Excellent Conceited Tragedie of Romeo and Juliet. As It Hath Been Often (with Great Applause) Plaid Publiquely, by the Right Honorable the L. of Hunsdon His Servants*. London: Printed by John Danter [and Edward Allde?], 1597.

———. *The Complete Pelican Shakespeare*. Edited by Stephen Orgel and A. R. Braunmuller. New York: Penguin Books, 2002.

Shannon, Laurie. *Sovereign Amity: Figures of Friendship in Shakespearean Contexts*. Chicago: University of Chicago Press, 2002.

Shapiro, James. *Shakespeare and the Jews*. New York: Columbia University Press, 1996.

Sheeha, Iman. "'I Have Paid the Duty to the Sonne, Which I Have Vowed to the Father': Serving the Father in John Ford's *'Tis Pity She's A Whore*." *Early Modern Literary Studies* Special Issue 26: John Ford (2017): 1–28.

Shell, Marc. *The End of Kinship: 'Measure for Measure' Incest, and the Ideal of Universal Siblinghood*. Baltimore: Johns Hopkins University Press, 1988.

Shurgot, Michael W. "Bertram's Scar and Courts of Healing in *All's Well That Ends Well*." *Shakespeare Bulletin* 37, no.3 (2019): 391–407.

Siegel, Paul N. "Christianity and the Religion of Love in *Romeo and Juliet*." *Shakespeare Quarterly* 12, no.4 (1961): 371–92.

Silverstone, Catherine. "New Directions: Fatal Attraction: Desire, Anatomy and Death in *'Tis Pity She's a Whore*." In *'Tis Pity She's a Whore: A Critical Guide*, edited by Lisa Hopkins, 77–93. London: Continuum, 2010.

Simonds, Peggy Muñoz. "Overlooked Sources of the Bed Trick." *Shakespeare Quarterly* 34, no.4 (1983): 433–34.

———. "Sacred and Sexual Motifs in *All's Well That Ends Well*." *Renaissance Quarterly* 42, no.1 (1989): 33–59.

Singer, P. N. *Galen: Psychological Writings*. Cambridge: Cambridge University Press, 2013.

Singleton, Charles S. "Machiavelli and the Spirit of Comedy." *Modern Language Notes* 57, no.7 (1942): 585–92.

Siraisi, Nancy G. "Vesalius and the Reading of Galen's Teleology." *Renaissance Quarterly* 50, no.1 (1997): 1–37.

Slights, Camille. "In Defense of Jessica: The Runaway Daughter in *The Merchant of Venice*." *Shakespeare Quarterly* 31, no.3 (1980): 357–68.

Slights, William W. E. *The Heart in the Age of Shakespeare*. Cambridge: Cambridge University Press, 2008.

Smith, Bruce R. "What Doing It in the Dark, without Words, Tells Us about Early Modern Sexuality." In *All's Well that Ends Well*, edited by Sylvan Barnet, 165–74. New York: Signet Classics Shakespeare, 2005.

Smith, Ian. "Othello's Black Handkerchief." *Shakespeare Quarterly* 64, no.1 (2013): 1–25.

———. *Race and Rhetoric in the Renaissance: Barbarian Errors*. New York: Palgrave Macmillan, 2009.

———. "The Textile Black Body: Race and 'Shadowed Livery' in *The Merchant of Venice*." In *The Oxford Handbook of Shakespeare and Embodiment: Gender, Sexuality, and Race*, edited by Valerie Traub, 170–85. Oxford University Press, 2016.

———. "We Are Othello: Speaking of Race in Early Modern Studies." *Shakespeare Quarterly* 67, no.1 (2016): 104–24.

Smith, Paul Edgar. "The Incest Motif and the Question of the Decadence of English Drama, 1603–1632." PhD diss., University of Pittsburgh, 1980.

Smith, Warren D. "Romeo's Final Dream." *Modern Language Review* 62, no.4 (1967): 579–83.

Snyder, Jon R. *Dissimulation and Culture of Secrecy in Early Modern Europe*. Berkeley: University of California Press, 2009.

Snyder, Susan. "*All's Well That Ends Well* and Shakespeare's Helens: Text and Subtext, Subject and Object." *English Literary Renaissance* 18, no.1 (1988): 66–77.

———. "Ideology and the Feud in *Romeo and Juliet*." *Shakespeare Survey* 49 (1996): 87–96.

———. "'The King's Not Here'": Displacement and Deferral in *All's Well That Ends Well*." *Shakespeare Quarterly* 43, no.1 (1992): 20–32.

———. "Naming Names in *All's Well That Ends Well*." *Shakespeare Quarterly* 43, no.3 (1992): 265–79.

Sofer, Andrew. *The Stage Life of Props*. Ann Arbor: University of Michigan Press, 2010.

Solomon, Julie R. "Morality as Matter of Mind: Towards a Politics of Problems in *All's Well That Ends Well*." *English Literary Renaissance* 23, no.1 (1993): 134–69.

———. "You've Got to Have Soul: Understanding the Passions in Early Modern Culture." In *Literary and Scientific Cultures in Early Modernity: Rhetoric and Medicine in Early Modern Europe*, edited by Stephen Pender and Nancy S. Struever, 195–228. Abingdon, UK: Ashgate Publishing, 2012.

Sorella, Antonio. *Magia, Lingua, e Commedia nel Machiavelli*. Firenze: L. S. Olschki, 1990.

Sparacio, Francesca R. "Il processo del comico ne *La Mandragola* di Niccolò Machiavelli." *Romance Review* 5, no.1 (Spring 1995): 73–85.

Spellberg, Matthew. "Feeling Dreams in *Romeo and Juliet*." *English Literary Renaissance* 43, no.1 (2013): 62–85.

Spiller, Elizabeth A. "From Imagination to Miscegenation: Race and Romance in Shakespeare's *The Merchant of Venice*." *Renaissance Drama* 29 (1998): 137–64.

———. *Reading and the History of Race in the Renaissance*. Cambridge: Cambridge University Press, 2011.

Stavig, Mark. *John Ford and the Traditional Moral Order*. Madison: University of Wisconsin Press, 1968.

Steinway, Elizabeth. "In Search of Alternative Kinship: Pregnancy without Proof in *All's Well That Ends Well*." *Medieval and Renaissance Drama in England* 33 (2020): 274–96.

Stensgaard, Richard K. "*All's Well That End's Well* and the Galenico-Paracelsian Controversy." *Renaissance Quarterly* 25, no.2 (1972): 173–88.

Stern, Tiffany. "'A Small-Beer Health to His Second Day': Playwrights, Prologues, and First Performances in Early Modern Theater." *Studies in Philology* 101, no.2 (2004): 172–99.

Stevens, Andrea Ria. *Inventions of the Skin: The Painted Body in Early English Drama, 1400–1642*. Edinburgh: Edinburgh University Press, 2013.

Stewart, Pamela D. "How to Get a Happy Ending: *Decameron* III.9 and Shakespeare's *All's Well*." *Studi sul Boccaccio* 20 (1991–92): 325–44.

Stone, Lawrence. *The Family, Sex and Marriage in England 1500–1800*. New York: Harper and Row, 1977.

Strier, Richard, and Carla Mazzio. "Two Responses to 'Shakespeare and Embodiment: An E-Conversation.'" *Literature Compass* 3, no. 1 (2005): 15–31.

Strong, David. "The Natural Rights Exerted in Shakespeare's Bed-Tricks." *Philosophy and Literature* 41 (2017): A76–A94.

Strout, Nathaniel. "The Tragedy of Annabella in *'Tis Pity She's a Whore*." In *Traditions and Innovations: Essays on British Literature of the Middle Ages and the Renaissance*, edited by David G. Allen and Robert A. White, 163–76. Newark: Delaware University Press, 1990.

Sugg, Richard. *The Smoke of the Soul: Medicine, Physiology and Religion in Early Modern England*. New York: Palgrave Macmillan, 2013.

Sullivan, Garrett A. "'Be This Sweet Helen's Knell, and Now Forget Her': Forgetting, Memory, and Identity in *All's Well That Ends Well*." *Shakespeare Quarterly* 50, no.1 (1999): 51–69.

Sumberg, Theodore A. "*La Mandragola*: An Interpretation." *Journal of Politics* 23, no.2 (1961): 320–40.

Tadmor, Naomi. "Early Modern English Kinship in the Long Run: Reflections on Continuity and Change." *Continuity and Change* 25, no.1 (2010): 15–48.

Targoff, Ramie. "Mortal Love: Shakespeare's *Romeo and Juliet* and the Practice of Joint Burial." *Representations* 120 (2012): 17–38.

Taylor, Thomas. *A Man in Christ; or, A New Creature*. London: H. Lownes, 1629.

Terpstra, Nicholas. *Lost Girls: Sex and Death in Renaissance Florence*. Baltimore: Johns Hopkins University Press, 2010.

Teuscher, Simon. "Flesh and Blood in the Treatises on the *Arbor Consanguinitatis* (Thirteenth to Sixteenth Centuries)." In Johnson et al., *Blood and Kinship*, 83–104.

Thomas, George. "The Parasite as Virtuoso: Sexual Desire and Political Order in Machiavelli's *Mandragola*." *Interpretation: A Journal of Political Philosophy* 30, no.2 (2003): 179–94.

Thompson, Ayanna, ed. *The Cambridge Companion to Shakespeare and Race*. Cambridge: Cambridge University Press, 2021.

———. *Passing Strange: Shakespeare, Race, and Contemporary America*. Oxford: Oxford University Press, 2011.

Tokson, Elliot H. *The Popular Image of the Black Man in English Drama, 1550–1688*. Boston: G. K. Hall, 1982.

Totaro, Rebecca. "Embedded in Shakespeare's 'Fair Verona.'" In *Contagion and the Shakespearean Stage*, edited by Darryl Chalk and Mary Floyd-Wilson, 255–76. New York: Palgrave Studies in Literature, Science and Medicine, 2019.

Traister, Barbara Howard. "'Doctor She': Healing and Sex in *All's Well That Ends Well*." In *A Companion to Shakespeare's Works*. Vol. 4, *The Poems, Problem Comedies, Late Plays*, edited by Richard Dutton and Jean E. Howard, 333–47. Malden, MA: Blackwell Publishing, 2003.

Traub, Valerie. "Mapping the Global Body." In *Early Modern Visual Culture: Representation, Race, and Empire in Renaissance England*, edited by Peter Erickson and Clark Hulse, 44–97. Philadelphia: University of Pennsylvania Press, 2000.

———. "The New Unhistoricism in Queer Studies." *PMLA* 128, no.1 (2013): 21–39.

———. *The Renaissance of Lesbianism in Early Modern England*. Cambridge: Cambridge University Press, 2002.

Trethewey, John. "Chimène's Dissembling and its Consequences." *Romance Studies* 17, no.2 (1999): 105–14.

Trevor, Douglas. "Love, Humoralism and 'Soft' Psychoanalysis." *Shakespeare Studies* 33 (2005): 87–94.

Tribble, Evelyn. "'The Raven O'er the Infectious House': Contagious Memory in *Romeo and Juliet* and *Othello*." In *The Routledge Handbook of Shakespeare and Memory*, edited by Andrew Hiscock and Lina Perkins Wilder, 105–15. New York: Routledge, 2017.

True, Micah. "King and Colony in Pierre Corneille's *Le Cid*." *French Studies: A Quarterly Review* 71, no.1 (2017): 1–14.

Turner, Robert Y. "Dramatic Conventions in *All's Well That Ends Well*." *PMLA* 75, no.5 (1960): 497–502.

Tylus, Jane. "Theater and Its Social Uses: Machiavelli's *Mandragola* and the Spectacle of Infamy." *Renaissance Quarterly* 53, no.3 (2000): 656–86.

Ultsch, Lori J. "*Epithalamium Interruptum*: Maddalena Campiglia's New Arcadia." *MLN* 120, no.1 (2005): 70–92.

———. "Maddalena Campiglia, 'Dimessa nel mondano cospetto?': Secular Celibacy, Devotional Communities, and Social Identity in Early Modern Vicenza." *Forum Italicum: A Journal of Italian Studies* 39, no.2 (2005): 350–77.

Valesio, Paolo. "The Practice of Literary Semiotics: A Theoretical Proposal." In *Centro Internazionale di Semiotica e di Linguistica: Working Papers and Pre-Publications*. Vol. 71, 1–24. Urbino: Università di Urbino, 1978.

Vaughan, Virginia Mason. *Othello: A Contextual History*. Cambridge: Cambridge University Press, 1994.

Vickers, Nancy J. "Diana Described: Scattered Woman and Scattered Rhyme." *Critical Inquiry* 8, no.2 (1981): 265–79.

Vilches, Patricia. "The Delegate Womb: Lucrezia's Body as Political Tool in Machiavelli's *La Mandragola*." *American Journal of Italian Studies* 22, no.60 (1999): 99–126.

Vitkus, Daniel J. *Turning Turk: English Theatre and the Multicultural Mediterranean, 1570–1630*. New York: Palgrave Macmillan, 2003.

———. "Turning Turk in *Othello*: The Conversion and Damnation of the Moor." *Shakespeare Quarterly* 48, no.2 (1997): 145–76.

Wack, Mary Frances. "From Mental Faculties to Magical Philters: The Entry of Magic into Academic Medical Writings on Lovesickness, 13th–17th Centuries." In *Eros and Anteros: The Medical Traditions of Love in the Renaissance*, edited by Donald A. Beecher and Massimo Ciavolella, 9–31. Ottawa: Dovehouse Editions, 1992.

Walker, Katherine. "'Daring to Pry into the Privy Chamber of Heaven': Early Modern Mock-Almanacs and the Virtues of Ignorance." *Studies in Philology* 115, no.1 (2018): 129–53.

Wall, Wendy. "De-generation: Editions, Offspring, and *Romeo and Juliet*." In *From Performance to Print in Shakespeare's England*, edited by Peter Holland and Stephen Orgel, 152–70. London: Palgrave Macmillan, 2006.

Walter, Melissa, and Sarah Klann. "Shakespeare Source Study in the Early Twenty-First Century: A Resurrection?" *Literature Compass* 15 (2018): 1–9.

Walter, Melissa Emerson. "Dramatic Bodies and Novellesque Spaces in Jacobean Tragedy and Tragicomedy." In *Transnational Exchange in Early Modern Theater*, edited by Robert Henke and Eric Nicholson, 63–80. New York: Routledge, 2008.

———. *The Italian Novella and Shakespeare's Comic Heroines*. Toronto: University of Toronto Press, 2019.

Waters, D. Douglas. "Fate and Fortune in *Romeo and Juliet*." *Upstart Crow* 12 (1992): 74–90.

Watson, Robert N. "Lord Capulet's Lost Compromise: A Tragic Emendation and the Binary Dynamics of *Romeo and Juliet*." *Renaissance Drama* 43, no.1 (2015): 53–84.

Webber, Edwin J. "The Dramatic Unities in the 'Mandragola.'" *Italica* 33, no.1 (1956): 20–21.

Welsh, Alexander. "The Loss of Men and Getting of Children: *All's Well That Ends Well* and *Measure for Measure*." *Modern Language Review* 73, no.1 (1978): 17–28.

Whigham, Frank. "Ideology and Class Conduct in *The Merchant of Venice*." *Renaissance Drama* 10 (1979): 93–115.

Wiesner-Hanks, Merry E. *Christianity and Sexuality in the Early Modern World: Regulating Desire, Reforming Practice*. New York: Routledge, 2000.

Wilcox, Helen. "Shakespeare's Miracle Play? Religion in *All's Well That Ends Well*." In *New Critical Essays*, edited by Gary Waller, 140–54. New York: Routledge, 2007.

Williams, Robert I. "Machiavelli's *Mandragola*, Touchwood Senior, and the Comedy of Middleton's *The Chaste Maid of Cheapside*." *Studies in English Literature, 1500–1900* 10, no.2 (1970): 385–96.

Wilson, Harold S. "Dramatic Emphasis in *All's Well That Ends Well*." *Huntington Library Quarterly* 13, no.3 (1950): 217–40.

Wiseman, Susan J. "'*Tis* Pity She's a Whore: Representing the Incestuous Body." In *Renaissance Bodies: The Human Figure in English Culture, c. 1540–1660*, edited by Lucy Gent and Nigel Llewellyn, 180–97. London: Reaktion, 1990.

Woshinsky, Barbara R. "Rhetorical Vision in *Le Cid*." *French Forum* 4, no.2 (1979): 147–59.

Wright, Heather Hadar. "Lucrezia in *Mandragola*: Machiavelli's New Prince." *Interpretation: A Journal of Political Philosophy* 36, no.2 (2009): 145–64.

Wyatt, Michael. *The Italian Encounter with Tudor England: A Cultural Poetics of Translation*. Cambridge: Cambridge University Press, 2005.

Younge, Richard. *The Cure of Misprision*. London: Thomas Paine, 1646.

Zorach, Rebecca. *Blood, Milk, Ink, Gold: Abundance and Excess in the French Renaissance*. Chicago: University of Chicago Press, 2005.

INDEX